Catalysis
by Gold

CATALYTIC SCIENCE SERIES

Series Editor: Graham J. Hutchings *(Cardiff University)*

*To view the complete list of the published volumes in the series, please visit:
http://www.worldscientific.com/series/css

CATALYTIC SCIENCE SERIES — VOL. 6

Series Editor: Graham J. Hutchings

Catalysis by Gold

Geoffrey C. Bond
Brunel University, UK

Catherine Louis
Université Pierre et Marie Curie, France

David T. Thompson
Consultant, World Gold Council, UK

Imperial College Press

ICP

Published by

Imperial College Press
57 Shelton Street
Covent Garden
London WC2H 9HE

Distributed by

World Scientific Publishing Co. Pte. Ltd.
5 Toh Tuck Link, Singapore 596224
USA office: 27 Warren Street, Suite 401-402, Hackensack, NJ 07601
UK office: 57 Shelton Street, Covent Garden, London WC2H 9HE

British Library Cataloguing-in-Publication Data
A catalogue record for this book is available from the British Library.

First published 2006 (Hardcover)
Reprinted 2016 (in paperback edition)
ISBN 978-1-911299-70-7

CATALYSIS BY GOLD
Catalytic Science Series — Vol. 6

ISBN-13 978-1-86094-658-5
ISBN-10 1-86094-658-5

Typeset by Stallion Press
Email: enquiries@stallionpress.com

Acknowledgements

We thank many researchers in gold catalysis for providing relevant manuscripts and papers, many of them before publication, and for their comments and summaries of recent developments and reviews. Those who have contributed in these ways include Alvaro Amieiro-Fonseca, Donka Andreeva, Antonio Arcadi, Valerie Caps, Silvio Carrettin, Avelino Corma, Hajo Freund, Maria Flytzani-Stephanopoulos, Wayne Goodman, Masatake Haruta, Stephen Hashmi, Graham Hutchings, Chang Hwan Kim, Harold and Mayfair Kung, Joszef Margitfalvi, Ben Nieuwenhuys, Vojtech Plzak, Francesca Porta, Laura Prati, Michele Rossi, Tatyana Tabakova and many others: we apologize for not mentioning everybody. In addition, we thank Alvaro Amieiro-Fonseca and Richard Holliday for valuable information on patents.

Much of the earlier literature for parts of this book was collected for a chapter entitled 'Catalytic Applications for Gold Nanotechnology' written by Sónia Carabineiro and one of us (DT) for a book on 'Nanocatalysis' recently published by Springer-Verlag, and the collaboration on identifying new gold catalysis literature has continued. Lina Mehta provided valuable assistance with drawing the organic equations and reaction schemes for Chapters 8 and 12, and Wendy Smith provided help with producing the typescript.

We are particularly grateful to Eric Short for a critical reading of Chapters 2 and 3, for the additional calculations contained in Chapters 5 and 10, and for the brief introduction to Density Functional Theory, which is inserted as an Appendix to Chapter 5.

The three figures incorporated into the cover design are reprinted with permissions from *Gold Bull.* **33** (2000) 41, Copyright 2000 World Gold Council; *J. Phys. Chem. B* **106** (2006) 7634, Copyright 2002 American Chemical Society; and *Chemical Record* **3** (2003) 75, Copyright 2003 John Wiley and Sons Inc.

Finally, we thank Graham Hutchings for the invitation to write this book and for his enthusiastic encouragement and stimulating Preface; and

Chris Corti and the World Gold Council for their interest and support. We feel confident that readers will find the contents useful in their research and in finding new applications for catalysis by gold.

Geoffrey Bond
Catherine Louis
David Thompson
August 2006

Preface

"Everything has its beauty but not everyone sees it"

Confucius, Analects, *ca.* 500BC

Gold is an element that has fascinated mankind for millennia. It is viewed as immutable, non-changing, the ultimate statement of wealth and beauty. Gold has been used by jewellers to create some of the most beautiful artefacts throughout history. Gold is invariably the metal selected by most couples as the outward sign of their love. The constancy of gold is born out of its chemical inertness when in a bulk form as it does not react with air and corrode. Gold has been a source of conflict, and it has also been fought over for millennia. Gold has been viewed as so important that alchemists tried to make it from less valuable base metals. Indeed, some of the most noted scientists in the seventeenth century formed a group called 'the mercuralists' who contended that gold was a particular combination of mercury and sulfur. However, for most people, the outward sign of beauty is obvious for gold, so why have I selected a relatively obscure quotation from Confucius for this Preface to a book on the topic of Catalysis by Gold. By the time you have finished the book, I hope you will have appreciated the statement. Gold has a hidden inner beauty for a scientist interested in catalysis, for it turns out that, when this immutable, lustrous metal is subdivided down to the nanoscale, it becomes an incredibly reactive material. In a nanocrystalline state, gold can activate carbon monoxide and dioxygen at temperatures as low as 197 K to form carbon dioxide. Such levels of activity are not replicated by other catalysts. It is an amazing discovery that gold is an active redox catalyst. Indeed, it can be the catalyst of choice. This hidden beauty had lain dormant for centuries, even though the catalytic activity of other nano-divided metals had been established. For example, Faraday demonstrated the catalytic activity of finely divided platinum for hydrogen reactions in the first half of the nineteenth century. Now the topic of catalysis by gold represents one of the fastest growing fields in science. Hundreds of scientific papers are appearing on the topic annually and the rate of growth of scientific discovery for catalysis by gold is currently exponential. New discoveries, particularly relating to the selective oxidation of alkenes, alcohols and even alkanes are being made with incredible speed. Against this background, it is timely that the authors have written a book bringing together these myriad of themes of catalysis by gold. It is a rich story and it is well told, it is a story you will enjoy reading.

Graham J. Hutchings
May 2006

Contents

CHAPTER 1

Introduction to Catalysis

1.1. The Phenomenon of Catalysis

In the early part of the 19th century, when the scientific study of chemistry was just beginning, it was observed that the occurrence of a number of chemical reactions was conditional upon the presence of trace amounts of substances that did not themselves take part in the reaction. In 1836 the Swedish Scientist J.J. Berzelius tried to bring these observations into the body of chemical knowledge by attributing their action to what he called their *catalytic power*: this action he named *catalysis* by analogy, he said, with *analysis*, which is *"the separation of the component parts of bodies by means of ordinary chemical forces. Catalytic power means that substances are able to awake affinities that are asleep at this temperature by their mere presence ..."* The word 'catalysis' comes from Greek words meaning 'a breaking down', and had been used from the time of Ancient Greece to signify the collapse of moral or ethical constraints, so Berzelius applied the term to those phenomena where the normal barriers to chemical reaction were removed. In journalistic use it has however come to mean 'a coming together', which at first sight is the opposite of breaking down, but 'a breaking down' of a barrier inevitably leads to 'a coming together', and it is significant that in the Chinese language the same word is used for both catalysis and marriage broker.

The phenomenon of catalysis occurs very widely. Our life and health, and that of all living things, depends upon the action of biological catalysts called *enzymes* that usually consist of proteins, which sometimes have a metal–atom-containing prosthetic group such as the chlorophyll or haem molecule. These remarkably effective biocatalysts are at the pinnacle of catalytic power and all synthetic catalysts strive to emulate them. The substances that are of use in chemical processing and in environmental control are, however, inorganic in nature, and can be classified into (i) metals, (ii) oxides, (iii) sulfides, and (iv) solid acids, although practical catalysts often contain components drawn from two of these categories. In particular, as we shall see shortly (Section 3.1), metals need to be employed as very

1

small particles in order to maximise their surface area, and because they are unstable in this state, it is necessary to separate them by attaching them to the surface of an oxide particle so that they are not in contact with each other. We then have a *supported metal catalyst*, and these materials will occupy our attention through much of this book (see Chapter 4 for ways of preparing them). In this form they occupy a phase that differs from the fluid phase in which the reactants exist, and they are therefore termed *heterogeneous catalysts*. However, there are many chemical species that can act catalytically when dissolved in a liquid phase in which one or more of the reactants are to be found: examples include the proton and hydroxyl ion, but of greater interest and importance to us are the salts and organometallic complexes of metals. These are termed *homogeneous catalysts*.

In the years following Berzelius, a number of further examples of catalytic action were discovered, but scientific appreciation of their mode of action had to await the arrival of experimental and theoretical techniques for the study of reaction rates. It then became possible for F.W. Ostwald to define a catalyst as *"a substance that increases the rate at which a chemical system approaches equilibrium, without being consumed in the process."* This handy form of words encapsulates the essential truth of the catalytic effect, and has stood the test of time; it carries with it a number of important implications that we should now explore. The first of these is that the position of equilibrium attained in a catalysed reaction is exactly the same as that which would ultimately be arrived at in its absence: this must be so because the equilibrium constant K is determined by the Gibbs free energy of the process, and this in turn is fixed by the enthalpy and entropy changes, thus:

$$\Delta G = \Delta H - T\Delta S, \tag{1.1}$$
$$K = \exp(-\Delta G/RT). \tag{1.2}$$

It is inconceivable that the same reaction could have two different sets of thermodynamic parameters, and this basic principle has been put to good use by using catalysts for determining heats of hydrogenation of alkenes at room temperature:[1] this would otherwise be impossible because reactions would be inordinately slow at all reasonable temperatures.

It is sometimes a source of confusion that different catalysts can effect different courses of reaction on the same molecule. A good example of this is the decomposition of ethanol, which when metal-catalysed undergoes dehydrogenation to ethanal and when oxide-catalysed is transformed by dehydration to ethene. The answer is, of course, that both reactions are thermodynamically favourable, but this example introduces the important concept of *catalytic specificity*, whereby a catalyst is able to select one particular route to the exclusion of others, through the kind of intermediate species that are formed on its surface. It is, however, important to appreciate that a catalyst can *only* assist a reaction which is thermodynamically allowed under the specified conditions, that is, for which the change in Gibbs free energy is negative. Considerable effort was spent in the last years of the 19th century attempting the catalysed synthesis of ammonia — under conditions where it later became obvious that the ammonia molecule was not stable.

There are a number of other qualities that catalysts possess which should be introduced at this point. It is not only different types of catalyst that afford different products: a single catalyst can also do this, so for example the hydrogenation of ethyne can lead with a platinum catalyst to a mixture of ethene and ethane, and ethene once formed can react further to ethane. The extent to which the intermediate product, which is often the desired one, is formed is measured by the *selectivity S*, where

$$S = r_{C_2H_4}/(r_{C_2H_4} + r_{C_2H_6}) \tag{1.3}$$

and the reaction scheme takes the form shown in Scheme 1.1. With certain metals such as palladium, nickel, copper (and gold, see Section 9.3.2), ethene is formed almost selectively. A second kind of selectivity is shown when two reactive molecules are present over a catalyst, and one of them reacts faster than the other because it is more strongly adsorbed on its surface. A further aspect of selectivity appears when there are two different reactive groups in the same molecule; thus for example styrene is easily reduced to ethylbenzene because the alkene side-chain is much more

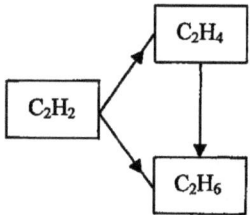

Scheme 1.1: Reaction pathways in the hydrogenation of ethyne.

reactive than the aromatic ring. This is an example of *regioselectivity*. If a reaction is capable of giving stereoisomeric products, a catalyst may exhibit *stereoselectivity*: thus 1,4-dimethylcyclohexenes can lead on hydrogenation to either *Z*- or *E*-dimethylcyclohexane ($Z \equiv cis$; $E \equiv trans$). Of particular current interest is the reduction of *prochiral* molecules, that is, those that develop centres of optical activity in the product, which therefore contains optical enantiomers. It is often desirable to create one of the products selectively, and a catalyst showing *enantiomeric selectivity* is therefore required.

1.2.   The Activation Energy of Catalysed Reactions

We now enquire how it is that a catalyst is able to accelerate the rate of a reaction. We may start with the concept proposed by Svante Arrhenius to describe the effect of temperature on a homogeneous (i.e. non-catalysed) gas-phase reaction: he stated that reaction rate r depended on the fraction of colliding molecules that between them had more than a critical amount of energy, which he called the *activation energy E*. This fraction increased exponentially with temperature in line with the Boltzmann distribution fraction, so that

$$r = Z \exp(-E/RT), \tag{1.4}$$

where Z is the collision number. The rate might be lower if collisions had to be orientationally acceptable, and so a *steric factor P* was later added to the right-hand side. It is not easy to compare a homogeneous gas-phase reaction proceeding in quite a large volume of space with a heterogeneous reaction occurring within a very much smaller volume at the surface of a solid, but, if the latter depends on the frequency of collisions of a reactant with the surface, this number expressed per cm^2 is typically about 10^{12} times smaller than the gas-phase collision frequency Z, and hence it has been concluded that to compensate for this the activation energy of a catalysed reaction has to be *at least* $65\,kJ\,mol^{-1}$ less than that of its homogeneous counterpart, and realistically must be $100\,kJ\,mol^{-1}$ less. This conclusion has been confirmed in cases where it has been possible to measure both. It has therefore become an article of faith in the theory of catalysis that *a catalyst acts by lowering the activation energy of the reaction*.

It must do this by creating a new and energetically more favourable reaction path, and we can visualise this by recalling that the activation energy can also be represented as the potential energy barrier that exists between reactants and products. This is the barrier that has to be broken down, so

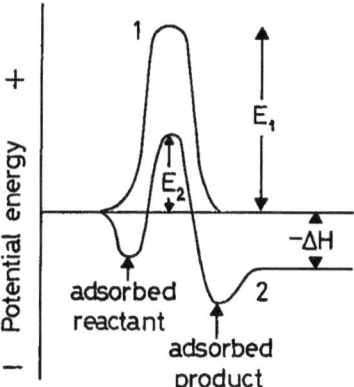

Figure 1.1: Potential energy profiles for (1) non-catalysed and (2) catalysed reactions.

the new reaction path is opened up as shown in Figure 1.1. This new path becomes possible because the reactants have first to be *chemisorbed* in the catalyst's surface, typically by the breaking of chemical bonds within the molecule and creating new bonds with the surface. A good example of this is the hydrogen molecule, which is quite stable and only dissociates into atoms at very high temperature; its dissociation energy is $410 \, \text{kJ} \, \text{mol}^{-1}$. However, in the presence of an active metal such as platinum, it is chemisorbed even at liquid hydrogen temperature by dissociation into two atoms: this process is exothermic, and can be depicted as

$$H_2 + 2^* \rightarrow 2H^*, \tag{1.A}$$

where the asterisk stands for a univalent adsorption site on the surface. A weakly held intermediate state of *physical adsorption* effectively eliminates the potential energy barrier by allowing close approach of the molecule to the surface (Figure 1.2). Thus the catalyst succeeds in accomplishing the difficult act of dissociating the molecule, which is the hardest part in the process of hydrogenation. Quite generally *the catalyst surface acts by preparing the reactants for reaction, by converting them into forms that will react with minimum energy input, that is, with a lower activation energy than would otherwise be needed.* This concept is readily extended to cover homogeneous catalysis by metal compounds, where coordination of the reactants at the metal centre replaces chemisorption at the surface. A principal concern of scientists working on catalysis is to identify these

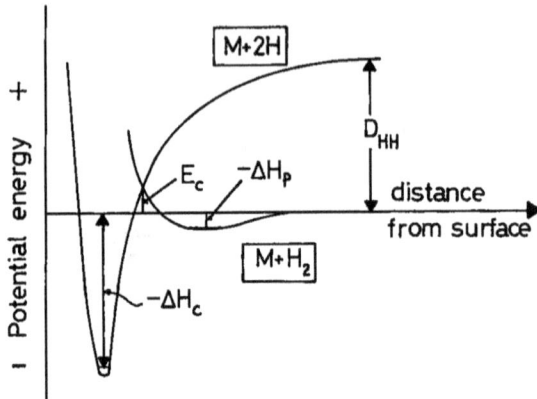

Figure 1.2: Potential energy curves for the approach of a hydrogen molecule and of two hydrogen atoms to a metal surface: E is the activation energy; $-\Delta H$ is the heat of adsorption; subscripts p and c are, respectively, physical adsorption and chemisorption.

new species and their modes of reaction, in other words, to establish the mechanism of the reaction.

If we display the temperature dependences of a reaction proceeding both homogeneously and heterogeneously catalysed as Arrhenius plots, using the equation in the form

$$\ln k = \ln A - E/RT, \tag{1.5}$$

k being the rate constant for the reaction, we see (Figure 1.3) that not only does the catalyst increase the rate, but also that it lowers the temperature at which the reaction achieves a useful rate, and it extends the temperature range in which these rates are available. In practice this is one of the most valuable attributes of a catalytic process, as it minimises the energy input needed, and hence the process costs.

1.3.   Ways of Using Heterogeneous Catalysts

To get the best out of a catalyst, it has to be deployed in the most appropriate way. The considerations that determine what this should be are shown in the Catalytic Cycle (Figure 1.4). Reactants in the fluid phase have first to be brought to the neighbourhood of the surface, where they must find an *active centre* where they can be chemisorbed in the right form and with

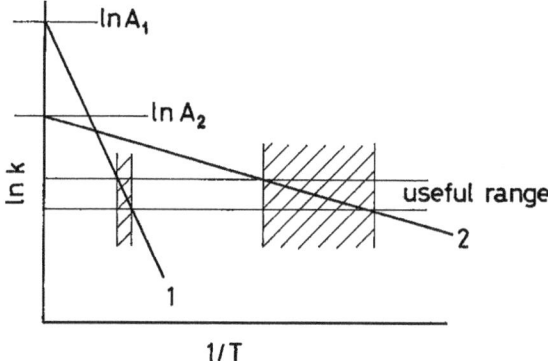

Figure 1.3: Arrhenius plots for (1) non-catalysed and (2) catalysed reactions: k is the rate constant; A is the pre-exponential factor.

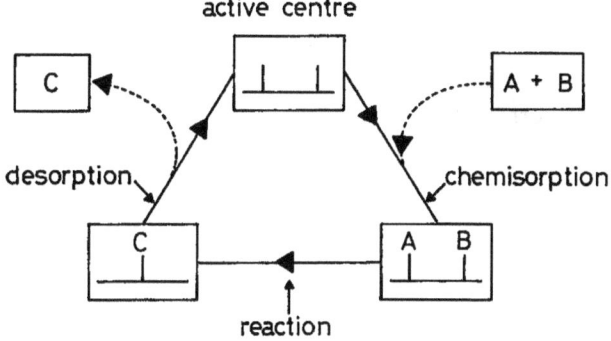

Figure 1.4: The catalytic cycle for the reaction of A + B → C.

the minimum energy, to ensure that their adsorbed states are sufficiently reactive. They react to form a product, which may itself be chemisorbed (Figure 1.1) and which must in that case desorb quickly, and diffuse away from the surface in order to recreate the vacant active centre. If the two diffusion steps are slower than the chemical reactions, the system is said to be under *mass-transport control*, and the catalyst is not being used efficiently, because the surface has to await the arrival of reactants or departure of products. For most purposes therefore it is better for the rate to be determined by the chemical steps at the surface, that is, to be under *kinetic control*. There are two simple ways of establishing which regime applies. (1) In a flow system, conversion should be a linear function of the

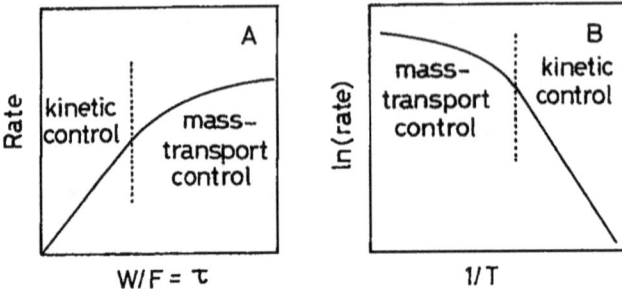

Figure 1.5: (A) Dependence of rate on contact time τ: W is the weight of catalyst; F is the flow-rate. (B) Arrhenius plot showing change from kinetic control to mass-transport control.

contact time τ, that is, to be inversely proportional to flow-rate F and proportional to catalyst weight (Figure 1.5(A)). In a stirred static reactor, it should be independent of speed of stirring, which affects the rate of diffusion of molecules close to the surface. (2) The temperature coefficient for mass-transport control is small and almost always lower than that for kinetic control: as temperature rises and the catalytic rate increases, a point is reached at which diffusion cannot keep pace, and mass-transport control sets in. This is most easily seen as a change in slope of the Arrhenius plot. (Figure 1.5(B)).

There are cases where it is actually desirable to operate under conditions of mass-transport control; this is so for example where it is an intermediate product that is wanted: fat-hardening is a case in point. More usually, however, one wishes to work under conditions where conversion is as close as possible to 100% and here it is inevitable that mass-transport control will apply, at least at the end of the catalyst bed. The physical structure of the catalyst then becomes of great importance, and much thought and skill is exercised in maximising access of reactants to the active centres. The form of reactor and the appropriate physical form of the catalyst have to be chosen with care.

If the reactants are in the gas phase, the reactor may be either *static* or *dynamic*. Small static reactors are convenient for basic research where either the reactants are expensive (e.g. isotopically labelled molecules) or the reaction slow. Dynamic reactors, where reactants flow through the catalyst bed, provide a better simulation of practical use: in a *fixed bed reactor* the catalyst remains in place, and it is in the form of large particles or pellets.

Alternatively it may take the form of a ceramic or metallic *monolith*, of which a variety of physical shapes is available; monoliths are now widely used as supports for the active catalyst, which lines the channels which permeate the structure. They find particular application for the control of exhaust from vehicles powered by internal combustion or diesel engines. If the catalyst particles are small enough, a fast flow of reactants causes the bed to expand and the particles to move about like molecules in a liquid. We then have *a fluidised bed reactor*, which affords a more uniform temperature profile than is possible in fixed bed reactors, and is therefore more apposite to strongly exothermic reactions.

In three-phase systems, where the solid catalyst is in contact with a liquid reactant or its solution plus a gaseous reactant, efficient agitation is required to effect dissolution of the gaseous molecule into the liquid and its transport to the catalyst surface. Such systems easily become mass-transport limited, especially when a very active catalyst is used. In *a batch reactor*, rapid shaking or stirring is needed, and catalyst particles must be small; it may operate at atmospheric pressure, or at superatmospheric pressure as an *autoclave*. Large catalyst particles can however be used with liquid reactants either in a *trickle-column reactor* or a *spinning-basket reactor*.

1.4. Understanding Catalysed Reactions

Under specified operating conditions, a reaction proceeds with a known amount of catalyst at a *rate* that can be expressed in units such as $mol\,s^{-1}$. It is better given as $mol\,g_{cat}^{-1}\,s^{-1}$, and if the fraction of the active component of the catalyst is known (and sometimes it is only guessed at) as the *specific rate*, in $mol\,g_M^{-1}\,s^{-1}$ (M = metal or other component). The ultimate step is taken if the area of the active phase is known: we can then use $mol\,m^{-2}\,s^{-1}$, which is the *areal rate*. For catalysts that adsorb carbon monoxide or hydrogen readily (see Sections 5.3 and 5.5), the number of surface metal atoms in a supported metal catalyst can be estimated by their chemisorption, obtained by measuring the amount needed to saturate the surface of the metal. Assuming each surface atom constitutes an active centre the rate can then be given as $mol\,mol_S^{-1}\,s^{-1}$ where S is the surface atom, i.e. as a *turnover frequency* (TOF). When this is not possible, the quantity 'mol$_S$' can also be derived through estimation of particle size by transmission electron microscopy (TEM) or X-ray absorption fine structure (XAFS). For many purposes however the areal rate (rate per unit area of active surface) is sufficient, because it can be changed into TOF by dividing by the calculated average number of surface atoms per m² (ca.10^{15}). Since the actual number of active centres is rarely known with precision, and is certainly often less than the number of surface atoms, TOF is a quantity that needs to be treated with great care, especially where there is some possibility of involvement of the support in the reaction. This difficulty does not however apply in the case of homogeneously catalysed reactions, however, because there the concentration of the active species is precisely known and so values of TOF are readily obtained.

In a static reactor the rate changes with time as the reactants are consumed, and the *initial rate* is often used. In a dynamic reactor under steady state conditions the rate is independent of time, and with a known flow of reactant into the reactor the observed fractional *conversion* is readily changed into a rate. What is of great interest in understanding a catalysed reaction is the response of the rate to variations in operating conditions, especially the concentrations or pressures of the reactants, and temperature. It is frequently observed that, at least over some limited range of temperature, the Arrhenius equation in the form

$$\text{rate} = A \exp(-E/RT) \qquad (1.6)$$

is obeyed, sometimes with great precision: if it is certain that the reaction is under kinetic control, E is the *apparent activation energy*.

For reactions of environmental interest, it has been customary to obtain the signature of a catalyst by a plot of conversion versus temperature, and unfortunately this procedure has been widely adopted for reactions catalysed by gold, especially the oxidation of carbon monoxide (Chapter 6) and the water-gas shift (Chapter 10). While it provides an easy means of ordering a series of catalysts into a qualitative hierarchy of activity, its limitations need to be stressed.

(1) 'Conversion' depends on the contact time or flow-rate, on the gas composition, on the amount of catalyst used and on the loading of the active component: changing any one of these could alter the relative activities of the members of the series.

(2) For highly exothermic reactions, such as carbon monoxide oxidation, it may be hard to keep the catalyst isothermal, and if the rate of heat dissipation no longer keeps pace with heat generation it becomes hotter than the temperature shown by the sensing device; the rate will then escalate

and quickly rise to 100% while the recorded temperature only rises by a few degrees. This phenomenon is known as *light-off* and the temperature at which this occurs is sometimes used as a measure of activity.

(3) At low conversions the rate is limited by the catalytic process on the surface (this is the *kinetic regime*), but when this becomes faster than the rate at which reactants can diffuse to the surface it becomes *diffusion-limited* (or *mass-transport-limited*). Its onset is gradual and can only be safely detected by changing conversion to a *rate* and then plotting its dependence on the measured temperature using the Arrhenius equation. When its slope starts to decrease, diffusion limitation is starting (Figure 1.5(B)): it may be noticeable at 50% conversion, and above 75% it will be dominant, so the comparison of high conversions only reveals aspects of the catalyst's *physical* structure and not of their true catalytic activity.

(4) It makes little sense to compare catalysts on the basis of the temperature of 50% conversion (or any other conversion) as is often done, because the statement that catalyst A is 10 degrees more active than catalyst B does not actually say by how much it is faster. Once again, comparison of *rates* at one or more temperatures is needed to show quantitatively how they compare. This does not work if these activities are vastly different, but other devices (such as showing the Arrhenius parameters as a compensation plot) can then be used.

(5) There is a danger that by forcing or allowing the catalyst to work at high temperature in order to obtain complete conversion, it may suffer structural damage such as sintering of the active component; this can be detected by following the reaction as the temperature is lowered back to its initial value, but this is rarely done. Deactivation due to structural change or any form of poisoning as the temperature is raised will clearly diminish the catalyst's ranking, and sometimes conversion fails to reach 100% for this reason. The activation energy will then also be false. The concept of *random* variation of temperature is hardly ever employed, although this is necessary to validate the activation energy. Running a catalyst to constant activity at low temperature before starting to raise it is no guarantee that further deactivation will not occur at higher temperature.

These cautionary words are needed because it is a matter for concern that conclusions about the merits of particular aspects of catalyst structure and composition are often based on very limited and sometimes unsatisfactory experimental work.

Deactivation of a catalyst is estimated by following the conversion as a function of time at a fixed temperature. It is of course pointless to do this when the conversion is close to 100%, as is sometimes done, because under such conditions the rate of the catalytic reaction can decline without affecting the observed conversion. Imagine a long catalyst bed in which initially all the reaction occurs in the first 10%; as this deactivates, the reactive zone moves further on, and this occurs progressively without apparent loss of activity, until all the catalyst is dead.

Consideration of how the rate of a catalysed reaction depends on the pressures or concentrations of the reactants in contact with it takes us into the realm of *chemical kinetics*. The simplest way of expressing this dependence for a reaction between two molecules is by an equation of the form

$$\text{rate} = kP_A^a P_B^b, \tag{1.7}$$

where k is a rate constant and the exponents a and b are the *orders of reaction*, respectively, of the reactants A and B, P being the pressure. This is an example of a *Power Rate Law*. Strictly speaking we should also provide for a possible effect of the pressure of the product on the rate, by including a term P_C^c, because occasionally the product is strongly adsorbed and inhibits the reaction. However, more often than not it is more weakly adsorbed than the reactants and does not interfere (i.e. $c = 0$), but all too frequently this is assumed without checking. It would be wise to check this by adding some product with the reactants to confirm that this is so. Now with catalysed reactions the values of a and b are often nonintegral, so that they do not allow of simple interpretation as is usually the case with non-catalysed reactions. The reason for this is that the rate is determined by the concentrations of the two reactants in chemisorbed states on the surface, usually expressed as their fractional surface coverages θ_A and θ_B, and these do not bear a straightforward relation to their concentrations in the fluid phase. They can however be connected by the *Langmuir adsorption equation* (sometimes called the Langmuir isotherm) which for reactant A takes the form

$$\theta_A = b_A P_A / (1 + b_A P_A + b_B P_B)^2, \tag{1.8}$$

where the b terms are the *adsorption coefficients* for each reactant, i.e. the equilibrium constants for their chemisorptions, and there is a corresponding equation for θ_B. The rate of reaction r is then proportional to the product of the two terms:

$$r = k\theta_A\theta_B = b_A P_A b_B P_B/(1 + b_A P_A + b_B P_B)^2. \tag{1.9}$$

This is the *Langmuir–Hinshelwood formalism*. We can see what this equation says by taking two extreme situations. (i) When the pressures of A and B are both low or when their adsorption coefficients are small, the rate is simply proportional to the product of the pressures, i.e. the reaction is first order in both reactants. (ii) If however θ_A is high because A is strongly adsorbed or its pressure is high, and if B is weakly adsorbed or its pressure is low, $b_A P_A \gg b_B P_B$, and then the reaction becomes first order in B and minus first order in A. Clearly there will be a range of intermediate conditions, and these can be summarised in Figure 1.6, which shows how the rate varies with P_A when P_B is held constant. It is however unusual for experimental measurements to reveal the whole of a curve of this type, and it is more usual for a section of it to be approximated by an exponent of the pressure, as in Equation (1.7). It is however important to appreciate that such an expression is only an approximation which applies over a limited range of pressure; however, it does provide a qualitative indication of the relative adsorption strengths of the reactants, but for quantitative work it is better to extract the adsorption coefficients from the appropriate form of the full rate equation (e.g. Equation (1.9)).

Although many reactions appear to proceed by a Langmuir–Hinshelwood mechanism, two other types are sometimes invoked and will receive passing mention in later chapters. In the *Rideal–Eley* (or

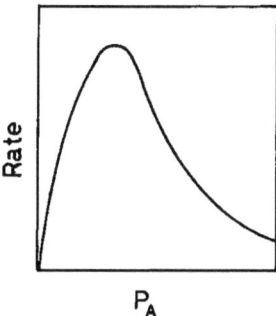

Figure 1.6: Langmuir–Hinshelwood formalism for a bimolecular reaction: dependence of rate on pressure of reactant A (see text).

Eley–Rideal) *mechanism*, one of the reactants comes directly from the fluid phase to react with the other, which is already chemisorbed. This procedure was devised to explain the kinetics of the hydrogen–deuterium reaction on certain metals (see Section 9.2), but has also been suggested for other reactions. The *Mars-van Krevelen mechanism* applies to oxidations catalysed by oxides that are easily reducible, and are therefore able to release their lattice oxide ions for the purpose of oxidising the other reactant; they are then replaced by the dissociation of molecular oxygen. With gold catalysts supported on such oxides, it is sometimes proposed that this mechanism plays a part in the total process.

Although orders of reaction for the oxidation of carbon monoxide in gold catalysts are sometimes reported,[2,3] application of rate equations based on Langmuir–Hinshelwood formalism (as Equation (1.9)) has only rarely been undertaken. In the case of this reaction, it is necessary to consider whether or not the reactants dissociate when they chemisorb and whether in either state they chemisorb competively (i.e. on the same site) or noncompetitively (i.e. on different sites). Results obtained on Au/TiO_2 at 273–313 K were tested[4] against each of the four possible rate equations, and that based on non-dissociative non-competitive chemisorption gave best fit; spectroscopic measurements strongly suggested that carbon monoxide is chemisorbed on the metal and the oxygen on the support (see Section 6.2.5). The orders of reaction were $\sim$0.4 for oxygen and 0.2–0.6 for carbon monoxide;[2] quite different values have been given for other catalysts and other experimental conditions,[2,3] but these could be accommodated by various values of the thermodynamic parameters for chemisorption.

Now the process of chemisorption is necessarily exothermic, so the operation of Le Chatelier's principle requires the surface coverages to decrease when temperature rises. This has important consequences for reaction kinetics, because the inhibition due to a strongly adsorbed reactant becomes less, and negative orders become more positive; the curves shown in Figure 1.7 illustrate this. A moment's consideration shows that the temperature coefficient of the rate must be a function of P_A; specifically, the activation energy will increase with P_A, thus justifying the term *apparent* activation energy. It is not always appreciated that activation energies derived from the temperature dependence of *rates* are not unique quantities, but depend on reactant pressures. The Arrhenius equation (1.5) describes the effect of temperature on the rate *constant*.

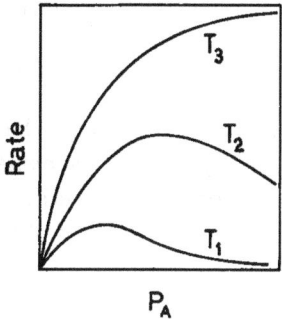

Figure 1.7: Dependence of rate on pressure of reactant B at three temperatures, $T_3 > T_2 > T_1$.

This matter can be resolved in the following way. The temperature dependence of an adsorption coefficient is given by

$$\mathrm{d}\ln b_\mathrm{A}/\mathrm{d}T = \Delta H_\mathrm{A}^\circ / RT^2, \tag{1.10}$$

where $\Delta H_\mathrm{A}^\circ$ is the standard heat of adsorption of A: thus is the *Van't Hoff isochore*. The observed effect of temperature on rate is therefore the consequence of two opposing effects, viz. (i) the positive effect on the rate of reaction of two adsorbed molecules assuming their surface coverages to be constant, and (ii) the negative effect of their decreasing coverages. We may associate (i) with the true activation energy E_t derived from the temperature-dependence of the rate constant k, so that

$$E_t = E_a - \Delta H_\mathrm{A}^\circ - \Delta H_\mathrm{B}^\circ. \tag{1.11}$$

Thus E_a will be less than E_t because the rate does not increase with temperature so fast as it would if surface coverages were constant; the heat terms are however negative from the system's point of view, so their values have to be *added* to E_a to obtain E_t. Of course the extent to which each heat term has to be taken into account depends upon how much the coverage term changes with temperature. If both reactants are very strongly adsorbed, the rate of coverage change will be small, in which case $E_a \approx E_t$; the 'orders' a and b will both be zero; but if both are weakly adsorbed, coverage change will be rapid, and the 'orders' both unity; both the heat terms then apply. In 1935, the Russian Scientist M.I. Temkin therefore devised the following equation to cover these and intermediate conditions:

$$E_t = E_a - a\Delta H_A^\circ - b\Delta H_B^\circ. \tag{1.12}$$

This analysis helps to explain one of the biggest mysteries of heterogeneous catalysis, namely *compensation phenomena*. It is often found that when the same reaction is followed over a series of different catalysts, or at different reactant pressures, or when a series of related reactions is used on the same catalyst, there is a correlation between the activation energy and the logarithm of the pre-exponential factor ln A (Equation (1.6)) of the form

$$\ln A = mE + c. \tag{1.13}$$

An example of this is shown in Figure 1.8. It means that if activation energy rises, and the rate in consequence of Equation (1.6) falls, the ln A term is increased in order to *compensate*. It is a simple algebraic consequence of this equation that there must be a temperature T_i at which all the rates in the series are the same, and only below this temperature does a lower activation energy betoken a faster rate. There has been very much discussion in the literature concerning the meaning and significance of this relation, but it has recently become clear that in every case compensation only occurs when *rate* measurements are used, and when therefore *apparent* activation energies are involved. What is of interest is the cause of the

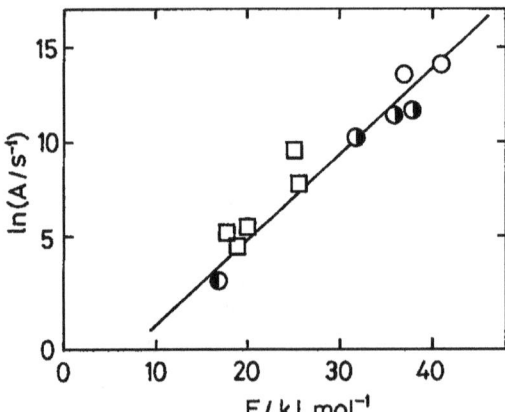

Figure 1.8: A 'compensation' plot showing a linear dependence of ln A upon E: □ Au/TiO$_2$ (*Catal. Today* **36** (2001) 153); ○ Au/TiO$_2$, ◖ Au/Al$_2$O$_3$, ◐ Au/SiO$_2$ (*J. Molec. Catal. A: Chem.* **199** (2003) 73).

variation of the activation energy, rather than the origin of the compensation, and the cause lies in the Temkin Equation (1.12). *Apparent* activation energies alter because of various inputs from heats of adsorption; the true activation energy remains the same.

The basic concepts of chemical kinetics as applied to heterogeneously catalysed reactions have been presented above because of their overriding importance in understanding how they proceed. There are however many other ways in which the structure and composition of adsorbed species can be explored and identified. The applicability of each method depends on the physical structure of the catalyst: for flat surfaces (single crystals, 'model' surfaces) low-energy electron diffraction (LEED) and sum-frequency generation (SFG) are appropriate, while for powdered materials infra-red spectroscopy (FTIR, RAIRS/IRAS) and X-ray absorption fine structure (XAFS) are suitable procedures. The last three methods can be used with dispersed systems such as supported metal catalysts. There is however one danger associated with all of them, namely, that they most easily notice the adsorbed species that are present in the greatest concentration, and since the key reactive species may only be a minor component great care has to be taken to ensure that these are accurately identified, and that their appearance correlates with the rate of reaction. Kinetic analysis is perhaps the only way of gaining direct access to the heart of the reaction, but a word of warning is still necessary; it is impossible to deduce a unique reaction mechanism simply from the reaction kinetic, because (as Karl Popper said) you cannot prove that all other mechanisms are excluded. However the converse is true: *no reaction mechanism can be valid that does not agree with the observed kinetics.*

1.5. The Catalytic Activities of Metals

The kinds of reactions catalysed by various types of solid are determined by the ability of the surface to convert the reactants into adsorbed forms that are conducive to making the desired product. So, for example, the metals of Groups 8–10 are particularly adept at reactions that require the dissociation of hydrogen molecules, i.e. hydrogenation and hydrogenolysis. Metals of Group 11 have the reputation of adsorbing hydrogen only weakly, and they are not therefore versatile catalysts for reactions needing hydrogen atoms. The base metals are useless for oxidations because they so readily become oxidised, and it is only the noble metals of these Groups that are useful oxidation catalysts, and then it is generally for non-selective or deep oxidation. Many transition metal oxides make splendid selective oxidation catalysts, and some of them, and particularly mixtures of them, are renowned for catalysing the selective oxidation of alkenes, alkanes and aromatic molecules. Acidic solids such as silica-aluminas, and especially zeolites, their crystalline analogues, are excellent for catalysing reactions of the carbocation type, which are initiated by protons.

With metals it is possible to drive our understanding further, and to see in a more quantitative way the principles that govern activity. Maximum rates will be found when the catalytic system, that is, the combination of reactants and catalyst, is such that the reactants (i) are so strongly adsorbed that the whole of the surface is utilised, but not so strongly that they are unreactive or poison the surface and (ii) are adsorbed in the forms that are appropriate for forming the desired product. The first consideration leads to the idea of the Volcano Curve (Figure 1.9), which exhibits a maximum rate when the two opposing needs are optimally balanced. This implies that strength of reactant adsorption enters into the picture *twice*, once to determine coverage and then to control reactivity. An additional requirement is for two reactants to be adsorbed with comparable strengths, since the rate depends on the product of their surface concentrations (Equation (1.9)), but this need cannot always be met.

Strength of adsorption is conveniently measured by the heat released when adsorption takes place, and for several molecules there is evidence to show that the strength decreases on passing from left to right across each of the three transition series; with some molecules (hydrogen, alkenes)

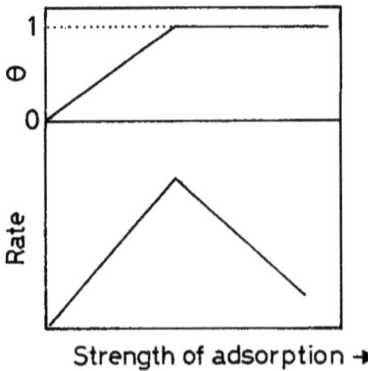

Figure 1.9: Volcano plot showing dependence of rate on strength of adsorption: the upper part shows the corresponding variation of surface coverage θ.

there is little variation within Groups 8–10, and with others (e.g. nitrogen) the ability to be chemisorbed cuts out at Group 8. These trends have been attributed to variation in the number of unpaired d-electrons or vacancies in the metal's d-band; the Group 11 metals, having filled d-levels, therefore fail to be very active in chemisorption. Drawing these concepts together, it is not surprising to find the most active metals for hydrogenation towards the end of each transition series (rhodium takes the prize for ethene hydrogenation but nickel is the most active base metal), with most metals occupying places on the right-hand side of the Volcano Curve (Figure 1.9) and only copper, and perhaps the other Group 11 metals (and possibly manganese) on the other side. Other classes of reaction however show somewhat different behaviours.

For example, maximum activity for alkane hydrogenolysis is to be found in Group 8 (Ru, Os) rather than in Group 9 or 10, because the hydrocarbon intermediates have to be multiply bonded to the surface, and metals in the later Groups have insufficient unpaired electrons for this purpose. Palladium is outstandingly the best metal for hydrogenating alkynes, but is of little use for hydrogenating aromatics.

There is one other important aspect of the catalytic activity of metals to introduce before concluding this brief survey; this is the concept of *structure sensitivity*, which will turn out to be relevant to much of the catalytic chemistry of gold. The idea of the active centre has already been noted (Section 1.3). For some reactions this comprises perhaps only one or two metal atoms, and it does not matter too much what their surroundings are; these reactions are termed *structure-insensitive*. Other reactions seem to require a larger assembly or ensemble of atoms arranged in a quite specific way; they are named *structure-sensitive*. Experimental evidence for the classification of reactions into these groups is of three kinds: (i) variation of specific rate with particle size, which alters the coordination number of surface atoms (particle size sensitivity, see Section 3.4); (ii) dependence of rate on the structure of single-crystal surfaces, including those having straight or kinked steps on them (surface structure sensitivity, see also Section 2.5.2); and (iii) dependence of rate on the composition of bimetallic particles containing one active and one inert metal (e.g. Pd–Ag, see Section 1.6).

1.6. Catalysis in Bimetallic Systems

In the wider field of heterogeneous catalysis, very much use has been made of catalysts containing two or more metals. Some of these have achieved industrial prominence, notably the platinum–iridium and platinum–rhenium combinations in petroleum reforming, and platinum–tin in alkane dehydrogenation, but much academic work has focused on catalysts containing an element of Groups 8–10 plus one of Group 11. The initial motivation for this work, which started in earnest in the middle of the last century, was to determine the importance of electronic structure of a metal in determining its activity,[5] but this was based on the mistaken belief that electrons from the two metals forming the bimetallic system were shared in a common pool. Although this work was misguided,[6,7] results of great interest were obtained, and the electronic theory came to be supplanted by concepts based on the size of the ensemble of the active metal and electronic modification of the active atoms by a ligand effect due to the vicinity of the other metal.[8] The probability of finding an active ensemble of a specified size is a function of the ratio in which the two metals are present, but note has to be taken of the tendency of the component of lower surface energy to segregate preferentially at the surface and liking best to occupy low coordination number sites. There has been much discussion over the years as to the relative importance of the ensemble and ligand effects; in the great majority of cases, the former is more usually predominant.

While in many cases addition of an inactive metal leads immediately to a decrease in activity, in some cases there is an initial increase. This has often been attributed to a decrease in the mean size of the active ensemble, which in turn, in the case of hydrocarbon reactions, minimises the formation of strongly-bonded dehydrogenated species that would lower the rate of the desired reaction, although sometimes improving its selectivity; a possible example of the effect of gold in doing this will be found in Section 13.5. There are, however, several instances of gold improving the activity of palladium in reactions involving only hydrogen (see Section 9.2). Bimetallic catalysts containing gold show activity that is superior to that of either component separately in the synthesis of hydrogen peroxide (Section 8.5), of vinyl acetate (ethenyl ethanoate) (Section 8.4), and in a number of other selective oxidations (Section 8.3). Sound explanations for these effects are not always available, but in some cases it is clear that the role of the gold is to modify favourably the performance of the palladium. It is not feasible to record all the many instances described in the literature[5,9,10] of gold acting purely or predominantly as an inert diluent, whatever benefits this may bring to the active component in terms of higher activity or better selectivity. In the following chapters, attention will to be largely confined to cases where the presence of gold leads to a significant and sustained improvement in performance.

For a further discussion of the structure and properties of bimetallic systems, see Sections 2.6 and 3.2.3; for the preparation of bimetallic catalysts, see Section 4.6; and for the mechanisms by which they work in oxidations, see Section 8.2.2. Most textbooks of physical chemistry have sections on adsorption and catalysis, but they frequently focus on studies made under ultra-high vacuum conditions with single crystal surfaces. While this work produces beautiful pictures, it has limited relevance to the more mundane world of practical catalysis. Other introductory treatments of about the level of this chapter, or slightly more advanced, are available,[5,7,11] as are deeper discussions of the kinetics of catalysed reactions.[12–14] Industrial processes using catalysts have also been described in detail.[15,16]

References

1. G.B. Kistiakowsky and A. Nickle, *Discuss. Faraday Soc.* **10** (1951) 175.
2. S.D. Lin, M. Bollinger and M.A. Vannice, *Catal. Lett.* **17** (1993) 245.
3. G.C. Bond and D.T. Thompson, *Catal. Rev.-Sci. Eng.* **41** (1999) 319.
4. M.A. Bollinger and M.A. Vannice, *Appl. Catal. B: Env.* **8** (1996) 417.
5. G.C. Bond, *Catalysis by Metals*, Academic Press, London, 1962.
6. V. Ponec and G.C. Bond, *Catalysis by Metals and Alloys*, Elsevier, Amsterdam, 1996.
7. G.C. Bond, *Metal-Catalysed Reactions of Hydrocarbons*, Springer, New York, 2005.
8. V. Ponec, *Appl. Catal. A: Gen.* **222** (2001) 31.
9. J. Schwank, *Gold Bull.* **18** (1985) 1.
10. D.T. Thompson, *Platinum Metals Rev.* **48** (2004) 169.
11. G.C. Bond, *Heterogeneous Catalysis: Principles and Applications*, Clarendon Press, Oxford, 2nd edition, 1987.
12. M. Boudart and G. Djéga-Mariadassou, *Kinetics of Heterogeneous Catalytic Reactions*, Princeton University Press, Princeton NJ, 1984.
13. R.A. van Santen and J.W. Niemantsverdriet, *Chemical Kinetics and Catalysis*, Plenum, New York, 1995.
14. K.J. Laidler, *Chemical Kinetics*, 3rd edition, Harper and Row, New York, 1987.
15. R.J. Farrauto and C.H. Bartholomew, *Fundamentals of Industrial Catalytic Processes*, Chapman and Hall, London, 1997.
16. *Catalyst Handbook*, M.V. Twigg, (ed.), 2nd edition, Wolfe, Frome, 1989.

CHAPTER 2

The Physical and Chemical Properties of Gold

2.1. Introduction

We must first explain why it is felt necessary to discuss the *physical and chemical properties of gold* before starting to consider its catalytic abilities. It would be a quite straightforward matter to list these properties without trying to understand how they arise, but if we seek to appreciate their full significance, we must think about how and why they differ from those of its neighbours, especially its antecedents in Group 11. Catalysis is a chemical phenomenon; it involves chemical reactions that proceed on surfaces, and before reaction can happen the molecules must chemisorb by what is to all intents and purposes another chemical reaction. It has long been known that there is a general parallelism between the strengths of chemisorption of simple molecules on metals and the stability of analogous bulk compounds where these exist.[1] This holds both in a qualitative sense, e.g. metals that form nitrides can also chemisorb nitrogen, and also quantitatively, e.g. heats of formation of oxides run parallel to heats of chemisorption of oxygen.[2] Both of these quantities are however in turn dependent on the same physical property of the element, namely, the latent heat of sublimation, modified by the differences in the electronegativities of the metal and the adsorbed atom. Heats of chemisorption are therefore easily estimated for systems where no bulk analogues exist, e.g. for hydrogen;[3] for carbon monoxide there are parallels between its heat of chemisorption and the M–CO bond strength in carbonyl complexes.

Since chemisorption must necessarily precede catalysis, the chemical and physical properties of gold ought therefore to help us to determine not only what reactions it can and cannot catalyse, but also in a quantitative sense what these levels of activity are likely to be. At various times catalytic activity has been associated with either geometric structure or electronic constitution or energetic parameters such as latent heat of sublimation, before it was finally appreciated that these and many other properties of metals are themselves intimately related. So, catalytic activity cannot be ascribed to a single metallic property: it is more likely that a number of factors act in concert to decide the types and strengths of bonds formed with molecules at the surface. *All* the properties of gold therefore become a proper subject for enquiry.

There is one major feature that appears to be of much greater importance with gold than with other metals: its catalytic ability in carbon monoxide oxidation and some other reactions is a steep function of the size of particle responsible. We shall therefore need to examine closely how the properties of gold depend on the size of the assembly of atoms. Fortunately, there is much relevant information to consider and to bear in mind when thinking about catalysis by gold: this is surveyed in the following chapter.

While every element is unique in some respects, gold occupies a position at one extreme of the range of metallic properties, and its legendary chemical inertness is attributable to chemical features that are not surpassed by any other metal. In order to appreciate exactly how outstanding these characteristics are, it will be necessary to contrast them especially with those of the metals that are its neighbours to the left, the right and above in the Periodic Classification. Inorganic chemists have long realised that trends between the elements in the First and Second Transition Series are not continued into the Third Series, and have sought explanations for this. The Lanthanide Contraction (see below) is in fact only partly responsible; of equal importance are the consequences of Einstein's Theory of Special Relativity, but these have only been recognised comparatively recently, and have not yet percolated into many chemistry textbooks. A short rehearsal of the origins of relativistic effects will make it easier to understand the reasons behind the physics and chemistry of gold, and why they differ from those of their neighbours.

2.2. The Origin of Relativistic Effects[4–14]

The Schrödinger wave equation[15] is generally thought to contain, at least in principle, the solution to all chemical problems: unfortunately it is wrong, or at least not universally valid,[16] in the sense that Newton's Laws of Motion are valid but are not applicable to electrons and protons. The problem is that it is non-relativistic, that is, it does not treat space and time in equivalent ways; another way of putting it is to say that it is not Lorentz–Fitzgerald invariant. A relativistic analogue was however devised by P.A.M. Dirac and published in 1928;[17] the Dutch physicist, H.A. Kramers, also developed a similar treatment at about the same time,[18] but his work is less well known.

It is unnecessary to enter very far into the complex mathematics that underlies the Dirac equation, but it is essential to grasp the ideas on which it is based and the conclusions it generates: this can be done in a perfectly straightforward way. For those interested to pursue its logical foundation, there are some semi-popular accounts,[19] one of the best being by F. Wilczek[20] who describes it as 'A Piece of Magic'. A good introduction to the mathematics appears in the 1958 edition of Chambers Encyclopaedia.[21] Dirac's relativistic wave equation is remarkable in many ways: it provides a logical explanation for the existence of both electrons and positrons having opposite spins, since the equation has four solutions.[20] It predicted the positron before it was discovered. It has areas of application far beyond those that concern us here; for example, in nuclear physics the equation is found to apply to all other fermions.[20,22] Its further modification would however lead us into quantum electrodynamics, but luckily these refinements have little *chemical* consequence.[13]

Dirac believed that changing to a relativistic wave equation would not significantly affect the calculated properties of the hydrogen atom. This is correct, but with heavier elements the inner electrons feel the large nuclear charge, and to maintain balance with the strong electrostatic field they must acquire speeds that are comparable with that of light:[7] according to Einstein's Theory of Special Relativity this causes their mass (M) to increase according to the equation

$$M = M_0/(1 - v^2/c^2)^{1/2},$$

where M_0 is their rest mass, v their speed and c is the speed of light. This is the basis of the *relativistic effect*, which exists in all atoms,[7,8] but only becoming significant when the atomic number Z exceeds about 50 (Sn). It increases roughly as Z^2, and for gold ($Z = 79$) and mercury ($Z = 80$) the $1s$ electrons have speeds of about 58% of that of light, and their mass is thereby increased by about 20%. The $1s$ orbital therefore shrinks, and the s orbitals of higher quantum number have to contract in sympathy, in order to maintain orthogonality; in fact the $6s$ orbital shrinks relatively more than the $1s$. The same effect also operates to a lesser extent on the p electrons, but d and f electrons are hardly affected, never coming close to the nucleus due to the centrifugal potential $l(l+1)/r^2$, l being the azimuthal quantum number and r the radius. In addition, their effective potential is more efficiently screened because of the relative contractions of the s and p shells: they therefore increase in energy and move outwards, this effect being called the *indirect relativistic orbital expansion*. This energetic stabilisation of the s and p electrons and destabilisation of the d and f electrons is accompanied by a splitting of those orbitals for which l is greater than zero, an effect that increases greatly with atomic number. This is illustrated by reference to the elements of Group 11 in Figure 2.1, where calculated relativistic orbital energies[23] are compared. The comparison cannot be exact, because element 111 (for which the name *röntgenium* (Rg) has been proposed for consideration by IUPAC), has the electronic structure $6d^9 7s^2$, whereas the earlier elements of this Group are all $nd^{10}(n+1)s^1$. However, the progressive raising of the energy of the $(n+1)s$ level and the increased splitting of the $nd_{3/2}$ and $nd_{5/2}$ levels is quite evident, the main chemical consequence being the greater ease with which d electrons can be engaged in chemical activity as atomic number increases.

The greater similarity between the elements in the Second and Third Transition Series compared to those in the First Series (Table 2.1) was formerly ascribed solely to the Lanthanide Contraction, caused by the failure of the $5d$ and $6s$ shells to occupy the expected space, because the $5f$ electrons do not adequately shield them from the increasing nuclear charge, by reason of the disposition of their orbitals: $5d$ and $6s$ electrons are therefore drawn

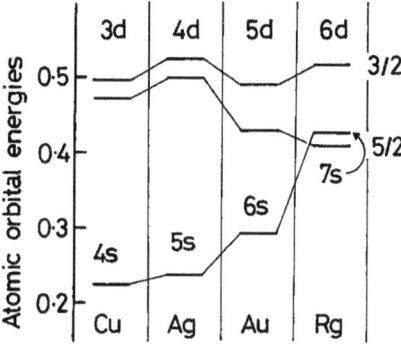

Figure 2.1: Atomic orbital energies for copper, silver, gold and röntgenium ($Z = 111$; Rg): note the electron configuration for Rg is $6d^9 7s^2$, for the others it is $n^{10}(n+1)s^1$.

Table 2.1: Physical properties of gold compared to those of copper and silver.

Property	Cu	Ag	Au
Atomic number	29	47	79
Atomic mass	63.55	107.868	196.9665
Electronic configuration	$[\text{Ar}]3d^{10}4s^1$	$[\text{Kr}]4d^{10}5s^1$	$[\text{Xe}]4f^{14}5d^{10}6s^1$
Structure	fcc	fcc	fcc
Lattice constant(nm)	0.361	0.409	0.408
Metallic radius(nm)	0.128	0.14447	0.14420
Density(g cm^{-3})	8.95	10.49	19.32
Melting temp. (K)	1356	1234	1337
Boiling temp. (K)	2843	2428	3081
Sublimation enthalpy (kJ mol^{-1})	337 ± 6	285 ± 4	343 ± 11
1st ionisation energy (kJ mol^{-1})	745	731	890

towards the nucleus. It is now thought however that the Lanthanide Contraction and the relativistic effect have approximately equal importance, but the latter leads to *selective* effects on the sizes and energies of the various electron shells, these accounting for chemical behaviour that is not otherwise explicable.

The size of the relativistic effect and its variation with nuclear charge has been calculated.[6,7] The relativistic contraction of the $6s$ shell (i.e. the fractional decrease in its actual size compared to its calculated nonrelativistic value) is shown for elements of nuclear charge 50–100 in Figure 2.2.[7,8] Most noticeable is the very sharp minimum at platinum and gold, where the contraction is about 17%; the greater rate of increase beyond $Z = 70$ is caused by the relativistically enhanced loss of nuclear screening ability of both the $5d$ and $4f$ electrons. The fall after gold is due to the decreasing importance of the $6s$ shell contraction in determining size.

The physical and chemical consequences of these effects will be discussed in the following sections. While they are well known to those working on the chemistry of the heavy elements, it is surprising and regrettable that they are not more widely known and appreciated. The Dirac equation is a thing of greater beauty and power, and provides an explanation for spin in a way that Schrödinger cannot do. In fact it has been said that *Spin is nature's way of signalling the correctness of Einstein's Theory of Special Relativity.*[22]

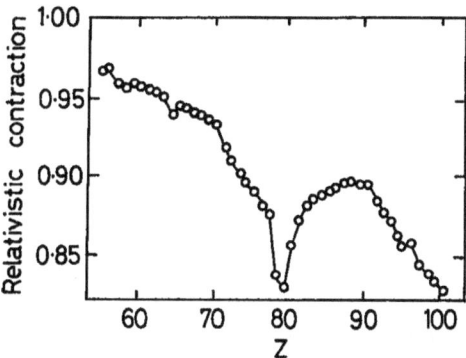

Figure 2.2: Relativistic contraction ($r_{\mathrm{rel}}/r_{\mathrm{nonrel}}$) for the $6s$ orbitals of the heavy elements as a function of the atomic number Z (based on Ref. 50).

2.3. Comparisons of the Chemistry of Gold with that of the Adjacent Elements[24]

This section concerns the chemistry of gold in its non-zero oxidation states,[25] and draws comparisons with adjacent elements, especially those of Group 11, in order to highlight the way in which relativistic effects manifest themselves, and to set the scene for what may be the tendency of metallic gold to form, or not to form, chemisorbed states at its surface. The physical character of metallic gold is considered later; organometallic complexes containing gold atoms or ions are also discussed in the following section.

The chemical properties of platinum, gold and their successors are dominated by the relativistic stabilisation of the $6s$ level.[6,7,26–32] The $6s^2$ 'inert pair effect' is well known to inorganic chemists, although many textbooks manage to discuss it without reference to its origin. It is not a result of the Lanthanide Contraction, which arrests but does not reverse the usual trends caused by increasing atomic mass. A further important consequence is that electrons in the $5d$ level become more easily mobilised for chemical reaction (see Figure 2.1): thus the electronic configuration of platinum is $5d^9 6s^1$, while that of palladium is $4d^{10}$, and this explains why the Pt^{IV} state is so much more easily available than Pd^{IV}. There are similar differences between rhodium and iridium. In Group 12 the Hg^I state in the form of the dimeric ion Hg_2^{2+} is not replicated in zinc or cadmium, and the lower oxidation states of thallium (Tl^I) and lead (Pb^{II}) also epitomise the stability of the $6s^2$ pair.

In the case of gold ($5d^{10} 6s^1$), its chemistry is determined by (i) the easy activation of the $5d$ electrons, as discussed above and (ii) its desire to acquire a further electron to complete the $6s^2$ level and not to lose the one it has. This latter effect awards it a much greater electron affinity and higher first ionisation potential than those of copper or silver (see Table 2.1), and accounts for the ready formation of the Au^{-I} state (see below).[4] The former effect obviously explains the predominance of the Au^{III} state, which has the $5d^8$ configuration (even the Au^V state $(5d^6)$ is accessible as in AuF_5[29]), the Au^I state being of lesser importance and the Au^{II} state being unknown except in a few unusual complexes.[33,34] Gold's electronegativity (2.4) equals that of selenium and approaches that of sulfur and iodine (2.5);[25] it is frequently said to have therefore some of the properties of a halogen. Its electrode potential $(E^0 = +1.691\,V)$ is also extremely high for a metal. Its electronic structure determines its nobility, and its inability in the massive form to interact with oxygen or sulfur compounds, i.e. to tarnish as silver and copper do, is in line with the instability of its oxide Au_2O_3, which decomposes at about 433 K and probably has a positive heat of formation; values between -3.7 and $+160\,kJ\,mol^{-1}$ are quoted in the literature.[35,36] The sulfides Au_2S and Au_2S_3 are known, but are of limited stability and importance.[25]

The electronic state of gold atoms in the massive state is not however exactly that of the free atom, because a weak white line on the leading edge of the L_{III} X-ray absorption edge (Section 3.3.2) signifies a small number of holes in the d-band caused by d–s hybridisation.[37,38]

Considerable interest has been shown in the Au^{-I} oxidation state.[4,14,33,39] Gold dissolves in solutions of the heavier alkali metals in liquid ammonia,[25] and the auride ion Au^- is formed; the electrical conductivity of caesium-gold alloys at 873 K shows a very sharp minimum at the 1:1 ratio, and the solid CsAu is regarded as a semiconductor.[4] It has the NaCl structure. Historically, the first auride to be prepared was $BaAu_2$ in 1938, although as long ago as 1923 Partington reported[24] that a eutectic point had been found in the phase diagram of the sodium–gold system at 1262 K, corresponding to the composition $NaAu_2$. Other alkali metal aurides are known, and the Au^- ion can be cryptated.[4] More recently, series of ternary oxides containing Au^{-I} have been discovered: these include M_3AuO (M = K, Rb, Cs), and compounds containing both Au^0 or Au^I and Au^{-I}, viz.

$$M_5Au_3O_2 = 4M^+[M^+Au^-][Au^0O]_2^{2-},$$
$$M_7Au_5O_2 = 3M^+[M^+Au^-]_4[Au^IO_2]^{3-}.$$

Their catalytic properties await investigation. Several other compounds containing Au^{-I} are known; tetramethylammonium auride is isostructural with the bromide,[40] and the deep blue addition compound $CsAu \cdot NH_3$ has recently been prepared and characterised.[4] The electron affinity of platinum is almost as large as that of gold, and the platinide ion Pt^{2-} has been formed as the compound Cs_2Pt, the structure of which has been established.[41]

Finally, we may note the existence of compounds of gold, which cannot be prepared and put in a bottle, but whose ephemeral character may imitate transient species formed in catalytic processes. These include the hydrides AuH_3 (i.e. $HAu(H_2)$) and AuH_5 (i.e. $H_3Au(H_2)$) which have been seen in low-temperature matrices, and $AuXe^+$ and $AuXe_2^+$ which have been detected by mass-spectrometry.[4] The compound $[AuXe_4][Sb_2F_{11}]_2$ has however actually been made.

2.4. The Aurophilic Bond[4, 14, 32–34, 42]

In numerous complexes containing two or more Au^I ions, it is generally observed that distances between pairs of such ions are unusually short (275–350 pm), and that some form of bonding must therefore exist between them. The effect is termed *aurophilic attraction* or *aurophilicity*. It is also observed when the gold ions are in different molecules that pack closely together in the solid state, and when they are located at opposite sides of a ring formed with bidendate ligands (transannular attraction). There is an enormous literature on gold complexes in which the effect occurs, and this has been reviewed.[34]

Aurophilicity has also been extensively studied by theoretical methods. It appears that the bond is due to dispersion forces of the type that hold molecules together in a liquid or solid, but very much stronger than normal van der Waals forces; it has the same kind of strength as the hydrogen bond in water and alcohols, and takes values between 10 and 100 kJ mol^{-1}, depending on the separation between the atoms. Other pairs of ions having a closed d^{10} or $d^{10}s^2$ configuration (e.g. Ag^I, Tl^I) show similar but smaller effects, and so one should perhaps employ *metallophilicity* as the generic term. The complex $CsAu \cdot NH_3$ also contains short Au–Au bonds (302 pm), which is not surprising because Au^{-I} also has a closed electron shell ($5d^{10}6s^2$).

2.5. Physical Properties of Gold and Adjacent Elements

2.5.1. Bulk properties

In continuance of our quest for the source of the surprising and unexpected catalytic activity shown by gold, we must now consider its physical properties in relation to those of its neighbours. We will deal first with its properties in the massive state, and later (in Chapter 3) consider how they change when the metal is in the form of small particles.

Gold crystallises in the face-centred cubic (fcc) habit, its lattice constant being fractionally smaller than that of silver[26,43,44] (Tables 2.1 and 2.2); in compounds and complexes, Au^I is smaller than Ag^I. This is in consequence of the relativistic contraction of the $6s$ level, and it is expected that it will be even greater with the $7s$ level; indeed the size of the next element of Group 11 (röntgenium) has been calculated to be no larger than that of copper. While in some respects the properties of gold reflect its greater atomic mass compared to copper and silver (e.g. density), in many the trend is reversed; thus its melting point and heat of sublimation are almost the same as that of copper (see Tables 2.1 and 2.2), the greater strength of the Au–Au bond being a consequence of its shorter than expected length. There is only one naturally occurring isotope of gold, so the atomic mass is known very precisely (196.9665). It has a non-zero nuclear spin quantum number ($I = 3/2$) and its nucleus is therefore 'magnetic', but its receptivity relative to the proton is only 2.77×10^{-5}, so it is a hard nucleus to study[45,46] (receptivity is proportional to the natural abundance of the active nuclide and the Larmor frequency, and to $I(I + 1)$). It also has a large quadrupole moment, which leads to line broadening, so very refined equipment is needed

Table 2.2: Physical properties of gold compared to those of platinum and mercury.

Property	Pt	Au	Hg
Atomic number	78	79	80
Atomic mass	195.08	196.9665	200.59
Electronic configuration	$[Xe]4f^{14}5d^96s^1$	$[Xe]4f^{14}5d^{10}6s^1$	$[Xe]4f^{14}5d^{10}6s^2$
Structure	fcc	fcc	A10
Lattice constant(nm)	0.392	0.408	0.299
Metallic radius(nm)[a]	0.1385	0.14420	0.151
Density($g\,cm^{-3}$)	21.41	19.32	13.53
Melting temp.(K)	2042	1337	234.1
Boiling temp.(K)	4443	3081	630
Sublimation enthalpy($kJ\,mol^{-1}$)	469 ± 25	343 ± 11	59.1 ± 0.4
First ionisation energy($kJ\,mol^{-1}$)	866	890	1007

[a]In 12-coordination.

for its study, and the consequential absence of hyperfine structure means that the NMR of gold is of limited diagnostic use to chemists.

Its optoelectronic properties are also unpredictable by extrapolation from its antecedents in Group 11. Its electrical resistivity is greater than that of silver (see Table 2.1), and its colour more closely resembles that of copper; its optical absorption in the visible region of the spectrum is due to the relativistic lowering of the gap between the 5d band and the Fermi level, without which it would be white like silver and have the same propensity to tarnish and corrode.[27] Polycrystalline gold surfaces have been characterised by Auger electron spectroscopy (AES).

Gold is extremely malleable: 1 g can be beaten into a foil of area $\sim 1\,m^2$, the thickness of which is less than 250 atomic diameters. The same amount can also be drawn into 165 m of wire that is 20 μm in diameter.[25] These characteristics, together with many others, were discussed in detail in a lengthy but fascinating paper by Michael Faraday in 1857.[47]

Gold forms alloys and intermetallic compounds with many other elements[48] (Section 2.6). It has no apparent ability to dissolve or occlude simple gases, although there is indirect evidence that hydrogen atoms can diffuse through it if formed on its surface by dissociation of molecules.[49]

2.5.2. The structure of single-crystal surfaces[50–55]

When a cut is made through a single crystal of a metal parallel to one of the layers of atoms, a surface is exposed that contains atoms, the location of which is defined by a *Miller index*. For a fcc metal such as gold, there are three low-index planes that are usually considered (see Figure 2.3). In the case of the (110) surface, the atoms in the trench have a coordination number of 11, so they make only a small contribution to surface properties. The arrangement of atoms produced instantaneously after the cut is not however necessarily the most stable one. Surface atoms experience a net force acting inwards, which gives rise to *surface tension* or *specific surface free energy*; this strain produces changes to the standard interatomic distance both parallel and normal to the surface, so it is not confined just to the surface layer. A 'surface' can therefore be several atom layers deep.

If the cut is made at a slight angle to the plane, the resulting surface comprises a series of flat terraces separated by steps of monatomic height.[51,52] These *stepped surfaces* have been widely investigated, because of their supposed closer resemblance to the small metal particles found in

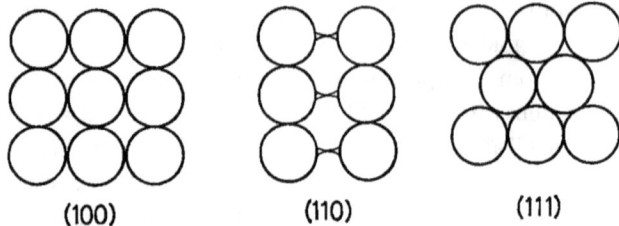

(100)   (110)   (111)

Figure 2.3: Structures of the surfaces of gold having the Miller indices (100), (111) and (110).

practical catalysts. If the cut is made at an angle to *two* planes, the steps are not straight but *kinked*. These surfaces contain atoms having a greater variety of coordination number, and sites on, above or below such steps have been advanced to explain aspects of catalytic behaviour.

In certain cases surfaces experience a more profound *reconstruction* than the quite modest modification to bond lengths referred to above.[50,56] The driving force for any change to the surface is the desire to minimise the total energy of the system, and sometimes there are alternative configurations that are much more stable than that exposed by the first cut. Most frequently such far-reaching rearrangements only occur after the chemisorption of molecules covering most of the surface; this is *adsorbate-induced reconstruction*. With iridium, platinum and gold, however, substantial changes take place spontaneously in the absence of an adsorbate: these are more marked with the somewhat open (100) and (110) planes, the (111) plane already being almost as stable as possible. The fact that only the $5d$ metals, and not those having $3d$ or $4d$ outer shells, reconstruct in this way suggests that relativistic effects are again at work:[13,50] the greater participation of the $5d$ electrons in interatomic bonding raises the latent heat of sublimation, so that surface restructuring gives a greater lowering of the surface energy.

We should briefly consider the changes that happen to the low Miller index planes of gold; these have been deeply researched, and are quite complex, so a simple summary must suffice. Gold is the only element the (111) surface of which reconstructs under UHV conditions.[50] The new structure is described by a complex stacking-fault-domain model in which there are areas (or domains) of both fcc and cph (close-packed hexagonal) structure; its Miller index is $(23 \times \sqrt{3})$, and 23 atoms occupy positions that would normally be taken by 22 atoms, and in consequence the new surface is

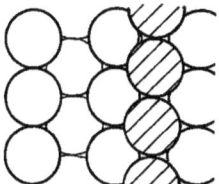

Figure 2.4: 'Missing row' reconstruction of the fcc(110) surface to make larger areas of (111) structure; the resulting form is designated (110) (2×1).

slightly corrugated. The net effect is therefore to lower the surface energy even further by squeezing in more atoms.

The Au(110) surface also reconstructs even under vacuum to form what is termed the *missing row structure* Au(110)(1 × 2). In the (110) surface the sides of the trenches contain triangles of atoms (Figure 2.3), so it can be regarded as a highly stepped (111) surface. The missing row structure is produced by moving one row to fill an adjacent trench (Figure 2.4), and this increases the extent of the (111) microfacets, and so lowers surface energy. The open (100) surface also reconstructs to give a surface layer of (111) geometry; this is denoted as Au(100)(5 × 1), although other designations have been suggested.

Chemisorption acts to relieve surface strain by utilising the free valences that emerge from the surface, and in the case of the 5*d* metals this can reverse the reconstruction either partially or completely. This is however much less evident with gold than with the other metals, due to its reluctance to participate in chemisorption.

Surface melting is observed close to the melting point; this understandably takes place at a lower temperature on the (100) surface than on the (111). Stepped surfaces rearrange to a more highly facetted and stable form before melting occurs.

The volume containing Ref. 56 should be consulted for other articles giving very detailed accounts of surface reconstruction.

2.6.   Bimetallic Systems Containing Gold

It is important to use a clear and consistent terminology in this area. The term 'alloy' is reserved for those cases, most commonly met when the metal is in the massive state, where the two components can be shown to be mixed

at the atomic level. Where this is not possible, as in the case of small particles, the combination is described as 'bimetallic'. A two-dimensional surface
'alloy' can be formed when the mutual solubility of the components is very
limited: a good example of this is the ruthenium–gold system, but osmium,
and the other Group 11 metals, form similar bimetallic systems.[53,57]

Gold in the massive state forms a continuous range of solid solutions
with nickel and with palladium,[58,59] but not with platinum, where there
is a misciblilty gap between 18 and 98% gold, wherein the intermetallic
compounds Pt_3Au, $PtAu$ and $PtAu_3$ are formed; there could be no clearer
expression of the influence of the relativistic effect on the behaviour of
two metals (Pd and Pt) that are in many ways so similar. Relatively little
use has been made of single crystals of alloys in catalytic studies, possibly
because the composition of the surface layer is not always the same as that
of the bulk, although sections cut in various ways through intermetallic
compounds reveal the components disposed in various ways, and these have
occasionally been used. Two factors are at work: (1) the equilibrium surface
in vacuo will be enriched with the component having the lower surface
energy (i.e. gold), as this minimises the energy of the whole system; on small
particles or rough surfaces, the gold atoms preferentially occupy sites of
low coordination number, because in those positions they have the greatest
effect on lowering the surface energy. (2) This effect is however counteracted
by the tendency of strongly adsorbed species to withdraw the active metal
to the surface;[60] the composition of the working surface cannot therefore
be predicted with precision.

In discussions of the behaviour of bimetallic systems it is generally
agreed that, provided the electronic structures of the two components are
not greatly different, mutual modification by a ligand effect is rarely important, and the ensemble size effect is usually dominant (see however Ref. 61);
this may not be true when the electronic structures are very different, as
for example with intermetallic compounds of the type $CrNi_2$, $HfIr_3$, Ce_2Ni
or $ZrPd_3$.[53,57] Results obtained by depositing thin films of gold onto the
surface of a single crystal metal of Groups 8–10 however cast some doubt
on this view; they imply some degree of electronic interaction, which may
originate in the epitaxial contact causing a lattice expansion or contraction,
with a consequent shift in the energies of the valence states.[50] A similar
effect may operate with massive alloys, where electron spectroscopy shows
that each element retains its identity to a large degree; nevertheless lattice parameters change in accordance with Vegard's law, namely, linearly
with composition. The effect on catalytic activity of subtle alterations to

the energies and occupancies of orbitals in which the evidence electrons are located is not however easily predicted.

The usefulness of bimetallic systems in catalytic studies was mentioned in Section 1.6, and their preparation is surveyed in Sections 3.2.3 and 4.6; their beneficial application to selective oxidation of organic molecules is particularly stressed in Sections 8.3–8.5.

References

1. K. Tanaka and K. Tamaru, *J. Catal.* **2** (1963) 366.
2. G.C. Bond, *Heterogeneous Catalysis: Principles and Applications*, 2nd edn., Clarendon Press, Oxford, 1987.
3. G.C. Bond, *Catalysis by Metals*, Academic Press, London, 1962.
4. P. Pyykkö, *Angew. Chem. Int. Ed.* **41** (2002) 3573.
5. K.S. Pitzer, *Acc. Chem. Res.* **12** (1979) 271.
6. P. Pyykkö and J.-P. Desclaux, *Acc. Chem. Res.* **12** (1979) 276.
7. P. Pyykkö, *Chem. Rev.* **88** (1988) 563.
8. K. Balasubramanian, *Relativistic Effects in Chemistry*, Wiley, New York, 1997.
9. P. Pyykkö, *J. Am. Chem. Soc.* **117** (1995) 2067.
10. N. Kaltsoyannis, *J. Chem. Soc. Dalton Trans.* (1997) 1.
11. L. Pisani, J.-M. André, M.-C. André and E. Clementi, *J. Chem. Educ.* **70** (1993) 894.
12. P. Strange, *Relativistic Quantum Mechanics*, Cambridge Univ. Press, Cambridge, 1998.
13. G.C. Bond, *J. Molec. Catal. A: Chem.* **156** (2000) 1; *Platinum Metals Rev.* **44** (2000) 146.
14. P. Pyykkö, *Angew. Chem. Int. Ed.* **43** (2004) 4412.
15. A.I. Miller, in *It Must be Beautiful: Great Equations of Modern Science*, Granta Books, London, 2002, p. 80.
16. G.C. Bond and E.L. Short, *Chem. and Ind.*, June 3 2002, p. 12.
17. P.A.M. Dirac, *Proc. Roy. Soc. A* **117** (1928) 610; **118** (1928) 351; **123** (1929) 714.
18. T. der Haar, *Masters of Modern Physics: The Scientific Contributions of H.A. Kramers*, Princeton Univ. Press, Princeton NJ, 1998.
19. B. Pippard, *Proc. Roy. Inst.* **69** (1998) 291.
20. F. Wilczek, in *It Must Be Beautiful: Great Equations of Modern Science*, G. Farmelo, (ed.), Granta Books, London, 2002, p. 102.
21. W. Wilson, in *Chambers Encyclopaedia*, New edn., Newnes, London, 1959, Vol. 11, p. 397. This article also presents the historical and philosophical background to quantum theory and wave mechanics in a readable way.
22. J. Maddox, *What Remains to be Discovered?* Macmillan, London, 1998, p. 73.
23. J.P. Desclaux, *Atom. Data Nucl. Data* **12** (1972) 311.
24. J.R. Partington, *A Comprehensive Treatise of Inorganic and Theoretical Chemistry*, Vol. III, Longmans Green, London, 1923, p. 491.
25. N.N. Greenwood and A. Earnshaw, *Chemistry of the Elements*, 2nd edn., Butterworth-Heinemann, Oxford, 1997.
26. G.C. Bond and D.T. Thompson, *Catal. Rev.-Sci. Eng.* **41** (1999) 319.
27. A.H. Guerrero, H.J. Fasoli and J.L. Costa, *J. Chem. Educ.* **76** (1999) 200.
28. M. Bardaji and A. Laguna, *J. Chem. Educ.* **76** (1999) 201.

29. N. Bartlett, *Gold Bull.* **31** (1998) 22.
30. P. Schwerdtfeger, M. Dolg, W.H.E. Schwarz, G.A. Bowmaker and P.D.W. Boyd, *J. Chem. Phys.* **91** (1989) 1762.
31. P. Pyykkö, *Inorg. Chim. Acta* **358** (2005) 4113.
32. P. Pyykkö, *Gold Bull.* **37** (2004) 136.
33. M.C. Gimeno and A. Laguna, *Gold Bull.* **36** (2003) 83.
34. H. Schmidbaur, *Gold Bull.* **33** (2000) 3; **23** (1990) 11.
35. S.J. Ashcroft and E. Schwarzmann, *J. Chem. Soc. Faraday Trans. I* **68** (1972) 1360.
36. J. Chevrier, L. Huang, P. Zeppenfeld and G. Comsa, *Surf. Sci.* **355** (1996) 1.
37. F.W. Lytle, P.S.P. Wei, R.B. Gregor, G.H. Via and J.H. Sinfelt, *J. Phys. Chem.* **70** (1979) 4849.
38. L.H. Matheiss and E. Dietz, *Phys. Rev. B* **22** (1980) 163.
39. A.-V. Mudring and M. Jansen, *Angew. Chem. Int. Ed.* **39** (2000) 3066.
40. P.D.C. Dietzel and M. Jansen, *Chem. Commun.* (2001) 2208.
41. A. Karpov, J. Nüss, V. Wedig and M. Jansen, *Angew. Chem. Int. Ed.* **42** (2003) 4818.
42. F. Mendizabal and P. Pyykkö, *Phys. Chem. Chem. Phys.* **6** (2004) 900.
43. G.C. Bond, *Catal. Today* **72** (2002) 5.
44. A. Bayler, A. Schier, G.A. Bowmaker and H. Schmidbaur, *J. Am. Chem. Soc.* **118** (1996) 7006.
45. M. Tokita and E. Haga, *J. Phys. Soc. Japan* **50** (1981) 482.
46. K. Zangger and I.M. Armitage, *Metal-Based Drugs* **6** (1999) 239.
47. M. Faraday, *Phil. Trans.* **147** (1857) 145; see also W.D. Mogerman, *Gold Bull.* **7** (1974) 22.
48. W. Rapson, *Gold Bull.* **29** (1996) 141.
49. G.C. Bond, *Surf. Sci.* **156** (1985) 966.
50. R. Meyer, C. Lemire, Sh.K. Shaikhutdinov and H.-J. Freund, *Gold Bull.* **37** (2004) 72.
51. R.I. Masel, *Principles of Adsorption and Reaction on Solid Surfaces*, Wiley, New York, 1996.
52. G.A. Somorjai, *Introduction to Surface Chemistry and Catalysis*, Wiley, New York, 1994.
53. G.C. Bond, *Metal-Catalysed Reactions of Hydrocarbons*, Springer, New York, 2005.
54. J.W. Niemantsverdriet, *Spectroscopy in Catalysis*, VCH, Weinheim, 1993.
55. K. Christmann, *Introduction to Surface Physical Chemistry*, Steinkopff, Darmstadt, 1991.
56. J.C. Campuzano, in *The Chemical Physics of Solid Surfaces*, D.A. King and D.P. Woodruff, (eds.), Elsevier, Amsterdam, 1994, Vol. 7, p. 75.
57. V. Ponec and G.C. Bond, *Catalysis by Metals and Alloys*, Elsevier, Amsterdam, 1996.
58. J. Schwank, *Gold Bull.* **18** (1985) 2.
59. E.G. Allison and G.C. Bond, *Catal. Rev.* **7** (1977) 233.
60. G. Maire, L. Hilaire, P. Legaré, F.G. Gault and A. O Cinneide, *J. Catal.* **44** (1976) 293.
61. Y.L. Lam, J. Criado and M. Boudart, *Nouv. J. Chem.* **1** (1977) 461.

CHAPTER 3

Physical Properties and Characterisation of Small Gold Particles

3.1. Overview

While the surface of massive gold exhibits some modest catalytic properties, and it is important to realise this, with highly dispersed forms the activity per unit mass of metal will be dependent *inter alia* upon the fraction of atoms at the surface, and this will clearly increase as the mean particle size is made smaller. This fraction is termed the *degree of dispersion*, and the way in which this varies with particle size can easily be calculated, assuming unit mass is converted into structureless spheres all of the same size.[1-3] The way in which dispersion, surface area and the number of atoms per particle depends on mean size is shown in Figure 3.1; this shows for example that a 2 nm particle has about 60% of its atoms at the surface. Essentially the same results are obtained if the particles are taken to be cubes exposing five faces (Figure 3.2). Such models cease to be applicable to very small particles, because the approximation to a sphere or cube is no longer valid. Instead it becomes of interest to see how the coordination number of surface atoms varies with size for particles of specified shape;[1,4] an example of this is shown in Figure 3.3, where results for cubo-octahedra are shown. Much use has been made of models of this kind for trying to understand how catalytic activity depends on particle size, but it has to be remembered that these calculations relate to *perfect* structures, i.e. to particles containing exactly the necessary number of atoms to complete the outer layer. They are thus statistically improbable, and it is far more likely that a real catalyst will contain particles in which the outside is to some extent incomplete. Further calculations have shown that the incidence of atoms with a chosen coordination number oscillates wildly as a further layer of atoms is added, and no two iterations of the procedure give the same results; moreover new coordination numbers appear and disappear. A better

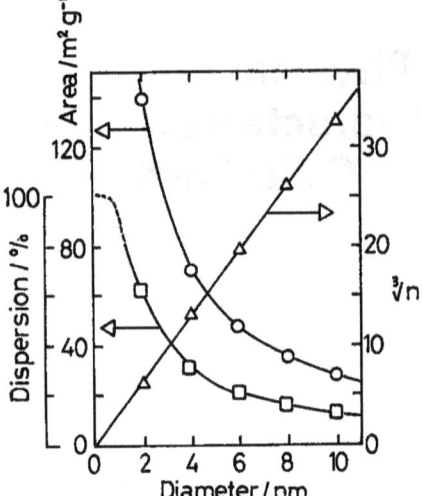

Figure 3.1: Dependence of surface area ($\bigcirc$), dispersion ($\square$) and $\sqrt[3]{n}$ (n is the number of atoms per particle) ($\triangle$) on particle diameter for uniform spheres of gold.

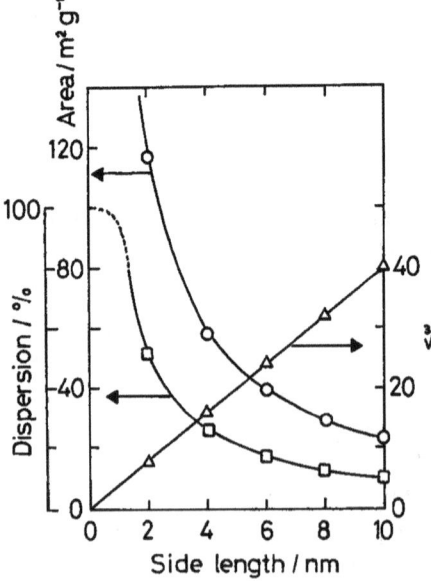

Figure 3.2: The same for uniform cubes exposing five faces.

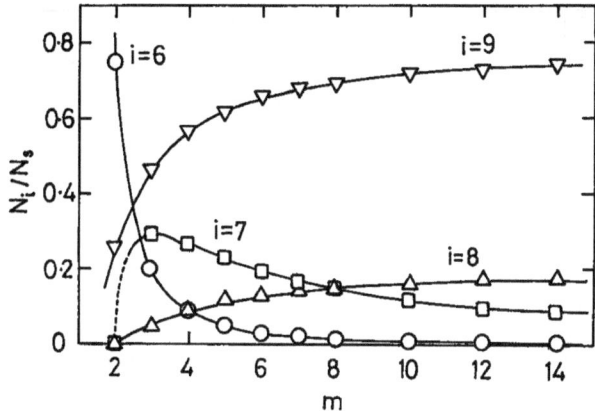

Figure 3.3: Fraction of surface atoms on perfect cubo-octahedra having the coordination numbers indicated as a function of the number of atoms m along each side.

alternative is to use the *free valence dispersion* D_{fv} defined as

$$D_{fv} = \sum (12 - c)/12n, \tag{3.1}$$

where c is the coordination number of each individual atom and n is their total number. This is less dependent on the exact sequence in which a new layer is constructed; it declines smoothly with increasing size for perfect cubo-octahedra as shown in Figure 3.4.

A word on *terminology* is in order. Physicists, and those that way inclined, use the term 'cluster' for small aggregates of atoms formed in the vapour phase, and generally having fewer than about 20 atoms: this term is retained. Organometallic chemists refer to polynuclear complexes stabilised by suitable ligands, often phosphines, as cluster compounds; these contain typically fewer than about 12 atoms, although stable symmetrical assemblies of 13, 39 and 55 atoms similarly stabilised, are also known (see below). Although it is swimming against the tide, the term 'small particle' is preferred to 'nanoparticle'; such particles have also been described as 'nanocrystal gold molecules'.[5]

Small particles are energy-rich, because the work that has to be done in subdividing a large piece of metal is used to break chemical bonds, and this energy ends up in the free valences or dangling bonds thereby created. As size is decreased, the proportion of atoms at or close to the surface that are incompletely bound to neighbours therefore increases; the lower the *mean*

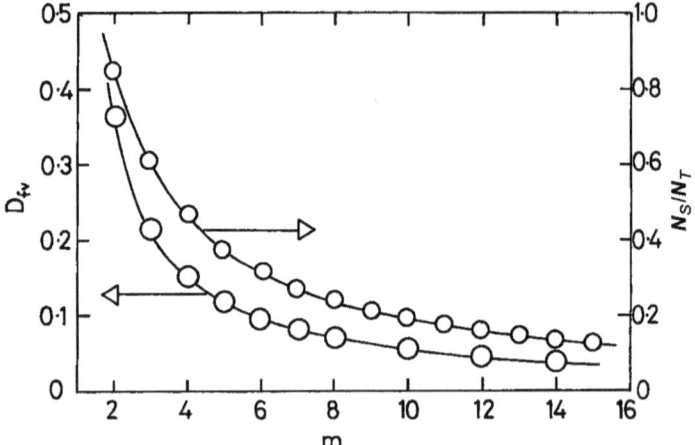

Figure 3.4: Dispersion (N_S/N_T) and free-valence dispersion (D_{fv}) for cubo-octahedra: dependence on number of atoms m per side (see text for definition of D_{fv}).

coordination number of atoms in a particle, the more unemployed orbitals there will be, and the greater will be their effect on the particles' properties (Figure 3.4). The change in a physical property is therefore often simply related to the inverse of the particle size.

There are many ways in which small metal particles can be created and examined (Section 3.2). When the gold particles are supported, the first step is to determine their mean size and size distribution; for this there is no real substitute for transmission electron microscopy (TEM). The various energetic and electronic properties then need to be examined, and the bases of the available experimental techniques will be briefly rehearsed in Section 3.3. Of particular interest is the point at which the change from metallic to nonmetallic behaviour occurs as size is decreased, because this corresponds very roughly to the point at which catalytic activity (at least for oxidation of carbon monoxide) starts to rise dramatically. Relevant experimental results and theoretical speculations are reviewed in Section 3.4.

3.2. Ways of Preparing Small Particles of Gold

3.2.1. Introduction[6]

The next three sections provide brief descriptions of ways of making clusters and small *unsupported* gold particles. Their physical and catalytic properties can sometimes be observed without their being supported, but for many purposes their deposition onto a support facilitates their examination; ways of doing this are considered further in Chapter 4. There are six main ways of preparing such particles. (1) Formation of 'clusters' in the gas phase, although what can be done with them is quite limited. (2) Deposition of such clusters onto a support, affording products of greater utility (Section 4.4.5). (3) Deposition of atoms onto a support, whereupon they aggregate and form a *model catalyst*. (4) Decomposition of a volatile compound at or near the surface of a support, this procedure being known as *chemical vapour deposition* (CVD) (Section 4.3.4). (5) Preparation of colloidal gold: this can also be made to adhere to a support (Section 4.3.6). (6) Traditional and novel methods for making supported metal catalysts (see Sections 4.1–4.4). These last methods start by creating chemical bonds between a gold precursor species and the support, and so the way of attachment of the metal may differ from that which applies when zero-valent atoms or particles are used.

3.2.2.   Gaseous clusters[7–9]

The preparation of gaseous metal clusters is a skilled occupation, requiring complex and expensive equipment; early work has been reviewed.[10] Originally the metal vapour was generated in a high-temperature oven, and expansion into high vacuum through a small orifice formed a supersonic, 'molecular beam'; cluster size declined exponentially towards higher size. This did not work for Transition Metals because their vapour pressure is too low: a high-power pulsed laser is now used to produce atomic metal vapour that can then be co-expanded into an inert carrier gas. A recent paper[11] summarises a procedure as follows: metal ions are transported by ion-optics through diffusion-pumped reaction chambers, deflected through $90°$ by a quadrupole deflector, and mass-selected by a quadrupole mass filter. They are then deposited onto a support, often an oxide single crystal at low kinetic energy ($< 0.2\,\text{eV atom}^{-1}$) under UHV conditions, a procedure termed 'soft landing'. Mass-selected ions have also been obtained by supersonic expansion of a cold ($40\,\text{K}$) laser-generated metal plasma, and then deposited gently on $MgO/Mo(100)$.[12]

3.2.3.   Colloidal gold

The notion of using the colloidal route to create small metal particles for catalytic use has considerable merit. In principle the mean size and size distribution can be controlled by appropriate choice of conditions independent of the influence of a surface, and depositing colloidal particles onto a support is not a problem,[13,14] unless it is necessary to remove a stabiliser by thermal treatment. A platinum colloid was first used to make supported catalysts in 1952.[15] Gold is outstanding in its ability to form colloidal dispersions having attractive colours and great stability:[16–23] one prepared by Michael Faraday[24] lasted until it was destroyed by thoughtless enemy action during World War II. Early attempts[25] to exploit the versatility of the preparation to study the effect of particle size on catalytic activity were not very successful, because the particles used were too large, and the reactions inappropriately chosen. The peculiar catalytic power of gold was not then appreciated.

Gold in its colloidal state has been the subject of intensive study that continues unabated, largely by materials scientists who find it a tractable material to test, both theoretically and experimentally, the properties of matter in a highly subdivided form.[26] A casual glance at the Abstracts section of any recent issue of Gold Bulletin will demonstrate the scale of the interest, but it is of little value to review the subject in depth, because of its limited relevance to catalysis, and because significant reviews are already available.[16–19,26] We concentrate therefore on those facets that have actual or potential bearing on the catalytic properties of gold.

The advantage of using the colloidal route for preparing supported gold catalysts (Section 4.3.6) lies in the way that conditions of preparation can be manipulated to give particles having a narrow size distribution about the desired mean, which can if wished be very small; particle shape can also be controlled in this way. The mean size and size distribution are governed by the relative rates of nucleation and growth.[27] The rapid creation of a large number of small nuclei, with termination of this process before growth starts, leads to a narrow size distribution of small particles.[28] The fewer the nuclei, the larger will be the mean size.[29] Initiation of growth before nucleation is finished leads to a broad size distribution.

The pioneering work by John Turkevich and his associates in the 1950s[25,30,31] used sodium citrate to reduce the $AuCl_4^-$ ion using the standard concentration (2.6×10^{-4} mol l^{-1}) led to particles of mean size about 20 nm, but doubling this gave much faster nucleation and more of them, with the expected consequences of obtaining smaller size. Halving the concentration had the opposite effect. Particles of this kind of size are not however single crystals, but are singly or multiply twinned. Nuclei are not formed by coalescence of single atoms formed by reduction of the $AuCl_4^-$

anion, but ill-defined complexes or polymers of gold atoms or ions with the reductant or its oxidation product (acetonedicarboxylic acid) are first made. Their reduction gives nuclei that are in the 1–2 nm range.[25]

Very many other reducing agents have been used.[1, 16, 21, 23, 32–34] Faraday used phosphorus,[20,24] and indeed reduction of the $AuCl_4^-$ ion by white phosphorus in diethylether gives 5 nm particles. Some are capable of making even smaller particles (e.g. 1–3 nm); these include sodium thiocyanate, poly(ethylene-imine), tetrakis[hydroxylmethyl]phosphonium chloride ($[(HOCH_2)_4P]^+Cl^-$)[1,19] and sodium borohydride (sodium tetrahydridoborate, $NaBH_4$).[35,36] Irradiation by X-rays or accelerated electron pulses generates solvated electrons, which easily reduce metal ions to atoms; reduction of $HAuCl_4$ in the presence of stabilising agents (CN^-, EDTA), or polymers that do not initiate reduction before the irradiation, gives particles the size of which becomes smaller and the distribution narrower as the dose rate is increased.[37,38] The Purple of Cassius produced by reduction of $HAuCl_4$ with stannous chloride has long been known.[20,39]

Stabilising agents are not essential for producing gold colloids,[40] although they are often used to promote longevity; commonly used materials include polyvinylalcohol (PVA),[41,42] polyvinylpyrrolidone (PVP), polydiallyldimethylammonium chloride (PDDA), and ethyltrimethylammonium bromide (CTAB).[27] Care in their use is necessary because they sometimes contain elements that are inimical to catalysis (S, P). With citrate reduction, the oxidation product acts as stabiliser. Oleylamine both reduces and stabilises.[34] Small particles are spherical, but larger particles (20–50 nm) acquire specific shapes and definite facets;[27,31] their shape can be controlled by varying the proportions of the ingredients to yields cubes, rods or shapes that in profile are triangular.[27] They have not yet been used as catalysts.

A unique feature of gold in its microparticulate state is the range of colours it can exhibit.[20,21,23] During citrate reduction the colour changes from grey through lavender to red;[19] suspensions of 1–3 nm particles are brown or dark orange-brown, and 5 nm particles are purple-brown or purple-red,[1] while large particles give a ruby red colour, and partially coagulated sols are blue, and cause scattering of incident light.[25] The optical absorption spectrum of the larger particles (< 40 nm) has a maximum at 520 nm, while the blue colour arises from absorption at 680 nm. Intermediate shades are therefore possible. The colour is due to a *plasmon resonance* initiated by the interaction of the electric field of visible light with the confined electron gas within the particle, causing collective oscillation of the conduction electrons with respect to the core.[16,22] The resonance weakens and shifts towards higher energy as size is decreased, and disappears below 2 nm.[5] The effect has excited the attention of theoreticians for many years, the first quantitative treatment having been given by Gustav Mie in 1908;[21−23,43] this accounts nicely for the change in colour that occurs as size is increased from 20 to 1600 nm.[5] A similar range of colours is observed when small gold particles are formed on colourless oxidic supports,[1,44−46] the colour, that is to say, the energy of the plasmon resonance band, depending on the dielectric constant of the support and the degree of intimacy of the gold particles with it, as well as on the fluid phase with which they are in contact. The colour may be recorded by measurement of the optical absorption.[47,48]

Bimetallic colloids may be made by certain of the methods used for gold alone (Section 4.6); the combinations of gold with platinum[41,49] and palladium[41,42] have received particular attention. Compounds of both have been reduced simultaneously with $HAuCl_4$ by sodium borohydride with PVA[42] to give in the latter case a dark brown colloid, but sequential reduction has also been used. With the copper–gold system, reduction by sodium borohydride with PVP gave particles that were annealed after separation by centrifuging;[50] below 448 K there was a single disordered solid solution, but at higher temperatures the ordered phases AuCu and $AuCu_3$ were obtained, depending on the ratio of the metals used, and could be redispersed. A trimetallic $AuCuSn_2$ phase has been made by reduction with stannous chloride and PVP in tetraethyleneglycol;[51] heating the dispersion at temperatures up to 473 K gave a product with a wide size distribution, the XRD pattern of which resembled that of the NiAs structure with copper and gold randomly occupying the arsenic site.

Although the mutual solubility of gold and platinum is limited (Section 2.6), there are many examples of small particles in which they appear to be in solid solution[41,49] (see also Section 4.4); this is a consequence of a difference in the way the valence electrons are used from the situation with massive alloys.

Bimetallic colloids containing gold and palladium or platinum have been deposited onto carbon or graphite for use as catalysts for the selective oxidation of organic compounds[52] (Section 8.3), in the same way as for pure gold colloids (Section 4.6).

3.2.4. Other methods

Particles within the colloidal size range can also be made by breaking down massive gold by the input of the necessary energy.[16] When this is mechanical, the process is termed *attrition*, but more useful is *metal vapour synthesis* where the energy input first produces atoms, which then coalesce on the surface of either an inorganic solid (typically a single crystal oxide surface such as $TiO_2(110)$[16,53,54]) or a cooled solid organic substance.[1,55,56] One procedure uses 'a commercial electrostatically focussed electron beam metal atom reactor'[57] — a compound noun of impressive length. The method involving the organic material has been developed over many years by Klabunde, and is termed *solvated metal atom dispersion* (SMAD) (see also Section 4.4.7). Deposition of gold atoms onto acetone at 77 K, followed by its removal, and introduction of dodecanethiol with toluene, gave 3–6 nm particles, the size distribution of which was made more uniform by 'digestive ripening' at 393 K.[56] Removal of the solvent gave 'soft shiny dark crystals' from which a wine-red colloid could be re-formed. *Laser ablation* has also been employed,[16] but does not give the desired control over size and size distribution.

In CVD, a volatile gold compound, usually $Me_2Au(acac)$ is caused to decompose on or near a supporting surface, with the formation of small particles that can be used as a model or conventional catalyst[58–60] (Section 4.3.4).

We should also note the availability of *cluster compounds* that contain assemblies of gold atoms numbering 13, 39 or 55;[7,16,19] they are stabilised by phosphine or other ligands and are quite readily prepared. Some remarks on the physical properties of the Au_{55} cluster will be made later (Section 3.4).

A number of other methods for preparing highly dispersed forms of unsupported gold and gold-containing bimetallic compositions may be briefly noted.[61,62] Blacks are made by chemical reduction of the appropriate solution of salts, using hydrogen, methanal (formaldehyde), methanoic (formic) acid or sodium borohydride; they are aggregated forms of transient colloids. Bimetallic oxides are made by fusion of mixed salts in sodium nitrate, and are recovered by leaching out the oxidant with water; this is the *Adams method*,[63] but has yet to be used with gold as a component. Oxides made in this way or by decomposition of mixed hydroxides or carbonates are generally reducible by hydrogen to give bimetallic powders. The *Raney method*[64] involves forming an alloy of the active metal with aluminium, followed by its leaching out with a strongly basic solution; the resulting powder, sometimes referred to as a skeletal metal, has a high surface area. The method has been used to make highly porous Raney gold *via* the alloy $AuAl_2$, and can be extended to make gold alloy powders.[65] It has also been applied to a great variety of combinations by A.B. Fas'man.[62]

Gold nanotubes have been made by electroless plating within the 220 nm diameter pores of a polycarbonate membrane.[66]

3.3. Techniques for the Study of Small Particles of Gold[3, 67–72]

The purpose of this section is to introduce the main methods that are used to study the physical properties of small gold particles, to give their acronyms for subsequent use, and where necessary to add a few explanatory or cautionary words. These methods are already well described in the literature, so only brief treatments are necessary. A small selection of references to their use in studying gold catalysts will be included as an introduction to the relevant literature.

3.3.1. Determination of size and structure

We consider first the methods that are used to determine their *structure* (size, size distribution, shape, composition, etc.). Because of the sensitivity of the physical and catalytic properties of small gold particles to their size, its determination becomes a central feature of efforts to characterise both practical and model catalysts. Because of the metal's chemical inertness, methods based on the chemisorption of simple molecules have in the past failed to give acceptable quantitative results, and they have not yet achieved routine status, but recently the chemisorption of hydrogen[73] and of oxygen on *highly dispersed* gold catalysts has afforded results that relate well to those given by the physical methods that were essential in the past (Sections 5.2 and 5.5.1), and continue to be of great benefit. *Scanning transmission electron microscopy* (STEM) gives a comparatively low-resolution image that shows the general features of a catalyst, and reveals for example the size and shape of support particles. By far the most useful technique for examining metal particles in the nanometer range is TEM,[74–76] which can also be used in a high-resolution mode (HRTEM) using a very energetic electron beam, to reveal the internal structure of metal particles. Much experience is needed to use these techniques with maximum efficiency and accuracy. With supported metal catalysts, sample preparation is critical; for example, sectioning to provide very thin slices of a material dispersed in a thermosetting resin (ultramicrotoming), which is used to determine the distribution of metal particles within porous supports, can cause mechanical damage to them. Gentler methods such as supporting the sample on a transparent membrane or carbon film are much to be preferred. To obtain a meaningful size distribution it is necessary to observe 500–1000 particles, either visually or instrumentally; one has also to ensure that the fields examined are truly representative of the whole sample, because the distribution of metal particles over a support can be distressingly inhomogeneous. The distribution curve will inevitably be skewed, because although there is a lower limit of zero (or whatever minimum size the instrument can reliably detect), there is no upper limit. On occasion one will see a bimodal distribution, i.e. one having two maxima, for example in partially sintered catalysts. Methods that rely only on an integrated signal from the whole sample will in such cases give totally misleading results. Detection of particles less than about 1 nm in size is difficult, and it is easier to see them when supports are used that contain only atoms of low atomic mass (Al_2O_3, SiO_2), because of the greater contrast they afford. The use of TEM on 'model' catalysts is usually more straightforward. HRTEM[77] will show the lattice planes of small particles, and this will confirm their identity; it will also show whether a particle is a single crystal or whether it is twinned or even multiply-twinned. The electron microscope can also be used in the electron diffraction mode, to give a diffraction pattern that identifies the object being viewed; this is most helpful when applied to the characterisation of bimetallic catalysts. *Scanning-tunnelling microscopy* and the associated *atomic-force microscopy* (AFM) also yield images of atomic resolution on flat specimens such as 'model' catalysts formed on single-crystal surfaces.[7,78–80]

Information on the size of metal particles in a catalyst is also given by *X-ray diffraction*;[71,75,81] the width of a peak at half-height (β) of a diffraction peak for the component of interest is related to its size (d) by the *Scherrer equation*

$$d = 0.9\,\lambda/\beta \cos\theta, \tag{3.2}$$

where λ is the wavelength of the X-rays and θ the angle of incidence. For small particles (less than about 5 nm) however the broadening of the peak is such that its width cannot be accurately estimated, and above 100 nm it is too narrow. The method inevitably gives an answer within these limits. The absence of an observable peak at the expected place is sometimes used as evidence that large particles are absent, but with very low loadings of metal ($< 1\%$) the signal would in any case be weak. Extremely small particles (≤ 1 nm) on titania-coated microporous silica have however been characterised by *powder XRD analysis*.[82] X-rays can also be scattered by a surface as well as reflected, and their analysis forms the basis of *small-angle X-ray scattering* (SAXS); its major advantage is that it detects particles as small as 1 nm. This method looks at X-rays scattered to within a few degrees of the primary beam, but those appearing at a larger angle also yield structural information: this technique is named *wide-angle X-ray scattering* (WAXS).[46] Use of an equation due to P. Debye gives a radial distribution function, and its comparison with calculated distribution functions derived from various models may enable the shape of the particles to be determined.[55,77,83] '*Anomalous' SAXS* employs synchrotron radiation to separate scattering by the support and its pores from that by the metal, by using the so-called anomalous or resonant behaviour of the atomic scattering amplitude of the metal close to its absorption edge. A large fraction of very small gold particles in 0.2% Au/C was detected in this way.[84]

When the energy of an incident X-ray photon hitting the surface exceeds the binding energy of the core level of an atom, the photon is absorbed and a photoelectron is emitted. Plotting the X-ray absorption coefficient *versus* incident photon energy shows extended sinusoidal fine structure above the absorption edge; this is the *extended X-ray absorption fine structure* (EXAFS, now usually called just XAFS) and is the basis of *X-ray absorption spectroscopy* (XAS).[55,73,85,86] The outgoing electron wave is backscattered by all atoms near the absorbing atom and, by means of Fourier analysis, the type, number, distance and thermal and static disorder of the backscatterers can be elucidated. Due to the fast attenuation of electron waves in matter, this is a local effect, and no long-range order is needed, as in XRD. The number of neighbours determines the intensity of the backscattering contribution, and for small metal particles having a narrow size distribution the XAFS technique supplies an estimate of their *mean* size. The method also informs on static and thermal disorder, because *Debye–Waller factors*[70] are part of the fitting procedure; increase in values obtained both at 77 and 300 K as size decreases indicates increasing structural disorder and greater mobility of surface atoms. If different phases or oxidation states are present, the XAFS spectrum will sum over all the contributions, and, providing results of good quality have been obtained, they can be deconvoluted and identified. This technique requires a source of synchrotron radiation, and needs a high level of skill and experience, both in performance and interpretation. The combination of XAFS with XRD and TEM is particularly powerful.

Examination of the low-frequency Raman modes is also claimed to provide a means of gold particle size determination.[87,88]

The nuclear and electronic structure of gold unfortunately rules out the use of several methods that have proved useful with other metals. Gold being diamagnetic does not lend itself to studies based on magnetisation, although the Au^{II} state is paramagnetic; gold has a non-zero nuclear spin quantum number, but the NMR of gold is only of limited diagnostic value to chemists[89,90] (Section 2.5.1).

3.3.2. Investigation of optoelectronic parameters

Having assessed the mean particle size and the shape of the size distribution, attention now turns to examining the optoelectronic properties of the gold particles; these reveal the way in which the electrons are employed, and in particular how their use varies with particle size. Perhaps the most useful and readily accessible methods are the *photelectron spectroscopies*. Electrons ejected by X-rays come from the more tightly held inner electron shells, and analysis of their energies identifies the level and spin state from which they come. The exact binding energy is however sensitive to the state of the element, and can reveal not only its oxidation state, but also finer nuances of its environment. This is *X-ray photoelectron spectroscopy*.[7,58,67,75,78,86,91-94] The technique is almost surface-specific, because the depth from which photoelectrons can escape is only a few nm. Valence electrons being less tightly held can be removed by UV radiation; such electrons are of course involved in chemical bonding and in forming conduction bands in metals. Any factor that affects electron density in this region will therefore be sensed by *ultraviolet photoelectron spectroscopy* (UPS).[76,78,95]

Mössbauer spectroscopy follows from the observation that a nucleus rigidly held in a lattice can undergo recoil-free emission and absorption of X-radiation. The separation of nuclear energy levels can be measured very accurately, and weak interactions between a nucleus and its electronic surroundings can be detected; this includes the element's oxidation state and its precise environment. Fortunately the ^{197}Au nucleus is suitable for Mössbauer spectroscopy;[7,17,89,97-101] the 77.3 keV excited state has spin and a half-life of 1.93 ns. In order to identify the oxidation state, however, both the quadrupole splitting and the isomer shift have to be used.[101] One of the earliest applications of Mössbauer spectroscopy in catalysis was to examine supported gold catalysts,[102] and subsequent studies have been reviewed;[101] some of their conclusions will be mentioned in Section 6.3.3.a. Simultaneous use of the ^{57}Fe nucleus with the ^{197}Au nucleus has permitted the study of Au/FeO_x catalysts, including identification of the support phases.[103–106] The technique is not, however, generally available as an X-ray source is required.

Whereas XAFS yields mainly structural information, analysis of the region close to the absorption edge informs chiefly about electronic structure. Ejected electrons of low energy ($\leq 30\,eV$) interact with the *valence* electrons of other atoms and this leads to an additional peak near the absorption edge; this is termed a *white line*, and the *X-ray absorption near-edge structure* (XANES or NEXAFS) is due to transitions of electrons in low-lying levels to vacant valence levels. The form of the spectrum in this region can be understood by considering the local density of states and symmetry, and oxidation state; the energy and intensity of the white line is therefore often helpful in identifying the oxidation state of the scatterer. This was recognised as long ago as 1976;[107] references to other early work have been collected.[108] The technique has been extensively applied to observing the transition from Au^{III} to Au^0 during the thermal treatment of catalyst precursors,[108–112] and has been used to detect the presence (or absence) of oxidised gold species in operating gold catalysts,[113] and to show the absence of Au^0 in catalysts treated with sodium cyanide.[114] Its applications have been reviewed.[109,115] The technique senses vacancies in the *d*-electron level or band, which in the case of gold *atoms* is filled, but where there is for example *d*-*s* hybridisation or where electron density is withdrawn (as with Au^I or Au^{III}), a XANES signal may appear.[7,85,115]

The low spatial resolution obtainable with electron spectroscopies makes the investigation of individual particles or clusters impossible, but with a remarkable new technique that combines STM with a spectroscopic component this can now be done. It is named *scanning-tunnelling spectroscopy* (STS), and its methodology merits a brief description.[16,75,116] By varying the applied voltage and measuring the tunnelling current, a map of the local density of states under the tip can be obtained. When the tip position and tunnelling gap are fixed, the resulting current-voltage spectrum holds information on the chemical situation of a single atom. Both occupied and unoccupied levels can be probed in this way: by varying the sample bias, the tunnelling current can be made to flow either from surface to tip (negative bias) or *vice versa* (positive bias). The first way senses occupied states and the second the unoccupied states. In this latter case the density of states in the tip can be assumed independent of voltage. The schematic Figure 3.5 shows how the method can be applied to measure the band gap in a semiconductor. Tunnelling occurs between the Fermi levels of sample

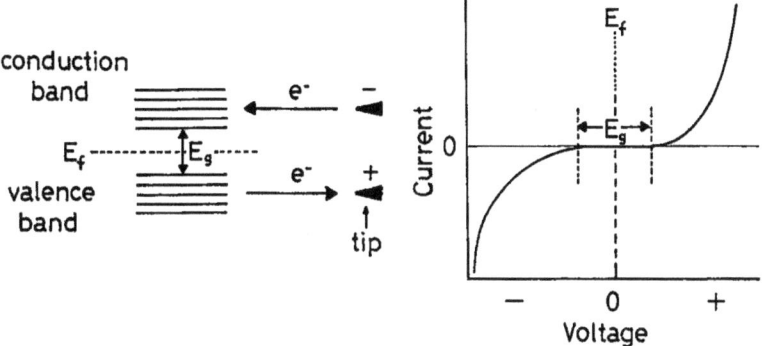

Figure 3.5: Schematic diagram showing the basis of scanning tunnelling spectroscopy as applied to a semiconductor. E_f: Fermi energy; E_g: width of band gap.

and tip, electrons emerging from whichever is negatively charged. STM has been applied to estimate the *local barrier height*, from which the local work function and energy gap may be deduced.[116,117] In *photoemission STM* the Mie plasmon resonance energy of a supported particle is probed by injecting an electron from an STM tip, and analysing the emitted photons.[7] *Electron holography*[80,92,119] has been used to provide values of the mean inner potential (i.e. the zero order Fourier coefficient of the crystal potential). Application of these techniques to small gold particles will be described in Section 3.4.

It should always be remembered that examination of easily reducible substances by high-energy radiation under high vacuum conditions is potentially able to effect their reduction, for example, by elimination of lattice oxygen. What is then observed does not necessarily reflect the composition of the original sample.

3.3.3. Other methods

Finally we should mention several techniques of limited applicability to the examination of small gold particles and their supports; these include:

- anion photoelectron spectroscopy,[7]
- time-of-flight secondary-ion mass-spectrometry (TOF-SIMS),[120]
- photoacoustic spectroscopy (PAS),[32]
- laser Raman spectroscopy (LRS),[35]

- ion-scattering spectroscopy (ISS; alternatively low-energy ion-scattering, LEIS),[69,80]
- UV/visible spectroscopy[121] and diffuse reflectance spectroscopy (DRS),[35]
- electron-spin resonance (ESR or electron paramagnetic resonance, EPR),[1,122]

this last method being useful in particular to detect electrons trapped at surface anion vacancies, and oxygen molecules adsorbed thereon as super-oxide (O_2^-) ions.[123,124]

3.4. Variation of Physical Properties with Size[6]

3.4.1. Introduction

The perceived importance of particle size for determining the catalytic activity of gold for some if not all reactions requires us to examine carefully how the physical properties vary with this parameter, since the strength of binding of reactants, and hence their reactivity towards one another, must inevitably depend on those characteristic features that such properties sense. In particular the manner of use of valence electrons in interatomic bonding will affect their availability for forming chemisorption bonds. These physical properties have been the subject of intense scrutiny, mainly by physicists, so their work has appeared mainly in physics journals, and is therefore not well known to the chemical/catalytic fraternity. The immense amount of information available, prompted by the ease with which particles of known size can be made by the methods just described, can only be briefly summarised: attention to the cited references is urged to obtain a greater depth of understanding.

From what has been said in Section 3.2 concerning methods of preparation it will be clear that the physical properties of small gold particles can be observed using a number of configurations. Their dependence on size most clearly appears with very small 'clusters' still in the vapour phase, because support interactions are absent, but the size range that can be examined is very limited. Interaction with the surface and consequential modification of properties has been studied using 'model' systems formed both by deposition of clusters and by coalescence of atoms on a flat surface. The properties observed using these systems, as well as those originating with colloids or supported metals can be subdivided into (i) structural, (ii) energetic and (iii) electronic, these last two being taken together under the term *optoelectronic*. We start by examining how all of them vary with size in gaseous and ligand-stabilised clusters, then we look at effects with supported particles in model and real catalysts that are essentially intrinsic to their size, and in Section 3.5 we consider metal-support interactions and their mutual influences.

3.4.2. Structure of gaseous and ligand-stabilised clusters

We may start by mentioning what is known about the structure of small gaseous clusters having n atoms.[7] Anionic clusters with eleven atoms or fewer are planar, the later change to 3D form when compared to copper and silver having a relativistic origin. Neutral complexes with six or fewer atoms are also thought to be planar. Larger clusters ($n > 30$) appear to be able to exist either in the fcc or icosahedral (five-fold cordinate) form, there being little difference between their stabilities. It may perhaps be wiser to inject a note of caution concerning the reliability of calculations that predict stablest structures or minimum energy routes for physical or chemical transformations. It does not automatically follow that these are the structures or procedures that will necessarily be adopted or followed, because they may not be available in practice, and metastable arrangements are not unknown.

Much attention has been given to large ligand-stabilised clusters containing 13, 39 or particularly 55 gold atoms;[7,16,125] relative ease of preparation, especially of the latter, has encouraged this. Little if any catalytic use appears to have been made of them, and the extensive reported work has been performed by physicists and materials scientists.[126] The Au_{55} cluster has atoms in the fcc arrangement, and is a cubo-octahedron having square (100) and triangular (111) faces;[127] interatomic distances are contracted by 7.5 pm compared to the normal bulk, and the Debye–Waller factor is lowered by 40% because there is in consequence less thermal vibration and a volume contraction of 7.6%. The bulk modulus rises by 63% because of the force acting inwards.[86] It has therefore been aptly described as *"a strained, squeezed piece of fcc gold with ligands on it"*.[127]

3.4.3. Structure of small supported gold particles

There have been numerous studies of the dependence of the shapes of small gold particles on the manner of their preparation and on the nature of

the support; titania surfaces have featured largely in this work, but first
we may note some early research on weakly interacting surfaces. Particles
of fivefold symmetry were exclusively present in gold evaporated onto an
organosilicon polymer solution; they were of course multiply twinned, and
transformed into single crystals by thermally induced growth.[7] There was a
1–3% contraction of the bulk lattice spacing, which had already been seen in
an early XAFS study of fcc gold particles on a Mylar film.[128] Gold particles
(~ 22 nm) on mica exposed (111) surfaces with (100) sides.[7] It appears that
the structure of small gold particles depends somewhat on the way they are
formed, and perhaps also on the support (Section 3.5). Analysis of X-ray
diffraction results by the Debye function method on $Au/Mg(OH)_2$ showed
both fcc cubo-octahedra and icosahedra, but interaction with the support
was weak and the latter sintered readily to give truncated dodecahedra.[55,83]
There is also an isolated mention of the bcc structure being seen in films.[129]
Based on XAFS observations, the Au–Au bond length in gold particles on
various supports[73] and stabilised by poly(N-vinyl-2-pyrrolidone)[130] short-
ened by several percent, and where this is due to intrinsic size effects it
can be rationalised by the liquid drop model,[130] that is, in consequence
of the operation of a surface energy akin to the surface tension of liquids.
Instances of bond-length changes where the support may play a role are
mentioned in Section 3.5.2.

Mössbauer spectroscopy also provides structural information.[97–100] In
particular it is possible to deduce from peak asymmetry the separate
appearance of surface atoms as size is decreased; they cause an increase
in intensity of quadrupole splitting, and two quadropole components indicate two kinds of surface atoms.[97] Au^I and Au^{III} species are also sometimes
recognised,[99] and sometimes not.[98]

The clearest evidence for the transition from nonmetallic to metallic
character should come from looking at gaseous anionic clusters.[7,131] Those
containing less than 20 (or 40) atoms exhibited odd-even oscillations in electron affinity up to 1 eV (96.5 kJ mol^{-1}); relativistic mixing of d and s states
was invoked to explain this.[7] Small clusters ($n < 20$) did not show a d-band,
but it appeared in larger clusters ($n < 233$) and grew with size.[132] Au_{55} has
been classified as a semiconductor on the basis of unrestricted hybrid DFT
calculation;[133] it was however resistant to oxidation.[134] Clusters having 70
or more atoms had a valence region quite that of bulk gold. Au_{20} clusters
have been examined by anion PES and appear to show great stability. If the
observations on the ligand-stabilised Au_{55} clusters are relevant, the region
between 55 and 70 atom sized particles would seem to be critical.

3.4.4. Optoelectronic properties of clusters and small supported particles

It is a well-known fact that, as the size of a metal particle is decreased, the overlap of the bands of valence electrons, with which we are mainly concerned, diminishes, and finally they are replaced by discrete energy levels characteristic of the isolated atom. This results in the loss of electrical conductivity and in the Mie plasmon resonance, an effect that has been noted with the Au_{55} and smaller clusters, on the basis of which they were described above as being 'molecular'. The extent of band overlap is temperature-sensitive because of thermal excitation, i.e. bands tend to convert to levels as temperature falls; thus 'metallic' properties may be seen at high temperature and insulator properties at low temperature. As an approximate guide we may take the relation

$$\delta = 4E_F/3n, \tag{3.3}$$

where δ is the separation between levels in a given band, E_F the Fermi energy and n is the number of atoms in the particle. This tells us that for a 2 nm particle the separation will be about 7×10^{-2} eV, which somewhat exceeds the thermal energy kT at room temperature, so that the band will have effectively transformed into a set of levels. The particle then behaves like a molecule rather than a metal.

It has been a major point of interest in the study of small metal particles to determine the precise point (if indeed there is a precise point) at which metallic character is lost. The broader context of this problem as it relates to other metals such as mercury and sodium has been discussed in a series of important papers by Peter Edwards and his associates.[135,136] The difficulty seems to be that there is no agreed criterion by which membership of the metallic state can be judged, and various physical techniques give somewhat different answers because they sense slightly different aspects of electron behaviour. The question has been earnestly addressed in the case of gold, partly because of the familiar sensitivity of catalytic activity to particle size; as noted above (Section 2.1) the way in which electrons are used to form metallic bonds determines the character of the free valences at the surface, and hence the kind of chemisorption bond that is formed with the reactants.

There have been quite a number of studies of small gold particles by XPS,[7,54,80,92,95,137–139] and it is almost always found that the binding energy of the $4f_{7/2}$ core level increases by 0.8–1 eV as the size falls below

about 5 nm. The precise point at which it begins to rise is difficult to determine because it is initially slow, but is rapid below about 3 nm. The effect
has been observed both with 'model' catalysts in which interaction with the
support may be slight (e.g. with graphite,[92] diamond[7] or NaCl[95]), as well
as with supports such as $TiO_2(110)(1 \times 1)$[80,138] and other oxides[54,58,59,75]
where interaction may or may not be expected, and indeed the form of the
dependence varies with the support,[54] for reasons that are not entirely clear
(see below). It is said also to depend somewhat on the form of pretreatment
and on particle shape.[7,58,59]

The reasons for this change in binding energy have been much discussed.
There seem to be two possible reasons:[138] (i) rehybridisation of valence
band energy levels, causing *initial state effects* that suppress photoemission
near the Fermi level and (ii) *final state effects* due to core-hole screening,[92]
caused by the positive charge left on the cluster, and strongly influenced
by the support. The arguments, which verge on the theological, are fully
rehearsed in Ref. 7: there is evidence to support both (i) and (ii), the latter
being supported by XANES measurements, where it has been thought that
final state effects predominate.[7]

When intelligent people look at the same results and draw opposite
conclusions, it is most probable that both are partly right and partly wrong,
that is to say, the truth lies somewhere in the middle. So we should not be
asking whether (i) or (ii) is right, but rather how much of each contributes
to the observed effect, and how their proportions change with the variables
of the system. In this case it is possible that the extents to which the
two possibilities operate may depend on the particle size; the initial state
effect is more likely to be significant with smaller particles, because strain
induced by lattice contraction will be greater, with consequential effects on
interatomic bond and the employment of electrons.[140]

There have been several investigations of supported gold particles by
STS, the theory of which was introduced in Section 3.3.2. It measures the
energy gap between the conduction and filled valence bands, and therefore
scans the region in which truly metallic properties are lost. Measurements
of the width of the band gap have been reported for gold particles of various size vacuum-deposited on $TiO_2(110)(1 \times 1)$[54,80,141] (see Table 3.1).
The quasi-2D particles are non-metallic, while the larger are metallic, but
interaction with the support dominates geometric and electronic structure: the size of the gap depends on intrinsic depletion of the density of
states near the Fermi surface. By electron holography[116,118] the mean inner

Table 3.1: Scanning-tunnelling spectroscopy of 'model' Au/TiO_2: dependence of band gap and cluster type on size.[54,80,141]

Diameter (nm)	Height (nm)	Layers	No. of atoms	Band gap (eV)	Type
1.5	0.6	~2	~35	>1.5	q-2D
2.5	1.0	~3	~170	0.6	3D
3.0	1.3	~4	~340	0.3	3D
4.0	1.5	~5	~600	0	3D

Note: The quasi-2D clusters are non-metallic, and the Au/TiO_2 interaction dominates the geometric and electronic structure.

potential for gold particles on titania started to become greater than the bulk value at 5 nm, but STM measurements of the local barrier height and band gap for $Au/TiO_2(110)(1 \times 2)$ in the range of particle heights from 0.1 to 1 nm suggested that metallic behaviour began to be seen at a particle *height* of only 0.4 nm.[116] The height of a single gold atom is merely 0.29 nm. This valuable paper[116] provides numerous other references to experimental work on the examination of supported gold particles by these and other techniques, as well as to relevant computational studies. Results have also been reported[119] for Au/graphite, but the band gaps quoted are much smaller (0.05 eV at 0.5 nm size): no gap was seen for particles above 2 nm.

Confirming our earlier conclusion, STS results show that the ligand-protected Au_{55} cluster has discrete energy levels, both occupied and unoccupied, inside the core, with a band gap of 0.17 eV. On the basis of electronic relaxation speed, however, it appears to be on the verge of being a metal.[7] There is therefore no unequivocal means of deciding the point at which metallic character is lost; indeed, its disappearance may not be sudden, and various techniques give different answers, but for most purposes we may say that its absence starts to be felt at about 5 nm and it is completely gone at 2 nm.

The overlap of *s*-, *p*- and *d*-electron bands close to the Fermi surface, which increases as particle size grows, allows hybridisation to occur,[142] so that the atomic structure $5d^{10}6s^1$ changes slightly to $5d^{10-x}6s^{1+x}$, as shown by the appearance of a weak white line in massive gold[143] (Section 2.6). This line has been found to be less intense in very small particles,[73,130] either because they become more atom-like[73] or because they acquire charge

from their surroundings[130] (e.g. Ti^{3+} ions). The first explanation does not
harmonise with DFT calculations (Section 5.3.3), which predicts that atoms
of low coordination number have some d-orbital vacancy, thus accounting
for the rise in chemisorption activity as particle size falls. The decrease in
the Au–Au bond length in small particles (Section 3.4.3) also reflects the
change in the way valence electrons are employed.

A further symptom of the change from metallic to non-metallic
behaviour is the fact that gold's redox potential decreased rapidly as particle size falls.[144]

The strength of interatomic bonding is also an electronic parameter that
is size-sensitive. It has long been known that the melting temperature of
small metal particles decreases with their size, and this has been clearly
established for gold. The extent of the decrease depends on the mean number of Au–Au bonds holding the particle together,[145] and this obviously
decreases as size goes down; the latent heat of fusion suffers a similar fall.
A quantitative relation between size and melting temperature has been
proposed:[146] it takes the form

$$(T_d - T_\infty)/T_\infty = -(4/\rho_s L d)[\gamma_s - \gamma_l(\rho_s/\rho_l)^{2/3}], \qquad (3.4)$$

where T = absolute temperature, ρ = density (in $kg\,cm^{-3}$), γ = surface
energy, L = latent heat of fusion (in $J\,kg^{-1}$), d = diameter; subscripts s and
l stand respectively for solid and liquid, and d and ∞ for the particle diameter and the massive metal. This predicts melting at 950 K for a 2 nm particle
and 850 K for a 1 nm particle, but experimentally 2 nm particles have been
shown to melt at about 500 K, and under the influence of an electron beam
at only 300 K.[147] Surface mobility will of course start well below the bulk
melting temperature (1336 K).[148] It is expected to start at the *Tammann
temperature*, which is half the bulk value expressed in K; a value of ~ 400 K
for 2.5 nm particles has been recorded,[149] compared to 690 K for massive
metal. So for particles of the size that are active in catalysis the surface
may well be fluid, remembering also that heat of reaction may raise the
particle temperature well above that felt by the thermocouple.

The variations with size in the colours exhibited by colloidal particles
and supported particles of similar dimensions have been discussed in Section 3.2.3; the plasmon resonance that is responsible is however influenced
by a number of extraneous factors, and it cannot be used quantitatively to
estimate size or degree of metallic character.

3.5. Metal–Support Interactions

3.5.1. Particle shape and bonding to support[150]

It is almost impossible to distinguish clearly between those physical or catalytic effects that are intrinsically dependent on particle size from those which are conditioned by contact with the support, because at least in the context of catalysis small particles are necessarily employed, and their utility depends on their being supported. Furthermore, the smaller the particle, the greater will be the fraction of atoms directly in contact with the support and therefore influenced by it, while at the same time the fraction of coordinatively unsaturated surface atoms also increases, and this changes the physical properties of the whole particle. It is therefore virtually impossible to draw a clear distinction between intrinsic particle size effects and those that are due to metal–support interactions. In Section 3.4.2 we noted some effects of size on structure in systems where the influence of the support was likely to be minimal; now we must examine effects of size in conjunction with the metal–support interaction.

The structure and stability of small gold particles is a function of the chemical and physical nature of the support on which they reside.[7,116,151,152] It is clear that the extent of the influence of the support on a metal particle will depend on the fraction of the metal atoms directly in contact with it; for particles of the same shape this will increase as the size decreases, but it will also depend on the shape of the particle, which is conditioned by the chemical forces at the interface. In principle the particle shape is determined by the *contact angle* θ defined by the equation

$$\cos\theta = (\gamma_{sg} - \gamma_{ms})/\gamma_{mg} \qquad (3.5)$$

where the γ's are the interfacial energies of the three interfaces (Figure 3.6).[150] When γ_{ms} is large, the metal will wet the oxide surface,

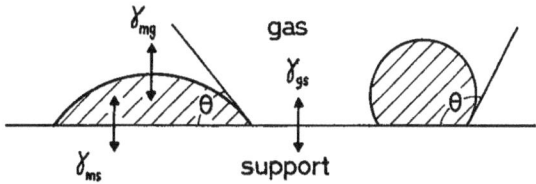

Figure 3.6: Interactions across the three interfaces around a supported gold particle, and the contact angle.

and particles will be hemispherical or truncated cubo-octahedral or in the limit thin islands, but when it is low the area of contact will be small, as for example with a sphere or cubo-octahedron. Increasing γ_{ms}, other things being equal, will increase the length of the perimeter of the metal particle, a region that has assumed importance for catalysis.[153] As size is increased, larger flat facets can develop and the mean surface coordination number will fall (Figure 3.4). Interfacial energies (γ_{ms}) have been measured for several gold/metal contacts (Table 3.2);[154–157] they vary considerably, but do not appear to be much influenced by particle size. The equilibrium shape should, however, depend on what is adsorbed at the gas-metal and gas-support interfaces; this will include the state of hydration of the support surface and any relics of the preparation such as sodium or chloride ions.

We may now review other experimental evidence, much of which has been obtained with the Au/TiO_2 system using TEM. Particle shape has been shown to depend on preparation method, deposition–precipitation (Sections 4.2.3 and 4.2.4) giving hemispherical particles, while impregnation (Section 4.2.1) and photodeposition (Section 4.4.2) gave spheres;[158,159] the latter were, however, much larger. Subsequent treatment has an effect: with Au/TiO_2 made by deposition–precipitation, hydrogen reduction at 473 K gave small particles that wetted the surface, but as temperature was increased they became smoother and facetted, and above 673 K they adopted the minimum energy shape of a truncated cubo-octahedron.[160,161] Au/TiO_2 made by the colloidal route contained spherical particles after calcination at 473 K, but larger hemispherical particles at 873 K.[162] The strength of bonding of gold *atoms* to the surface of $TiO_2(110)$ is considerably less ($210\,kJ\,mol^{-1}$) than that between the atoms themselves. With Au/TiO_2 similarly made, particles were more flat after drying at 323 K, but after calcination at 673 K the larger particles were facetted.[14,163] No evidence has been found for the existence of a layer of gold cations

Table 3.2: Interfacial energies between gold particles and oxide surfaces.

Oxide	d_{Au} (nm)	Interfacial energy ($J\,m^{-2}$)	Reference
Al_2O_3	1	0.27	154
MgO(100)	5	0.43	155
$TiO_2(110)$A	4	0.75–1.21	156
$TiO_2(110)$R	200	0.9–1.0	157

A: anatase; R: rutile.

between the metallic phase and the support, although its existence has been postulated,[164] and is very well established for supported noble metals of Groups 8–10.

Detailed structural information is obtained by HRTEM on both model and practical supported gold catalysts. Particles formed by vacuum evaporation of metal atoms have long been known to decorate steps and point defects such as oxide ion vacancies on nearly flat surfaces;[53,79] this implies that adsorbed atoms are quite mobile. Particles made in this way are preferentially located at anion defects,[165] and their quenching by oxygen treatment of $Au/TiO_2(110)$ increases particle mobility at high temperature.[137] DFT calculations suggest there is no adhesion of gold to a perfect titania surface,[111,116] but that with Au/TiO_2 vacancies beneath particles should lead to higher interaction energies and hence to flatter particles.[111] Reducible oxide supports should therefore be better at forming and stabilising small particles than ceramic oxides. In the case of $Au/CeO_2–Al_2O_3$, gold particles are preferentially attached to ceria sites.[166] DFT calculations have also been carried out on a number of rutile-type oxides as supports, but the conclusions are as yet unconfirmed by experiment.[167]

Extensive HRTEM studies of gold supported on anatase (A) and on rutile (R) have been reported.[168] With $Au/TiO_2(A)$ made by deposition–precipitation, particles were chiefly hemispherical and fcc single crystals, their preferred orientation being the epitaxial $(111)[011]_{Au}//(112)$ $[02-1]_{TiO_2}$.[168] A better match was obtained when the atoms of the $Au(111)$ plane were in contact with the anatase oxide ions than with its titanium ions. Electron holography with HRTEM showed small ($< 2\,nm$) particles made contact angles less than $90°$, while with larger ($5\,nm$) particles it was more than $90°$.[165] This reflects the decrease in the fraction of gold atoms at the interface and under the direct influence of the support, as noted above. With $Au/TiO_2(R)$, the particles mostly had their $Au(111)$ plane parallel to the rutile(110) plane, with no preferred orientations. Similar studies have also been reported for gold on titania single crystals ((110)(A), (110)R). Au/ZrO_2 prepared by DP contained mainly multiply twinned particles (14 nm) in contrast to Au/TiO_2, but larger particles (8 nm) formed by reduction at 573 K had even more of them.[169]

Very small gold particles can also be formed on magnesia and brucite $(Mg(OH)_2)$.[55,83] Au/MgO made by deposition–precipitation contained particles smaller than 1 nm that were claimed to show icosahedral and fcc cubooctahedral structures,[170] but this is hard to believe as the diameter of a single gold atom is already 0.29 nm, and 1 nm particles affixed at the steps of MgO(100) suffer structural and orientational fluctuation at room temperature under the influence of the radiation used to observe them.[7] Thus by a kind of uncertainty principle we cannot tell what their undisturbed structure is. The support tends to stabilise these motions because a high interfacial energy has to be overcome to secure the rearrangement, but exact epitaxy with the support is lost with large particles ($n \geq 1500$). With smaller particles in epitaxial registry with the support, there are however interesting mutual disturbances: with Au/MgO(100) there is strain in *both* the gold *and* the magnesia as effort is made to match lattice parameters at the interface, while the Au–Au distances at the edges of the particle were expanded to compensate for compression in the middle (see Figure 3.7). A 10% lattice expansion for the first three layers of a small gold particle in Au/TiO$_2$(P-25) has been observed, decreasing to 3% for the outermost layers;[160] similar results have been found with model systems (Au/MgO(100),[155] Au/TiO$_2$(110)[156]), a conclusion also reached by DFT calculations.[28] Paradoxically, however, XAFS analysis has usually shown a slight lattice *contraction* in small (< 3 nm) particles[108,112,160] (see however Ref. 130).

In the Au/Al$_2$O$_3$/NiAl(100) system, hemispherical particles occur even at low coverage,[7] unlike the situation with titania; size distribution was narrow, and particles were stable to 600 K, implying low mobility of adsorbed atoms. Paradoxically, on alumina large particles migrate and coalesce faster than small ones, presumably because the metal–support interaction is weaker; but with Au/FeO the diffusivity of atoms is higher due to a lower concentration of surface defects.

One cannot be certain that the effects observed with model systems, especially those concerning strength of adhesion and contact angle, are necessarily reproduced in catalysts made by more conventional methods or

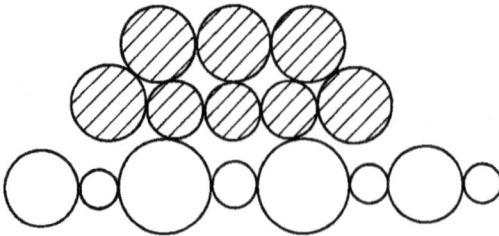

Figure 3.7: Illustration of the mutual disturbance of the gold and MgO lattices at the interface of Au/MgO(100).

on hydroxylated supports, but they serve to remind us of factors that quite probably are imitated by catalysts put to more practical use.

It will be clear from the work on the Au/MgO system mentioned above that there can be mutual effects at the metal–support interface; in the following sections an effort is made to separate the effects working in each direction.

3.5.2. Influence of the support on gold particles

Not many techniques provide direct evidence for an effect of the support on gold particles, and exchange of charge in either direction would probably be limited by the resulting image charge created. However there is evidence that the difference in binding energy (BE) of the $Au4f_{7/2}$ level from that of bulk gold varies with the support for gold particles of about the same size. The increase of BE with Au/SiO_2 (1.6 eV) was greater, and started at a larger size, than with $Au/TiO_2(110)$[54] (0.8 eV, see Figure 3.8), while in other work[171] the sequence with 3 nm particles was

$$Au/ZrO_2 > Au/TiO_2 > Au/Al_2O_3$$

but no correlation with activity for oxidation of carbon monoxide was found. The IR spectrum of adsorbed carbon monoxide is often a sensitive indicator of changes of electron density, but its adsorption on gold particles deposited

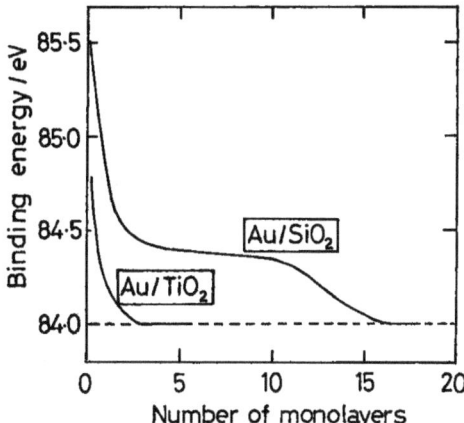

Figure 3.8: Dependence of binding energy of the $4f_{7/2}$ electrons on gold coverage (number of monolayers) for 'model' Au/TiO_2 and Au/SiO_2 systems.[54]

on FeO(111), Fe_3O_4(111) and thin alumina film showed no significant differences.[172] There are good indications that gold particles are located over anion vacancies on the support,[116] but theoretical calculations suggest that any transferred electron would stay near the vacancy, the conduction electrons effectively screening the bulk of the particle from the interface, and hence limiting the region affected. Calculations on Au/TiO_2(110) showed[173] that the shift of the $5d$ band towards the Fermi level increased with gold coverage, and the density of states for only two monolayers was almost the same as for bulk gold. The evidence for electronic interaction of the support on gold particles in systems of catalytic interest is therefore slight.

There is, however, one sense in which the support may influence the state of the gold. Ceria and zirconia appear to stabilise ionic states of gold,[75] and to increase their charge, by a process such as

$$Au^0 + Ce^{4+} \rightarrow Au^+ + Ce^{3+} \tag{3.A}$$

utilising perhaps gold atoms at the periphery of particles. This is probably the route whereby gold atoms dissolve into other supports having reducible cations, such as titania, ferric oxide (see Sections 6.5.2.a and f and the following section). Basic oxides (La_2O_3, Cs_2O) decrease the effective ionic charge.[75]

3.5.3.   Influence of gold particles on the support

Reciprocally the presence of gold particles may influence the properties of the support. The reduction of reducible supports (TiO_2, CeO_2, Fe_2O_3) by hydrogen is catalysed by gold;[174–176] this could be caused by hydrogen spillover from the metal or by changes induced in the electronic properties of the support. For example, gold alters the band gap of titania[176] and ceria, and causes a shift in the $Ce^{4+} \leftarrow O^{2-}$ charge-transfer band in the UV. With Au/Fe_2O_3, reduction of the support occurs during carbon monoxide oxidation,[177] and causes phase transformation of the support,[178] but gold retards the anatase $\rightarrow$ rutile phase change,[179] normally occurring at about 970 K and notoriously sensitive to the presence of impurities. Using laser Raman spectroscopy on Au/TiO_2 (anatase), the intensity of the E_{1g} mode at 145 cm^{-1} arising from the extension vibration of the anatase structure was drastically reduced by the presence of gold,[180] which also increased the lattice constant of ceria, although this could also have been due to a decrease in its particle size.

Titanium ions in the neighbourhood of gold particles have exhibited electronic changes, the Ti2p level suffering a band-bending of 0.15 eV. Argon ion bombardment of TiO$_2$(110) created oxygen vacancies and Ti^{3+} ions;[59] deposition of gold atoms on these defects led to a charge transfer from gold to the Ti^{3+} centres and this caused a *negative* shift in binding energy that cancelled out the *positive* shift seen on a defect-free surface. At the same time the signal due to the Ti^{3+} ions disappeared and the transfer produced a strong depletion of charge in the valence band close to the Fermi energy, i.e. it was an initial state effect. The negative charge left by the oxygen removal was calculated not to remain just on adjacent Ti^{3+} ions but to be diffused over a wider region.

Oxygen vacancies are expected to be more abundant near to gold particles in consequence of the Schottky junction at a metal–semiconductor interface.[181] The charge polarisation or Coulombic barrier caused by the hetero-junction between metal and oxide leads to a transfer of charge from the oxide to the Fermi level of the metal, and the oxide vacancies thus created may have relevance in catalysis[182] (see Chapter 6). Charge transfer in this sense is possible with gold particles large enough to exhibit high electronegativity, but is less likely with small particles where transfer in the opposite sense is more likely.

References

1. G.C. Bond and D.T. Thompson, *Catal. Rev.-Sci. Eng.* **41** (1999) 319.
2. G.C. Bond, *Surf. Sci.* **156** (1985) 966.
3. J.R. Anderson, *Structure of Metallic Catalysts*, Academic Press, London, 1975.
4. R. van Hardeveld and F. Hartog, *Surf. Sci.* **15** (1969) 189.
5. M.M. Alvarez, J.T. Khoury, T.G. Schaaff, M.N. Shafigullin, I. Vezmar and R.L. Whetten, *J. Phys. Chem. B.* **101** (1997) 3706.
6. W. Romanowski, *Highly Dispersed Metals*, Ellis Horwood, Chichester, 1987.
7. R. Meyer, C. Lemire, Sh.K. Shakhutdinov and H.-J. Freund, *Gold Bull.* **37** (2004) 72.
8. M.B. Knicelbein, *Ann. Rev. Phys. Chem.* **50** (1999) 79.
9. D. Stolcic, M. Fischer, G. Ganteför, Young Dok Kim, Q. Sun and P. Jena, *J. Am. Chem. Soc.* **125** (2003) 2848.
10. M.M. Kappes and E. Schumacher, *Surf. Sci.* **156** (1985) 1.
11. U. Heiz, A. Sanchez, S. Abbet and W.-D. Scheider, *Chem. Phys.* **262** (2000) 189.
12. A. Sanchez, S. Abbet, U. Heiz, W.-D. Scheider, H. Häkkinen, R.N. Barnett and U. Landman, *J. Phys. Chem. A.* **103** (1999) 9573.
13. G. Martra, L. Prati, C. Manfredotti, S. Biella, M. Rossi and S. Coluccia, *J. Phys. Chem. B* **107** (2003) 5433.
14. J.D. Grunwaldt, C. Kiener, C. Wögenbauer and A. Baiker, *J. Catal.* **181** (1999) 223.

15. G.C. Bond and J. Turkevich, *Trans. Faraday Soc.* **52** (1956) 1235.
16. G. Schmid and B. Corain, *Eur. J. Inorg. Chem.* (2003) 3081.
17. G. Schmid, *Clusters and Colloids, from Theory to Application*, VCH, Weinheim, 1994.
18. V.H. Perez-Luna, K. Aslan and P. Betala in *Encyclopaedia of Nanoscience and Nanotechnology*, H.S. Nalwa, (ed.), Amer. Sci. Publ., Stevenson Ranch, Calif., 2004, Vol. 2, p. 27.
19. P.J. Dyson and D.M.P. Mingos in *Gold: Progress in Chemistry, Biochemistry and Technology*, H. Schmidbaur, (ed.) Wiley, Chichester, 1999, p. 512.
20. J.R. Partington, *A Comprehensive Treatise of Inorganic and Theoretical Chemistry*, Vol. III, Longmans Green, London, 1923, pp. 491–618.
21. S. Link and M.A. El-Sayed, *Internat. Rev. Phys. Chem.* **19** (2000) 409.
22. M.-C. Daniel and D. Astruc, *Chem. Rev.* **104** (2004) 293–346.
23. J.P. Wilcoxon, R.L. Williamson and R. Baughman, *J. Chem. Phys.* **98** (1993) 9933.
24. M. Faraday, *Phil. Trans.* **147** (1857) 145; see also W.D. Mogerman, *Gold Bull.* **7** (1974) 22.
25. J. Turkevich, *Gold Bull.* **18** (1985) 86, 123.
26. *Colloidal Gold — Principles, Methods and Applications*, M.A. Hayat, (ed.), Academic Press, San Diego, Vols. 1 and 2, 1989; Vol. 3, 1991.
27. T.K. Sau and C.J. Murphy, *J. Am. Chem. Soc.* **126** (2004) 8648.
28. S.S. Saraiva and J.F. de Oliveira, *J. Disper. Sci. Technol.* **23** (2002) 837.
29. K. Takynama, *Bull. Chem. Soc.* Japan **31** (1958) 544.
30. J. Turkevich, J. Hillier and P.C. Stevenson, *Discuss. Faraday Soc.* **11** (1951) 55.
31. J. Turkevich, G. Garton and P.C. Stevenson, *J. Colloid Sci., Suppl.1* (1954) 26.
32. G.V. Hartland, M. Hu, O. Wilson, P. Mulraney and J.E. Sader, *J. Phys. Chem. B* **106** (2002) 743.
33. K. Hayakawa, T. Yoshimura and K. Esumi, *Langmuir* **19** (2003) 5517.
34. M. Aslam, Lei Fu, Ming Su, K. Vijayamohanan and V.P. Dravid, *J. Mater. Chem.* **14** (2004) 1795.
35. K. Mallich and M.S. Scurrell, *Appl. Catal. A: Gen.* **253** (2003) 527; K. Mallick, M.J. Witcomb and M.S. Scurrell, *Appl. Catal. A: Gen.* **259** (2004) 163.
36. H. Tsunoyama, H. Sakurai, N. Ichikuni, Y. Negishi and T. Tsukuda, *Langmuir* **20** (2004) 11293.
37. E. Gachard, H. Remita, J. Khatouri, B. Keita, L. Nadjo and J. Belloni, *New J. Chem.* **22** (1998) 1257.
38. J. Belloni, M. Mostafavi, H. Remita, J.-L. Maringier and M.-O. Delcourt, *New J. Chem.* **22** (1998) 1239.
39. L.B. Hunt, *Gold Bull.* **9** (1976) 134.
40. P. Beltrame, M. Comotti, C. Della Pina and M. Rossi, *Appl. Catal. A: Gen.* **297** (2005) 1.
41. N. Dimitratos, F. Porta, L. Prati and A. Villa, *Catal. Lett.* **99** (2005) 181.
42. C.L. Bianchi, P. Canton, N. Dimitratos, F. Porta and L. Prati, *Catal. Today* **102–103** (2005) 203.
43. G. Mie, *Annal. Phys. Lpz.* **5** (1908) 377.
44. P.A. Sermon, G.C. Bond and P.B. Wells, *J. Chem. Soc. Faraday Trans. I* **75** (1979) 385.
45. G.C. Bond and P.A. Sermon, *Gold Bull.* **6** (1973) 102.
46. S. Galvagno and G. Parravano, *J. Catal.* **55** (1978) 178.
47. M. Lee, L. Chae and K.C. Lee, *Nanostr. Mater.* **11** (1999) 195.

48. M. Arai, M. Mitsui, J.-I. Ozaki and Y. Nishiyama, *J. Coll. Interfac. Sci.* **168** (1994) 473.
49. P.A. Sermon, J.M. Thomas, K. Keryou and G.R. Millward, *Angew. Chem. Int. Edn.* **26** (1987) 918.
50. A. Sra and R.E. Schaak, *J. Am. Chem. Soc.* **126** (2004) 6667.
51. B.M. Leonard, N.S.P. Bhuvanesh and R.E. Schaak, *J. Am. Chem. Soc.* **127** (2005) 7326.
52. C.L. Bianchi, S. Biella, A. Gervasini, L. Prati and M. Rossi, *Catal. Lett.* **85** (2003) 91.
53. A. Vijay, G. Mills and H. Metiu, *J. Chem. Phys.* **118** (2003) 6536.
54. C.C. Chusei, X. Lai, K. Luo and D.W. Goodman, *Topics in Catal.* **14** (2001) 71.
55. W. Vogel, J. Bradley, O. Vollmer and I. Abraham, *J. Phys. Chem. B* **102** (1998) 10853.
56. S. Stoeva, K.J. Klabunde, C.M. Sorensen and I. Dragieva, *J. Am. Chem. Soc.* **124** (2002) 2305.
57. R.W. Devenish, T. Goulding, B.T. Heaton and R. Whyman, *J. Chem. Soc. Dalton Trans.* (1996) 673.
58. J. Radaik, C. Mohr and P. Claus, *Phys. Chem. Chem. Phys.* **5** (2003) 172.
59. S. Schimpf, M. Lucas, C. Mohr, U. Rodemerck, A. Brückner, J. Radnik, H. Hofmeister and P. Claus, *Catal. Today* **72** (2002) 63.
60. M. Okumura, S. Tsubota, M. Iwamoto and M. Haruta, *Chem. Lett.* (1998) 315.
61. G.C. Bond, *Metal-Catalysed Reactions of Hydrocarbons*, Springer, New York, 2005.
62. V. Ponec and G.C. Bond, *Catalysis by Metals and Alloys*, Elsevier, Amsterdam, 1996.
63. R. Adams and V. Voorhees, *J. Am. Chem. Soc.* **44** (1922) 183.
64. M. Raney, *Ind. Eng. Chem.* **32** (1940) 1199; *US Patent* 1628190 (1927).
65. M.B. Cortie, A.I. Maarhoof and G.B. Smith, *Gold Bull.* **38** (2005) 1.
66. M.A. Sanchez-Castillo, C. Couto, Won Bae Kim and J.A. Dumesic, *Angew. Chem. Int. Ed.* **43** (2004) 1140.
67. J.R. Anderson, *Structure of Metallic Catalysts*, Academic Press, London, 1975.
68. M. Che and C.O. Bennett, *Adv. Catal.* **36** (1989) 55.
69. J.W. Niemantsverdriet, *Spectroscopy in Catalysis,* VCH, Weinheim, 1993.
70. K. Christmann, *Introduction to Surface Physical Chemistry*, Steinkopff, Darmstadt, 1991.
71. *Handbook of Heterogeneous Catalysis,* Vol. 2, G. Ertl, H. Knözinger and J. Weitkamp, (eds.), Wiley-VCH, Weinheim, 1997.
72. G. Cocco, S. Enzo, G. Fagherazzi, L. Schiffino, I.W. Bassi, G. Vlaic, S. Galvagno and G. Parravano, *J. Phys. Chem.* **83** (1979) 2527.
73. J.T. Miller, A.J. Kropf, Y. Zha, J.R. Regalbuto, L. Delannoy, C. Louis, E. Bus and J.A. van Bokhoven, *J. Catal.* **240** (2006) 222.
74. S. Giorgi, C. Chapon, C.R. Henry, G. Nihoul and J.M. Penisson, *Phil. Mag.* **64** (1991) 87.
75. D.C. Meier, X.-F. Lai and D.W. Goodman in: *Surface Chemistry and Catalysis,* A.F. Carley, P.R. Davies, G.J. Hutchings and M.S. Spencer, (eds.), Kluwer Academic/Plenum, New York, 2002, p. 147.
76. L. Guczi, D. Horváth, Z. Pászti, L. Toth, Z.E. Horváth. A. Kavacs and G. Petõ, *J. Phys. Chem. B* **104** (2000) 3183.
77. D.A.H. Cunningham, W. Vogel, R.M. Torres-Sanchez, K. Tanaka and M. Haruta, *J. Catal.* **183** (1999) 24.

78. L. Guczi, G. Petö, A. Beck, K. Frey, O. Geszti, G. Molnár and C. Daróczi, *J. Am. Chem. Soc.* **125** (2003) 4332.

79. E. Wahlstrom, N. Lopez, R. Schaub, B. Thostrap, A. Ronnau, C. Africh, E. Laesgaard, J.K. Nørskov and F. Besenbacher, *Phys. Rev. Lett.* **90** (2003) 026101/1-4.

80. C. Xu, S. Oh, G. Liu, D.Y. Kim and D.W. Goodman, *J. Vac. Sci. Technol. A* **15** (1997) 1261.

81. A.M. Venezia, G. Pantaleo, A. Longo, G. Di Carlo, M.P. Casaletto, F.L. Liotta and G. Deganello, *J. Phys. Chem. B.* **109** (2005) 2821.

82. W.-F. Yan, V. Petkov, S.M. Mahurin, S.H. Overbury and S. Dai, *Catal. Comm.* **6** (2005) 404.

83. W. Vogel, D.A.H. Cunningham, K. Tanaka and M. Haruta, *Catal. Lett.* **40** (1996) 175.

84. A. Benedetti, L. Bertoldo, P. Canton, G. Goerigk, F. Pinna, P. Riello and S. Polizzi, *Catal. Today* **49** (1999) 485.

85. J. Guzman and B.C. Gates, *J. Phys. Chem. B* **107** (2003) 2242.

86. M.A. Marcus, M.P. Andrews, J. Zegenhagen, A.S. Bommannavar and P. Montano, *Phys. Rev. B.* **42** (1990) 3312.

87. G. Compagnini, *Mater. Sci. Eng. C: Biomim. Supramol. Systems* **C19** (2002) 181.

88. R.S. Cataliotti, G. Compagnini, C. Crisafulli, S. Minicò, B. Pignataro, P. Sassi and S. Scirè, *Surf. Sci.* **494** (2001) 75.

89. M. Tokita and E. Haga, *J. Phys. Soc. Japan* **50** (1981) 482–489.

90. K. Zangger and I.M. Armitage, *Metal-Based Drugs* **6** (1999) 239.

91. G.U. Kulkarni, C.P. Vinod and C.N.R. Rao in: *Surface Chemistry and Catalysis*, A.F. Carley, P.R. Davies, G.J. Hutchings and M.S. Spencer, (eds.), Kluwer Academic/Plenum, New York, 2002, p. 191.

92. C.N.R. Rao, V. Vijayakrishnan, H.N. Aiyer, G.U. Kulkarni and G.N. Subbanna, *J. Phys. Chem.* **97** (1993) 11157.

93. H. Tsai, E. Hu, K. Perng, M. Chen, J.-C. Wu and Y.-S. Chang, *Surf. Sci.* **537** (2003) L447.

94. A.M. Venezia, G. Pantaleo, A. Longo, G. di Carlo, M.P. Casaletto, F.L. Liotta and G. Deganello, *J. Phys. Chem. B* **109** (2005) 2821.

95. H. Roulet, J.M. Mariot, G. Dufour and C.F. Hague, *J. Phys. F: Metal Phys.* **10** (1980) 1025.

96. J.W. Niemantsverdriet and T. Butz in: *Handbook of Heterogeneous Catalysis*, G. Ertl, H. Knözinger and J. Weitkamp, (eds.), Wiley-VCH, Weinheim, 1997, Vol. 2, p. 512.

97. L. Stievano, S. Santucci, L. Lozzi, S. Calozero and F.E. Wagner, *J. Non-Cryst. Solids* **234** (1998) 644.

98. A. Goosens, M.W.J. Crajé, A.M. van der Kraan, A. Zurijnenburg, M. Makee, J.A. Moulijn, R.J.H. Grisel, B.E. Nieuwenhuys and L.J. de Jongh, *Hyperfine Interact.* **139/140** (2002) 59; *Catal. Today* **72** (2002) 95

99. Y. Kobayashi, S. Nasu, S. Tsubota and M. Haruta, *Hyperfine Interact.* **126** (2000) 95.

100. P.M. Paulus, A. Goossens, R.C. Thiel, G. Schmid, A.M. van der Kraan and L.J. de Johngh, *Hyperfine Interact.* **126** (2000) 194.

101. L. Stievano and F.E. Wagner, *AIP Conference Proc.* **765** (2005) 3.

102. W.N. Delgass, M. Boudart and G. Parravano, *J. Phys. Chem.* **72** (1968) 3563.

103. N.A. Hodge, C.J. Kiely, R. Whyman, M.R.H. Siddiqui, G.J. Hutchings, Q.A. Pankhurst, F.E. Wagner, R.R. Rajaram and S.E. Golunski, *Catal. Today* **72** (2002) 133.

104. R.M. Finch, N.A. Hodge, G.J. Hutchings, A. Meagher, Q.A. Pankhurst, M.R.H. Siddiqi, F.E. Wagner and R. Whyman, *Phys. Chem. Chem. Phys.* **1** (1999) 485.

105. F.E. Wagner, S. Galvagno, C. Milone, A.M. Visco, L. Stievano and S. Calogero, *J. Chem. Soc. Faraday Trans.* **93** (1997) 3403.

106. S.T. Daniells, A.R. Overweg, M. Makkee and J.A. Moulijn, *J. Catal.* **230** (2005) 52.

107. I.W. Bassi, F.W. Lytle and G. Parravano, *J. Catal.* **42** (1976) 139.

108. Eun Duck Park and Jae Sung Lee, *J. Catal.* **186** (1999) 1.

109. J. Guzman and B.C. Gates, *J. Am. Chem. Soc.* **126** (2004) 2672.

110. J.C. Fierro-Gonzalez and B.C. Gates, *J. Phys. Chem. B* **109** (2005) 7275.

111. N. Lopez, J.K. Nørskov, T.V.W. Janssens, A. Carlson, A. Puig-Molina, B.S. Clausen and J.-D. Grunwaldt, *J. Catal.* **225** (2004) 86.

112. J.H. Yang, J.D. Henno, M.P. Raphulu, Y.-M. Yang, T. Caputo, A.J. Groszek, M.C. Kung, M.S. Scurrell, J.T. Miller and H.H. Kung, *J. Phys. Chem. B* **109** (2005) 10319.

113. V. Schwartz, D.R. Mullins, W.F. Yan, B. Chem, S. Dai and S.H. Overbury, *J. Phys. Chem. B.* **108** (2004) 15782.

114. J.T. Calla and R.J. Davis, *Catal. Lett.* **99** (2005) 21.

115. M.C. Kung, C.K. Costello and H.H. Kung, in *Specialist Periodical Reports: Catalysis*, J.J. Spivey and G.W. Roberts, (eds.), Roy. Soc. Chem., London **17** (2004) 152.

116. K. Okazaki, S. Ichikawa, Y. Maeda, M. Haruta and M. Kohyama, *Appl. Catal. A: Gen.* **291** (2005) 45.

117. Y. Maeda, T. Akita, H. Okumura and M. Kohyama, *Jap. J. Appl. Phys.* **43** (2004) (7B) 4595.

118. S. Khikawa, T. Akita, M. Okumura, M. Haruta and K. Tanaka, *Mater. Res. Soc. Symp. Proc.* **727** (2002) 11.

119. C.P. Vinod, G.U. Kulkarni and C.N.R. Rao, *Chem. Phys. Lett.* **289** (1998) 329.

120. L. Fu, N.Q. Wu, J.H. Yang, F. Qu, D.L. Johnson, M.C. Kung, H.H. Kung and V.P. Dravid, *J. Phys. Chem. B* **109** (2005) 3704.

121. J. Margitfalvi and S. Göbölös, in *Specialist Periodical Reports: Catalysis*, J.J. Spivey and G.W. Roberts, (eds.), Roy. Soc. Chem., London **17** (2004) 1.

122. S. Schimpf, M. Lucas, C. Mohr, U. Rodemerck, A. Brückner, J. Radnik, H. Hofmeister and P. Claus, *Catal. Today* **72** (2002) 63.

123. H. Liu, A.I. Kozlov, A.P. Kozlova, T. Shido, K. Akasura and Y. Iwasawa, *J. Catal.* **185** (1999) 252.

124. M. Okumura, J.M. Coronado, J. Soria, M. Haruta and J.C. Conesa, *J. Catal.* **203** (2001) 168.

125. M.C. Fairbanks, R.E. Benfield, R.J. Newport and G. Schmid, *Solid State Commun.* **73** (1990) 431.

126. P. Schwerdtfeger, *Angew. Chem. Int. Ed.* **42** (2003) 1892.

127. L.R. Wallenberg, J.-O. Bovin and G. Schmid, *Surf. Sci.* **56** (1985) 256.

128. A. Balerna, E. Bernieri, P. Picozzi, A. Reale, S. Santucci, E. Burattini and S. Mobilio, *Surf. Sci.* **156** (1985) 206.

129. R. Burch in *Specialist Periodical Reports: Catalysis*, G.C. Bond and G. Webb, (eds.), Roy. Soc. Chem., London, 1985, Vol. 7, p. 149.

130. H. Tsuboyama, H. Sakurai, N. Ichikuni, Y. Negishi and T. Tsukuda, *Langmuir* **20** (2004) 11293.

131. A. Franceschetti, S.J. Pennycook and S.T. Pantelides, *Chem. Phys. Lett.* **374** (2003) 471.

132. K.J. Taylor, C. L. Pettiette-Hall, O. Chesnovsky and R.E. Smalley, *J. Chem. Phys.* **96** (1992) 3319.

133. M. Okumura, Y. Kitagawa, M. Haruta and K. Yamaguchi, *Appl. Catal. A: Gen.* **291** (2005) 37.

134. H.-G. Boyen, G. Kästle, F. Weigl, B. Kostowski, C. Dietrich, P. Ziemann, J.P. Spatz, S. Riethmüller, C. Hartman, M. Möller, G. Schmid, M.G. Garnier and P. Oelhafen, *Science* **297** (2002) 1533.

135. P.P. Edwards and M.J. Sienko, *Internat. Rev. Phys. Chem.* **3** (1983) 83; *Chem. Britain*, Jan. 1983, p. 39; *J. Chem. Educ.* **60** (1983) 691; *Acc. Chem. Res.* **15** (1982) 87.

136. P.P. Edwards, R.L. Johnston and C.N.R. Rao in: *Metal Clusters in Chemistry*, P. Braunstein, L.A. Oro and P.R. Raithby, (eds.), Wiley-VCH, Weinheim, 1999, Vol. 3.

137. A.N. Pestryakov, V.V. Lunin, A.N. Kharlanov, N.N. Bogdanchikova and I.V. Tuzovskaya, *Eur. Phys. J. D: Atom., Molec. and Opt. Physics* **24** (2003) 307.

138. A. Howard, D.N.S. Clark, C.E.J. Mitchell, R.G. Egdell and V.R. Dhanak, *Surf. Sci.* **518** (2002) 210.

139. I. Coulthard, S. Degen, Y.-J. Zhu and T.K. Sham, *Canad. J. Chem.* **76** (1998) 1707.

140. B. Richter, H. Kuhlenbeck, P.S. Bagus and H.-J. Freund, *Phys. Rev. Lett.* **93** (2004) 026805.

141. M. Valden, X. Lai and D.W. Goodman, *Science* **281** (1998) 1647.

142. L.F. Mattheiss and R.E. Dietz, *Phys. Rev. B* **22** (1980) 1663.

143. P.K. Jain, *Structural Chem.* **16** (2005) 421.

144. L.D. Burke, A.J. Ahern and A.P. O'Mullane,*Gold Bull.* **35** (2002) 3.

145. T. Castro, R. Reifenberger, E. Choi and R.P. Andres, *Phys, Rev. B* **13** (1990) 8548.

146. M.M. Maye, W. Zheng, F.L. Leibowitz, N.K. Ly and C.-J. Zhong, *Langmuir* **16** (2000) 490.

147. Ph. Buffat and J.-P. Morel, *Phys. Rev. A* **13** (1976) 2287.

148. M.A. Listvag, *J. Molec. Catal.* **20** (1983) 265.

149. J. Ross and R.P. Andres, *Surf. Sci.* **106** (1983) 11.

150. G.C. Bond in *Handbook of Heterogeneous Catalysis*, G. Ertl, H. Knözinger and J. Weitkamp, (eds.) VCH, Weinheim, 1997, Vol. 2, p. 752.

151. M. Jakob, H. Levanon and P.K. Kamat, *Nano Lett.* **3** (2003) 353.

152. H. Häkkinen, S. Abbet, A. Sanchez, U. Heiz and U. Landman, *Angew. Chem. Int. Ed.* **42** (2003) 1297.

153. M. Haruta, *Catal. Today* **36** (1997) 153.

154. D. Chatain, F. Chabert, V. Ghetta and J. Fouletier, *J. Am. Ceram. Soc.* **73** (1996) 1568.

155. S. Giorgio, C. Chapon, C.R. Henry, G. Nihoul and J.M. Penisson, *Phil. Mag. A.* **64** (1991) 87.

156. S. Giorgio, C.R. Henry, B. Pauwels and G.P. Tendeloo, *Mater. Sci. Eng. A* **297** (2001) 197.

157. F. Cosandey and T.E. Madey, *Surf. Rev. Lett.* **8** (2001) 73.

158. M. Haruta, *CATTECH* **6** (2002) 102.

159. D.A.H. Cunningham, W. Vogel, H. Kageyama, S. Tsubota and M. Haruta, *J. Catal.* **177** (1998) 1.

160. R. Zanella, S. Giorgio, C.-H. Shin, C.R. Henry and C. Louis, *J. Catal.* **222** (2004) 357.

161. R. Zanella, C. Louis, S. Giorgio and R. Touroude, *J. Catal.* **223** (2004) 328.

162. S. Tsubota, T. Nakamura, K. Tanaka and M. Haruta, *Catal. Lett.* **56** (1998) 131.
163. J.-D. Grunwaldt, M. Maciejewski, O.S. Becker, A. Fabrizioli and A. Baiker, *J. Catal.* **186** (1999) 458.
164. G.C. Bond and D.T. Thompson, *Gold Bull.* **33** (2000) 41.
165. S.T. Lee, G. Apai, M.G. Mason, R. Benbow and Z. Hurych, *Phys. Rev. B* **23** (1997) 505.
166. M.A. Centeno, C. Portales, I. Carrizosa and J.A. Odriozola, *Catal. Lett.* **102** (2005) 289.
167. Z.-P. Liu, S.J. Jenkins and D.A. King, *Phys. Rev. Lett.* **93** (2004) 156102.
168. T. Akita, K. Tanaka, S. Tsubota and M. Haruta, *J. Electron Microsc.* **49** (2000) 657.
169. C. Mohr, H. Hofmeister and P. Claus, *J. Catal.* **213** (2003) 86.
170. D.A.H. Cunningham, W. Vogel, H. Kagayama, S. Tsubota and M. Haruta, *J. Catal.* **177** (1998) 1.
171. S. Arü, F. Mortin, A.J. Renouprez and J.C. Rousset, *J. Am. Chem. Soc.* **126** (2004) 1199.
172. S. Shaikhutdinov, R. Meyer, M. Naschitzki, M. Bäumer and H.-J. Freund, *Catal. Lett.* **86** (2003) 211.
173. N. Lopez and J.K. Nørskov, *Surf. Sci.* **515** (2002) 175.
174. M. Khoudiakov, M.-C. Gupta and S. Deevi, *Appl. Catal. A: Gen.* **291** (2005) 151.
175. L.I. Llieva, D.H. Andreeva and A.A. Andreev, *Thermochim. Acta* **292** (1997) 169.
176. M.A. Dabeila, M.C. Raphulu, E. Mokoena, M. Avalos, V. Petranovskü, N.J. Coville and M.S. Scurrell, *Mater. Sci. Eng. A* **396** (2005) 70.
177. M.M. Schubert, S. Hackenberg, A.C. van Veen, M. Muhler, V. Plzak and R.J. Behm, *J. Catal.* **197** (2001) 113.
178. F.E. Wagner, S. Galvagno, C. Milone, A.M. Visco, L. Stievano and S. Calogero, *J. Chem. Soc. Faraday Trans. I* **93** (1997) 3403.
179. M.A. Debeila, M.C. Raphulu, E. Mokoena, M. Avalos, V. Petranovskü, N.J. Coville and M.S. Scurrell, *Mater. Sci. Eng. A* **396** (2005) 61.
180. W.-L. Deng, J. De Jesus, H. Saltsburg and M. Flytzani-Stephanopoulos, *Appl. Catal. A: Gen.* **291** (2005) 126.
181. J.C. Frost, *Nature* **334** (1988) 577.
182. M. Okumura, Y. Kitagawa, M. Haruta and K. Yamaguchi, *Appl. Catal. A: Gen* **291** (2005) 37.

CHAPTER 4

Preparation of Supported Gold Catalysts

4.1. Introduction

4.1.1. Principles of the preparation methods and definitions

It is desirable to begin this chapter with some general remarks about of use of terms to describe methods of preparation, because they are sometimes used loosely and inaccurately. The methods fall into two classes: (i) the support and the gold precursor are formed at the same time and (ii) the gold precursor is applied to the preformed support. The procedure for method (i) is termed *coprecipitation* (Section 4.2.2), and although capable of many minor variations (e.g. as in the sol–gel method, Section 4.4.1), it does not have the same wealth of alternative procedures as does the second method. These may be listed as follows.

In *impregnation* the pores of the support are filled with a solution of the gold precursor. Where just the needful volume is used, the method is *impregnation to incipient wetness*, but sometimes an excess of solution is used and the solvent removed by evaporation, so that concentrated solution finally resides in the pores (Sections 4.2.1 and 4.3.1). In these methods, the metal dispersion ultimately obtained depends critically on the conditions of drying.[1]

In *ion adsorption*, either cationic gold species are adsorbed through electrostatic interaction on surface O^- groups of the support when the solution pH is higher than the point of zero charge (PZC) (Section 4.2.5), or less commonly anionic gold precursors are adsorbed on surface OH_2^+ groups when the solution pH is lower than the PZC of the support (Section 4.3.5). Hydroxyls on oxide surface may be protonated or deprotonated depending on the solution pH, and the value of pH for which the total electric charge of the surface is zero is the PZC. These processes of *ion adsorption* have to be distinguished from those in which precursor species interact with the support surface through a chemical process that can be termed *grafting*, which may take the form of an hydrolysis of the precursor by hydroxyl groups on the surface. The term is generally appropriate for *chemical vapour deposition* (Section 4.3.4) and for methods involving organogold complexes (Sections 4.3.2 and 4.3.3), but we will see that it also occurs during other types of preparation.

The term deposition–precipitation (DP) is often employed in the preparation of gold catalysts (Sections 4.2.3 and 4.2.4), but most often incorrectly. Strictly speaking, it implies a process whereby the hydroxide or hydrated oxide is deposited onto the surface of the support as the result of gradually raising the pH of the solution in which the support is suspended. The precipitate may be nucleated by the support surface, and when properly performed, all the active phase becomes attached to the support and none is left floating by itself. This procedure was originally developed by Geus and co-workers,[2,3] for making supported nickel and copper catalysts, and the somewhat complex physical chemistry involved has been fully discussed.[2,4] It would seem that for preparation of gold catalysts, the term DP has been applied when the reactions occurring might be more accurately described as grafting or ion adsorption as it will be shown in Section 4.2.3.

Other manifestations of the second class of method include: deposition of colloidal gold onto a support (Section 4.3.6), and photochemical or sonochemical activation of the precursor to encourage its interaction with the support.

The size of gold particles is a very important parameter in obtaining active catalysts[5–8] for many reactions. For most of the reactions, only the catalysts with gold particles smaller than 5 nm lead to high activity;[9,10] this is especially true for the oxidation of carbon monoxide[11–13] (see Chapter 6). The method of preparation strongly influences the particle size.[9,13–15] The goal of this chapter is to gather together the methods of preparation used for gold supported on powder oxides and other supports. The preparation and use of massive metal surfaces (single crystals, film, foil, etc.) have been described elsewhere,[16] and the structures of the surfaces of gold single crystals were introduced in Section 2.5.2. Methods for making small unsupported gold particles (gaseous clusters, colloids, etc.) were treated in Section 3.2, and the procedure for making 'model' catalysts by depositing atoms or clusters onto flat oxide surfaces was also mentioned there. In this chapter, only the main methods for making gold catalysts on powder supports will be described with their advantages and disadvantages. One must note that each method admits of numerous variations, many of which have been used in various laboratories. Each possible variation is a potential source of influence on the composition and structure of the finished catalyst, but unfortunately the methods used are often not described in adequate detail, many important aspects being either not noted, or totally ignored. It is therefore very difficult to relate performance to the procedure used in the preparation.

We will try to separate as far as possible the procedures of preparation from the chemical reactions that occur during these preparations. For instance in the case of impregnation to incipient wetness, both the metal ions and the counter-ions are deposited on the support, but the speciation of the metal-containing ions may vary with the pH of the solution in contact with the support, and the way in which they interact with the support will also depend on the PZC of the support.

The precursors used in preparing supported gold catalysts are salts or complexes where gold is usually in the $+3$ oxidation state; a few of them are in the $+1$ oxidation state, which is more unstable. The Au^{III} precursor most often used is chloroauric acid ($HAuCl_4 \cdot 3H_2O$), which is a commercially available orange solid; in aqueous solution it is a strong acid, quite capable of dissolving alumina and magnesia. Moreover, the speciation of the gold ions depends strongly on the concentration, pH and temperature of the solution (Section 4.1.2). Gold chloride ($AuCl_3$) is also occasionally used.

Thus, for most of the preparation methods the gold is in $+3$ oxidation state after drying, but it is easily reduced to the zero state by thermal treatments, which can be performed with any gas, reducing or oxidising. Indeed, oxidic forms of gold are unstable in air, since auric oxide (Au_2O_3) is formed endothermically. Thermal treatments are often carried out in air, but it is important that the metal is formed under controlled conditions because variables such as the nature of the gas, the flow rate, the heating rate and the final temperature, all influence the particle size (see Section 4.7). The preparation method, the thermal treatment and the nature of the oxide support may also affect the morphology of the particles. Precursors to gold catalysts are not easy to handle because they are very sensitive to ambient conditions (light and air), and recommendations for the conditions of preparation and storage are made in Section 4.8.

4.1.2. Gold speciation

When $HAuCl_4$ is dissolved in water, the chloroauric anion hydrolyses to form anionic hydroxychloro-gold(III) complexes $[Au(OH)_xCl_{4-x}]^-$. The increase in pH induces changes of colour of the solution from yellow to colourless, indicating changes in gold speciation. Thermodynamic calculations point out that three kinds of reactions occur as the pH is raised:[17] (i) hydrolysis by replacement of chloride ion by hydroxyl ion; (ii) displacement of chloride ion from a complex anion by water, giving a neutral species; and (iii) loss of a proton from a neutral hydrated ion, Several groups[18-20] characterised the gold speciation as a function of pH by various techniques. As shown in Table 4.1, the increase of pH results in deeper hydrolysis of the $[AuCl_4]^-$ complex. The results are not all in good agreement, but they were not performed under the same conditions, and the extent of hydrolysis also depends on the gold and chlorine concentrations, i.e. on the ionic strength. The sequence of speciation does not depend on the gold concentration, but is shifted to lower pH.[20,21] It is noteworthy that these descriptions do not report the presence of neutral gold species in contrast with the thermodynamic predictions [reaction (ii)]. Hydroxychloro-gold(III) complexes were found more stable than predicted by the equilibrium constants arising from thermodynamic calculation or potentiometric constants.[19] As a consequence, the formation of $Au(OH)_3$ predicted by these calculations[22-24] was not observed; it was even highlighted that a solution of $HAuCl_4$ (10^{-2} M) and NaCl (1 M) at pH 9.2 does not lead to the formation of gold hydroxide in contrast to the predictions.[19]

The ageing time of a gold solution at fixed pH and at room temperature also influences the concentrations of the gold complexes, because equilibration of the speciation is established slowly.[21,25-27] The increase of temperature of $HAuCl_4$ solution increases the hydrolysis of gold species. For instance, $[AuCl_4]^-$ at 294 K transforms into $[Au(OH)Cl_3]^-$ at 323 or 463 K.[28] Solutions at pH values of 5.75 and 6.5, which contained different hydroxychloro-gold(III) complexes at ambient temperature (Table 4.1, Ref. 3), showed changes in speciation with increasing temperature.[29] In the solution at pH 5.75, hydroxyl groups were replaced by chloride ligands on heating, resulting in a transformation from $[AuCl_3(OH)]^-$ to $[AuCl_4]^-$, and in the solution at pH 6.5, there was an increase in the number of hydroxyl ligands, from $[Au(OH)_2Cl_2]^-$ to a mixture with $[Au(OH)_3Cl]^-$ between 323 and 373 K.

4.1.3. Fulminating gold

The question of the hazards of explosive fulminating gold must be addressed. Fulminating gold is a family of gold compounds containing

Table 4.1: Gold speciation in $HAuCl_4$ solutions at room temperature as a function of pH.

HAuCl₄ solution	pH or range of pH of predominance of the various Au^{III} speciation					Characterisation technique	References
	$[AuCl_4]^-$	$[Au(OH)Cl_3]^-$	$[Au(OH)_2Cl_2]^-$	$[Au(OH)_3Cl]^-$	$[Au(OH)_4]^-$		
10^{-2} M in NaCl (1M)[a]	pH 1.4–6.2	pH 6.2–8.1	pH 8.1–11	pH 11–12	pH 12?	Raman, UV-visible	18
10^{-1}–10^{-3} M[a]	pH 2	pH 7.5	pH 9.2			XANES, EXAFS	19
2×10^{-2} M[b]	pH 1–3.8	pH 3.8–5.2	pH 5.2–6.6	pH 8.2–11.2	pH 11.2–12	Raman	20
$1 - 2.5 \times 10^{-3}$ M	pH ≤ 4				pH 7	EXAFS	30

[a]pH range of predominance of the various Au^{III} speciation.

[b]Each pH range corresponds to a single species except at pH 6.6–8.2: mixture of Au(OH)$_2$Cl$_2^-$ and Au(OH)$_3$Cl$^-$.

nitrogen. The dried compound *explodes very violently*, upon a mere touch.[31] It was reported that *"supported metal catalysts which contain gold should never be prepared by impregnation of a support with solutions which contain both gold salts and NH_4OH. The dried catalysts contain extremely shock sensitive gold–nitrogen compounds which may explode with the lightest touch"*.[32] This problem was also reported by Johnson Matthey.[33] However, gold catalysts have been prepared this way using careful and thorough washing without explosions being experienced,[26,34–37] but such good fortune cannot be guaranteed.

The composition of fulminating gold is not well formulated and depends on the preparative route, but it contains Au–N bonds. It usually arises from a reaction between gold oxide, hydroxide or chloride and aqueous ammonia or an ammonium salt.[31,38] All detonate to give gold, nitrogen, ammonia and water. Fulminating gold is variously described as a dirty olive-green powder when it is prepared from auric hydroxide and ammonia, or as a yellow precipitate or a black powder when it is prepared from auric oxide and concentrated ammonia. Several formulae have been proposed: $AuN \cdot NH_3 \cdot 1.5H_2O$, $Au_2O_3 \cdot 4NH_3$, $Au_2O_3 \cdot xNH_3$ + $NH(AuNH_2Cl)_2$ or gold hydrazide $(AuNH \cdot NH_2) \cdot 3H_2O$, but it could be also an ammine $Au[(NH_3)_2(OH)_2]OH$.[39,40]

In conclusion, we recommend that if ammonia is used to prepare gold catalysts, great care must be exercised since the chemistry of fulminating gold is not well known. For instance, the $[Au(NH_3)_4]^{3+}$ complex can only be obtained by very slow addition of concentrated ammonia to a solution of $HAuCl_4$ while keeping the pH lower than 5.[39] However, if the ammonia solution is added too fast or a large excess is used, an explosive fulminating gold compound is formed.[39,40] In fact, it would be better to find alternative preparative routes, which do not involve the use of ammonia since there are too many unknowns involved in the explosive hazards; the use of ammonia is unsuitable for scaled-up manufacturing of gold catalysts.

4.2. Methods of Preparation using Gold Chloride Precursors

4.2.1. Impregnation

The very first supported gold catalysts were prepared by *impregnation* (IMP)[9,41] since this is the simplest method, and can be used with any support. The precursors most often used were chloroauric acid

(HAuCl$_4$)[41-45] and gold chloride (AuCl$_3$);[46] the ethylenediamine complex ([Au(en)$_2$]Cl$_3$)[44,47] has also been used. Silica, alumina and magnesia were the first oxide supports used, but catalysts were also prepared with titania and ferric oxide. Most of these preparations led after thermal treatment to large gold particles (10–35 nm) even for low gold loading (1–2 wt.%), and to catalysts with poor activity.

Large particles may arise because the presence of chloride ion promotes mobility and agglomeration of gold species during thermal treatment.[10,48,49] In addition, it is known to be a poison for several catalytic reactions. It is still present on the support even after calcination to 873 K,[50] but reduction with hydrogen seems to remove it as HCl.[10] A steam treatment would be even more effective. Nevertheless, Au/Fe$_2$O$_3$ prepared by impregnation and reduced in hydrogen still exhibited large gold particles (10–30 nm).[51]

Some of the AuCl$_4^-$ ions (or species derived from it) interact with titania during incipient wetness impregnation at the natural pH of the solution ($<$1), since not all of it is removed by washing with water (remaining gold loading 0.6–0.9 wt.%).[52] Then after calcination the mean particle size was much smaller (3 nm) than for an unwashed sample ($>$10 nm). This is not however a practicable method for obtaining well-dispersed gold particles, because of the necessity to recover and re-use the gold lost in the washing.

4.2.2. Coprecipitation

Small gold particles supported on various oxides such as α-Fe$_2$O$_3$, NiO and Co$_3$O$_4$, were first obtained by *coprecipitation* (COPPT) in 1987.[15,53,54] This was performed by adding sodium carbonate to aqueous solutions of HAuCl$_4$ and the nitrate of the metal that will lead to production of the support. The coprecipitates were washed, dried, and calcined in air; the method led to high gold dispersion ($\bar{d}_{Au} <$10 nm) (Table 4.2), and to oxide surface areas larger than in the absence of gold. It is a single step method, easy to carry out. However, some of the gold particles could be embedded in the bulk of the support.[55] Au/TiO$_2$ can apparently only be successfully prepared in the presence of magnesium citrate.[56] Possible reasons are proposed in Section 4.2.3. One of the most widely studied materials made by this method is gold on iron oxide, because of the many possible structures that the support can adopt.[15,57,58] This will be discussed in Chapter 6 (Section 6.3.3.1).

Some characterisation of catalysts prepared in this way has been reported.[59,60] It is not however clear whether all the gold in solution is

Table 4.2: Characteristics of Au/oxide samples prepared by coprecipitation.[15]

Oxide support	[Au] (wt.%)	$T_{calc.}$ (K)	$\bar{d}_{Au}$ (nm)
MgO	2	473	<2
TiO$_2$	5	873	5
Fe$_2$O$_3$	5	673	4
Co$_3$O$_4$	5	673	6
NiO	10	673	8
ZnO	5	673	5
Al$_2$O$_3$	5	573	5
In$_2$O$_3$	5	673	5
SnO$_2$	5	673	3
SiO$_2$	5	573	20
Cr$_2$O$_3$	5	673	>30
CdO	5	295	21

precipitated, and what is the nature of the precipitate. No detailed study has been reported of the physical chemistry associated with the precipitation of the gold precursor and the hydroxides. The method is said only to work for certain metal oxides, because the rates of precipitation of the two hydroxides and their affinity determine the gold particle size.[61]

4.2.3. Deposition–precipitation

The method is commonly referred to as deposition–precipitation (DP), because metal hydroxide is supposed to precipitate on the oxide support[11,62] (see Section 4.1.1). It has been widely used for preparing oxide-supported gold catalysts having small particle sizes; Table 4.3 summarises what has been found with three supports. A typical preparation method is as follows. After adding the support to an aqueous solution of HAuCl$_4$, the pH of the suspension is raised to a fixed value, usually 7 or 8, by adding sodium hydroxide or carbonate, after which it is heated at 343 or 353 K with stirring for ~1 h. After thorough washing with water usually at ~323 K to remove as much of the sodium and chlorine as possible, the product is dried under vacuum at 373 K, and often calcined in air at higher temperature.

This basic method has numerous variations, such as pH, temperatures of preparation and washing, for instance at room temperature instead of higher temperature,[25,63,64] and use of other bases, such as ammonia.[65,66] Some of the parameters have been systematically investigated in a study

Table 4.3: Characteristics of Au/oxide catalysts prepared by DP.

Oxide support	pH	[Au] (wt.%)	$T_{calc.}$ (K)	$\bar{d}_{Au}$ (nm)	References
TiO$_2$	7 to 10	0.5–3.1	673	3.5–2.5	73
	~8	~1	n.r.	2–5[a]	65
	7	3.3	573	1.4 (0.7–2.4)	52
	8	1.8	573	1.8 (0.7–4.3)	52
	8	3.1	573	1.4 (0.7–2.7)	52
MgO	9.6[b]	5	553	0.7–4[a]	74
Al$_2$O$_3$	n.r.	0.94	473	2.4 ± 1.1	75
	7	1.1	623	3–5[a]	76

[a]size range; [b]addition of Mg citrate; n.r.: not reported.

using an automated dispenser and various supports.[65] No variation in the results was found whether sodium hydroxide or carbonate or ammonia was used as base. Changes in the ageing period from 2 to 12 h or in temperature (298 and 343 K) did not have any influence either, but there were problems of reproducibility in the preparations that were not found in another study.[27] Whether sodium carbonate was added to the gold solution before or after addition of titania did not produce significant changes in the results.[65] This result is surprising since in another study it was reported that the pH of the solution significantly decreased when titania was added, and also during the DP.[67]

The method works well with supports having an PZC greater than five, such as magnesia, titania (usually Degussa P25, ~70% anatase and ~30% rutile), alumina, zirconia and ceria,[10,37,65] but it is not suitable for silica (PZC ~ 2), silica–alumina (PZC ~ 1), tungsta (PZC ~ 1),[10] or for supports such as activated carbon[68] (see Section 4.5.4). For zeolites, surprisingly, it does seem to work (see Section 4.5.1).

There are only a few studies reporting the influence of the pH of the gold solution on the gold loading on titania.[62,65,67] Figures 4.1–4.3 show that there is general agreement that the gold loading decreases at high pH. However, the results are divergent for the lower pH's since either it is maximum at pH 6, or it continues increasing as the pH decreases. Except for the lowest nominal gold loading of 1 wt.%, all the gold is not deposited on the support, and the yield of DP does not exceed 60%. The size of gold particles decreases with pH (Figure 4.4). The optimal conditions are proposed to be those corresponding to pH about 7–8,[27,62,67] because it

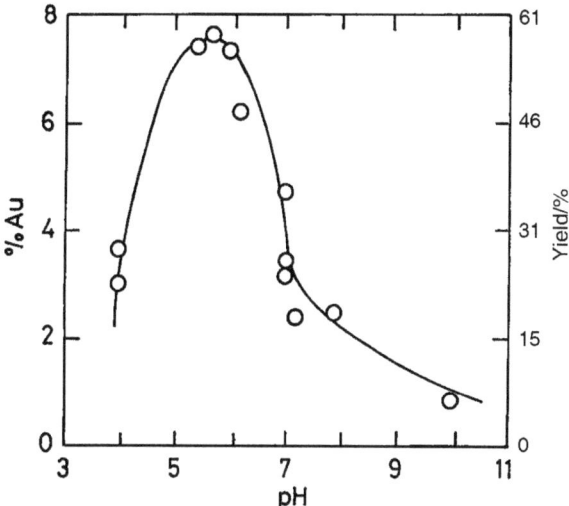

Figure 4.1: Gold loading and yield of DP versus solution pH for Au/TiO_2 (Degussa P-25) catalysts prepared by DP (NaOH added to the suspension, preparation at 343 K, nominal Au loading 13 wt.%).[62]

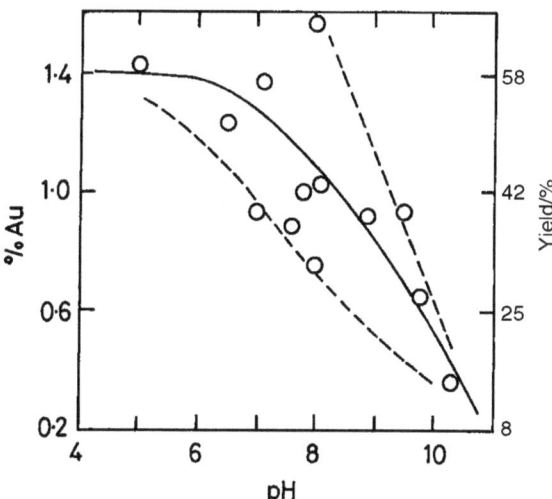

Figure 4.2: Gold loading and yield of DP versus solution pH for Au/TiO_2 (Degussa P-25) catalysts prepared by DP (Na_2CO_3, no indication of when it is added, preparation at 300 K, nominal Au loading 2.4 wt.%).[65]

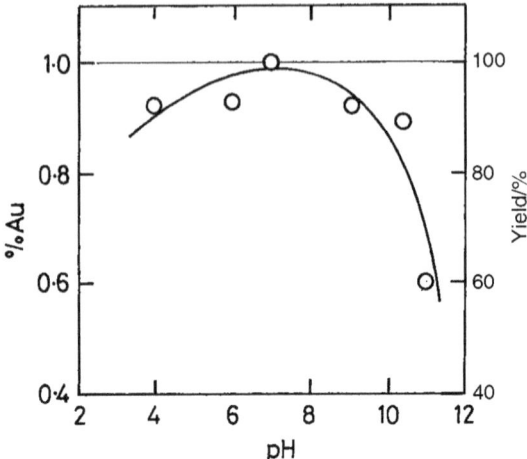

Figure 4.3: Gold loading and yield of DP versus solution pH for Au/TiO$_2$ (Degussa P-25) catalysts prepared by DP (titania added to the NaOH solution, preparation at 343 K, nominal Au loading 1 wt.%).[67]

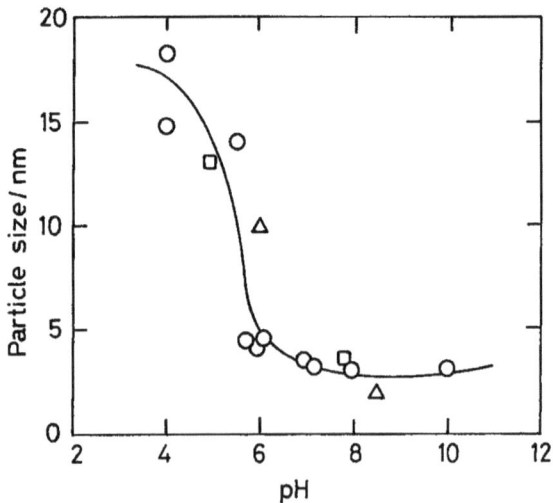

Figure 4.4: Average gold particle size versus solution pH in Au/TiO$_2$ catalysts prepared by DP and after calcination at 673 K in air; circles;[62] squares;[65] triangles.[67]

corresponds to the best compromise between the gold loading and the gold particle size.

When the pH is above the PZC, both the titania surface and the gold species are negatively charged. It was first proposed[11,62] that at such pH there is precipitation of $Au(OH)_3$ on nucleation sites consisting of $AuCl(OH)_3^-$ adsorbed on specific surface sites, the nature of which was not discussed in the paper. The decreasing gold loading when pH increased from 6 to 10 (Figures 4.1–4.3) was explained by the increasing solubility of $Au(OH)_3$. However, under the conditions of gold concentration and pH used, there is obviously no precipitation of gold hydroxide if one refers to Section 4.1.2. Moreover, in contrast to the principle of the DP as described in Section 4.1.1, all the gold in solution cannot deposit onto the support.

An alternative explanation combining interpretations found in two separate studies on Au/TiO_2[37,67] could be that when pH is above but close to 6, which is the PZC of titania, there is surface complex formation:

$$TiOH + [AuCl(OH)_3]^- \rightarrow Ti\text{-}[O\text{-}Au(OH)_3]^- + H^+ + Cl^-. \qquad (4.1)$$

This is consistent with a XAFS study performed after gold deposition at pH 8 but before washing, which showed that Au^{III} is tetracoordinated and that there is no chloride left in the first coordination sphere, but only oxygen atoms.[37] Characterisation of the deposited gold phase is not an easy task since it is amorphous, and since the EXAFS does not 'see' second neighbours, it is difficult to fully identify the gold species. A surface complex of this kind (Equation (4.1)) has already been proposed by geochemists for Au/Al_2O_3; they observed that (i) the maximum of gold adsorption on alumina occurs at pH close to the PZC of γ-alumina ($\sim$8), i.e. where the number of neutral hydroxyl groups is maximum;[69] (ii) the adsorption of anionic hydroxychloro-gold(III) complex on goethite (FeO(OH), PZC = 8.1) increases as pH increases from 4 to 7,[70] which is opposite to the typical behaviour for anion adsorption on positively charged oxide surfaces (Section 4.2.5). It may also be noted that, for the three studies reported above, two of them showed that the maximum gold loading is obtained when the pH is close to the PZC of titania ($\sim$6) (Figures 4.1 and 4.2), which is also consistent with Equation (4.1). Surface complex formation is also consistent with the fact that, at higher pH, the amount of gold decreases because of the decreasing number of hydroxyl groups.[37] The HCl released according to Equation (4.1) could also explain the decrease in pH observed with the time of DP.[67]

On the other hand, the decreasing amount of chloride ion in the catalysts, due to the hydrolysis of the gold complex when pH increases, is consistent with decreasing gold particle size observed for three sets of samples
(Figure 4.4). Indeed as mentioned above (Section 4.2.1) chloride encourages
mobility of gold and particle aggregation. Hence, there would be a narrow
range of pH where sufficient gold could be deposited with a minimal chloride
in the gold complex, leading to small metal particles.

This method called DP, which consists of fixing the pH of the HAuCl$_4$
solution by addition of a base, obviously does not correspond to that
of the principle of deposition–precipitation, which consists of a gradual
increase of the pH to avoid precipitation in solution (Section 4.1.1). In the
present method, there is no Au(OH)$_3$ precipitation, but only grafting of
gold complexes.

Unfortunately, the interpretation proposed above does not seem to apply
straightforwardly to the case of the Au/Al$_2$O$_3$ system. Three different studies on Au/γ–Al$_2$O$_3$ catalysts prepared at various pH values by addition
of sodium carbonate showed that both the gold loading and the yield of
deposited gold decrease as pH increased from 5 to 8–9, then dropped at
higher pH (Figures 4.5–4.7).[25,63,71] No maximum was observed as expected
at a pH near that of the PZC of γ-alumina ($\sim$8). An explanation for the

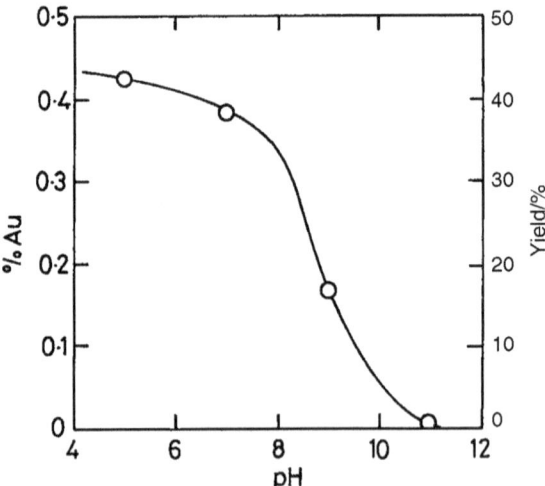

Figure 4.5: Gold loading and yield of DP versus solution pH for Au/Al$_2$O$_3$
catalysts prepared by DP by addition of Na$_2$CO$_3$ (preparation at 300 K,
nominal Au loading 1 wt.%).[25]

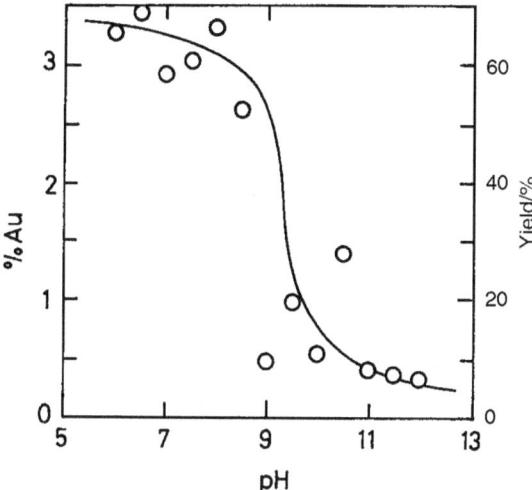

Figure 4.6: Gold loading and yield of DP versus solution pH for Au/Al_2O_3 catalysts prepared by DP by addition of Na_2CO_3 (preparation at 343 K, nominal Au loading 5 wt.%).[71]

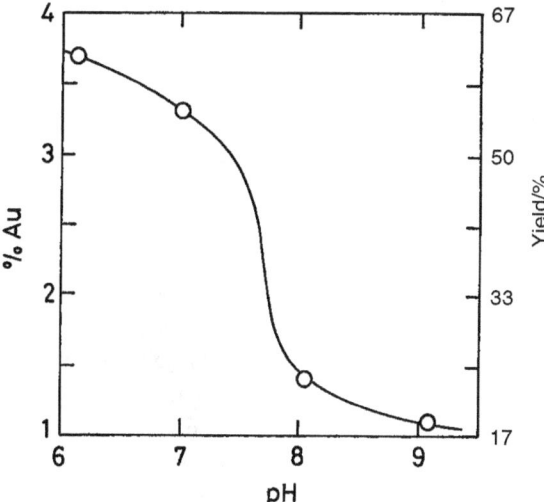

Figure 4.7: Gold loading and yield of DP versus solution pH for Au/Al_2O_3 catalysts prepared by DP by addition of Na_2CO_3 (preparation at 300 K, nominal Au loading 6 wt.%).[63]

somewhat high gold loading at lower pH could be that the number of surface hydroxyl groups is larger for alumina ($\sim$200 m^2 g^{-1}) than for titania
($\sim$50 m^2 g^{-1}) because of its larger surface area, and that the fraction that
is positively charged is higher for alumina than for titania at low pH. As
a consequence, anion adsorption could be the predominant mechanism at
pH below $\sim$8. Gold particle size was not systematically measured in these
studies, but it seems that, as in the case of titania, they are smaller at
higher pH.

In some cases, magnesium citrate has been added to the preparative
solutions to get smaller gold particles. The role of the citrate anions,
$C_3H_7O_7^{3-}$, is not clear. Citrate ion is a reducing agent, so it may reduce
Au^{III} to Au^0 in solution, and could also adsorb on the oxide surface, and
'act as a sticking agent, which blocks the coagulation of gold particles'.[61]
It is also a strong ligand, and may compete with the hydroxyl ligands of
the anionic hydroxychloro-gold(III) complexes, displace them, and avoid
the formation of gold clusters after drying and before calcinations.[49,72] It
was also proposed that magnesium citrate was effective in removing the
remaining chloride ion from the catalyst.

4.2.4. Deposition–precipitation with urea (DPU)

Using urea ($CO(NH_2)_2$) makes it possible to employ the real procedure of
DP as first described.[2,3] Urea acts as a 'delay base' since there is no reaction
when it is dissolved in a suspension of the support in the aqueous metal salt
solution at room temperature; hydrolysis only occurs when this is heated
above 333 K, according to the equation:

$$CO(NH_2)_2 + 3H_2O \rightarrow CO_2 + 2NH_4^+ + 2OH^-, \qquad (4.2)$$

whereby there is a gradual and homogeneous release of hydroxyl ions and
increase of pH throughout the solution.

In the first attempt to prepare supported gold catalysts in this way,[77,78]
HAuCl$_4$ was added to a suspension of the support in urea solution heated
at 353 K, but after calcination at 673 K, quite large gold particles were
obtained on titania, silica and alumina. In other work,[79] complete deposition of gold onto alumina (5 wt.%) was achieved, but particles were again
large after calcination.

In a more extensive study,[37,52] it was found that small gold particles
could be obtained on titania, alumina and ceria (Table 4.4) providing that
the DP time at 353 K was long enough (at least 4 h). Moreover, it was found

Table 4.4: Au/oxide samples prepared by DP with urea at 353 K with a nominal gold loading of 8 wt.%.[37,52]

Oxide support	BET surface $(m^2 g^{-1})$	PZC	DP time (h)	Final pH	[Au] (wt.%)	[Cl] (wt.%)	$\bar{d}_{Au}$ (nm)
TiO$_2$	45	~6	1	3.0	7.8	0.041	5.6
			2	6.3	6.5	0.122	5.2
			4	7.0	7.7	<0.03	2.7
			16	7.3	6.8	<0.03	2.5
			90	7.8	7.4	<0.03	2.4
CeO$_2$	260	~6	1	4.3	7.9	—	8.1
			16	6.6	8.2	—	<5[a]
γ-Al$_2$O$_3$	100	~7.5	1	4.3	6.9	—	6.9
			16	7.1	7.2	—	2.3
SiO$_2$	250	~2	1	5.2	2.9	—	>20
			16	7.0	3.7	—	>20

[a]Estimated by XRD (poor contrast between gold particles and CeO$_2$ by TEM).

that all the gold in solution (8 wt.%) was deposited onto these supports within the first hour, while the pH of the suspension was still acidic (pH ~ 3) (Table 4.4). Thereafter samples 'matured' while the pH continued to rise, reaching a plateau at pH around 7 after 4 h. After careful washing with water, drying at 300 or 363 K under vacuum, and calcination in air at 573 K, the gold particles were found to decrease in size as the DP time was increased (Table 4.4).

The chemical processes occurring during these preparations have been explored[37]. The gold phase that precipitates onto the support was orange in colour, and its chemical composition was $AuN_{2.2}O_{1.2}C_{0.9}H_{4.2}Cl_{0.1}$. The nature of the amorphous precipitate has not been elucidated, but one can note that the ratio of N:O:C:H$_2$ is close to that of urea (2:1:1:4). What is proposed is that it arises from a reaction between anionic hydroxychloro-gold(III) complex and the products of hydrolysis of urea since there is no precipitation as long as the suspension is not heated, as attested by Raman spectroscopy.[37] It was therefore proposed that deposition starts when anionic hydroxychloro-gold(III) species present in the solution at pH between ~2 (initial pH) and ~3 (pH of precipitation) begins to interact with the positively charged oxide surface. They act as nucleation sites for the precipitation of the orange compound. This interpretation is consistent with the fact that for a silica support, which has a PZC ~2, not all of

the gold is deposited and the gold particles are large (Table 4.4). The fact
that the gold particle size decreases with the time taken for DP (Table 4.4)
was attributed to a phenomenon of peptisation, i.e. of redispersion of the
supported gold phase.[37]

This DPU method is applicable to the same supports as for DP at
a fixed pH (Section 4.2.3). It also leads to small gold particles, but with
a longer preparation time (at least $\sim$4 h instead of 1 h). However, it has
the advantage that all the gold in solution is deposited onto the support;
therefore there is no loss of gold in solution, and the loading can be easily
controlled *a priori*. With this method, it is easy to prepare a set of samples
with the same gold loading but different particle sizes, by systematically
varying the time allowed.

In addition to titania, alumina and ceria supports (Table 4.4), the DPU
method has been applied to other supports. Gold particles supported on
ferric oxide were quite small (3–7 nm) after thermal treatment is static air
at 623 K.[80] Another study reported the use of the DPU method to deposit
gold onto other supports (MgO, CaO, SrO_2 and BaO).[81] After calcination
at 673 K, gold particles of moderate size were obtained on magnesia (8 nm)
and on calcia (6 nm). In all cases, all the gold in solution was deposited on
the support.

4.2.5.   Anion adsorption

Since $HAuCl_4$ gives anionic complexes in aqueous solution (Section 4.1.2),
anion adsorption is in principle possible providing that the pH of the solution is lower than the point of zero charge of the support; the support
surface is then positively charged. This method was studied for the preparation of Au/TiO_2 at a fixed pH 2, under various conditions of time and
temperature, 293 or 353 K;[52] equilibrium was reached quickly (<15 min),
but the gold loading never exceeded 1.5 wt.%, which corresponds to about
20% of the nominal quantity of gold in solution: values of $\bar{d}_{Au}$ were always
between 4 and 6 nm. Under these conditions, the main species in solution at
293 and 353 K are $AuCl_3(OH)^-$ and $AuCl_4^-$ (see Section 4.1.2), and these
can interact electrostatically with the titania surface (PZC $\sim$ 6). Moreover,
if one refers to the results of impregnation followed by washing with water
reported in Section 4.2.1,[52] where the main gold species in the impregnation solution at pH <1 is $AuCl_4^-$, after washing with water, some gold still
remains, so electrostatic interaction is also occurring with this method of

preparation:

$$TiOH_2^+ + AuCl_4^- \rightarrow [TiOH_2^+, AuCl_4^-]. \qquad (4.3)$$

4.2.6. Removal of chloride when using gold chloride precursors

Most of the methods described so far involve $HAuCl_4$ as precursor, but the presence of chloride is detrimental to obtaining small gold particle size and good catalytic activity. DP is usually performed at quite high pH, so that most of the Au–Cl bonds are hydrolysed, and thorough washing with water at the end of the preparation removes the remaining chloride. With impregnation and anion adsorption, much lower pH is used, so that the complex still contains chloride in its coordination sphere; this results in sintering to form rather large gold particles during calcination.[26]

However, there are effective methods for removing chloride after gold deposition. Two groups[26,36,82] have reported that after preparation of Au/Al_2O_3 using anion adsorption, washing with ammonia solution led to the elimination of the chloride (but see safety note above in Section 4.1.3). After calcination, much smaller gold particles (4 nm) were obtained than after washing with water (10–20 nm), and no large change in the gold loading was observed (1.5 instead of 2 wt.%). Based on these results, another group has recently washed catalyst precursors with a solution of ammonia just after impregnation of various supports (silica, alumina and titania) with $HAuCl_4$.[34] Surprisingly, gold was not eliminated, and small gold particles (~3 nm) could be obtained after calcination at 573 K, and apparently on any types of supports including silica. Another group proposes a treatment under gaseous ammonia followed by washing with water to remove the chloride after impregnation and get small gold particles on titania.[83] In contrast to the warning reported in Section 4.1.3 regarding using gold and ammonia, none of these groups reported that gold was explosive.

4.3. Methods of Preparation using Chloride-Free Gold Precursors

4.3.1. Impregnation in the aqueous phase

Methods of preparation that do not involve a chloride precursor are definitely more desirable. The use of gold acetate $(Au(OAc)_3)$ for impregnation has led to Au/Al_2O_3 having small gold particles (1.1 wt.% Au,

$\bar{d}_{Au} = 4.8\,\text{nm}$),[49] but the solubility of this salt in water is very low ($<10^{-5}\,\text{M}$) and varies from one batch of the commercial compound to another. Potassium aurocyanide ($KAu^I(CN)_2$) has given small particles when used with several supports ($<1\,\text{wt.\%}$ Au, $\bar{d}_{Au} = 3\,\text{nm}$) but much larger particles were found at higher loading.[44,84]

4.3.2. Deposition of gold–phosphine complexes

The gold–phosphine complex $[Au^I(PPh_3)]NO_3$ (**Au1**) and the cluster $[Au_3^I Au_6^0(PPh_3)_8](NO_3)_3$ (**Au2**) have been used to prepare supported gold catalysts.[85] The supports were impregnated with a solution of the precursor in dichloromethane followed by evaporation of the solvent. Thermal treatment in air at 673 K gave somewhat large gold particles, but much smaller particles were obtained by impregnation of freshly precipitated 'hydroxide' with acetone solutions of the precursors[86] (see also Table 4.5). More recently a solution of $[Au_2^I Au_4^0(PPh_3)_6](BF_4)_2$ in dichloromethane was used to prepare Au/TiO$_2$ having small gold particles (1 wt.\% Au, $\bar{d}_{Au} = 4.7\,\text{nm}$).[87]

Extensive characterisation studies have been performed with various techniques (XAFS, XPS, ^{31}P CP/MAS–NMR, IR and Raman spectroscopies)[13] to characterise **Au1** after deposition, and its decomposition and conversion into metal particles upon thermal treatment. The transformation of hydroxide supports into oxides was also followed. The amorphous hydroxides contain a larger number of surface hydroxyl groups and surface defects than do the corresponding oxides, and they interact more efficiently with **Au1**, which dissociates on the surface, releasing the

Table 4.5: Characteristics of Au/oxide catalysts prepared by deposition of phosphine gold complexes.

Catalyst	$T_{\text{treatment}}$	$\bar{d}_{Au}$ (nm)	Technique	References
Au1/SiO$_2$	673 K in air	16	XRD Au(111)	85
Au1/TiO$_2$	773 K in 5\% H$_2$/Ar	16	XRD Au(200)	85
Au2/TiO$_2$	773 K in 5\% H$_2$/Ar	18	XRD Au(200)	85
Au1/α-Fe$_2$O$_3$	673 K in air	12	XRD Au(111)	85
Au1/α-Fe$_2$O$_3$	673 K in air	30	TEM	86
Au1/Fe(OH)$_3$	673 K in air	3	TEM	86
Au1/TiO$_2$	673 K in air	30	XRD Au(200)	88
Au1/Ti(OH)$_4$	673 K in air	$\sim$3	XRD Au(200)	88

nitrate ion; the $[Au(PPh_3)]^+$ cation is retained by interaction with hydroxyl groups, which induces a shift of the ν_{OH} IR band. During thermal treatment, the precursor ion and the hydroxide decompose simultaneously, and this facilitates the gold-support interaction, which reduces mobility of the gold species during calcination, preventing sintering. This explains why the particle sizes obtained with hydroxides are smaller than with oxides (Table 4.5).

The advantage of using a phosphine complex is that it contains no chloride, and the counter-ion is easily decomposed or eliminated. The complex has to be synthesised (not easy or cheap), and nonaqueous solutions are needed, which means that the support must be dehydrated, the solvent dried, and the finished catalyst stored in ampoules sealed under vacuum.

4.3.3. Deposition of other organogold complexes

The isonitrile gold(I) nitrate ($[Au^I(NO_3)(CN-Bu^t)]$) has been used to prepare Au/Fe_2O_3;[89] the complex was added to a slurry of freshly precipitated ferric hydroxide in acetone. After removal of solvent and calcination (673 K), small gold particles (<4 nm) were obtained with a loading of 3 wt.% Au.

The dimethylgold acetylacetonate complex ($Me_2Au(acac)$, $(CH_3)_2$ $Au^{III}(OCCH_3)_2CH$)) is an attractive chloride-free precursor, but its use requires the absence of water and oxygen. After adding a dehydrated oxide support to a solution in acetone and residence overnight in a refrigerator, filtration and calcination led to the results indicated in Table 4.6.[90] The low gold loading on silica was ascribed to competitive adsorption of the solvent, and the large particle size to a weak interaction between precursor and support.

Table 4.6: Characteristics of Au/oxide catalysts prepared by deposition of $Me_2Au(acac)$ in acetone (nominal Au loading in solution: 3 wt.%).[90]

Oxide support	[Au] (wt.%)	$T_{calc.}$ (K)	$\bar{d}_{Au}$ (nm)
Al_2O_3	2.84	673	6.6
TiO_2	2.28	673	2.9
SiO_2	0.18	673	16.0

The Me$_2$Au(acac) precursor has also been used to deposit gold on magnesia,[91] titania,[92] γ-alumina[93] and NaY zeolite.[94,95] Using the Schlenk tube technique and a glove box, deoxygenated hexane or n-pentane was added to a mixture of partially dehydroxylated support and the precursor, and after stirring for one day, the solvent was removed by evacuation. XAFS characterisation showed that Me$_2$Au(acac) reacts with two surface hydroxyl groups and oxide cations to form $\{SO\}_2Au^{III}(CH_3)_2$ and $\{S\}$(acac) species. Treatment from 323 to 573 K led to the removal of the ligands, reduction to Au^0 and formation of small gold particles (3 nm for Au loadings of about 1 wt.%).

The great advantage of using Me$_2$Au(acac) is that it contains no chloride and that it is commercially available (but expensive). Any support can be used, but the preparation must be conducted in the absence of moisture and air. Catalysts made in this way must therefore necessarily remain laboratory curiosities, as the method is quite unsuited to manufacture on a large scale.

4.3.4. Chemical vapour deposition

In chemical vapour deposition (CVD), a volatile organogold compound reacts with the surface of a support, on which it decomposes to zero-valent particles of gold, or if the support is a flat surface, to a coherent film. The process, which can be operated in a variety of ways, can also be used for the preparation of supported gold catalysts.[96] Dimethylgold acetylacetonate is the only compound that has proved suitable for this purpose. Adsorption of its vapour at 306 K on dried oxides, including silica, mesoporous MCM-41, and carbon supports, gave after calcination satisfactorily small gold particles (Table 4.7),[97] but with a broader size distribution than is given

Table 4.7: Characteristics of gold on various supports prepared by chemical vapour deposition of Me$_2$Au(acac).[97]

Support	[Au] (wt.%)	$T_{calc.}$ (K)	$\bar{d}_{Au}$ (nm)
Al$_2$O$_3$	5.3	573	3.5 ± 2.7
TiO$_2$	4.7	573	3.8 ± 2.7
SiO$_2$	6.6	673	6.6 ±3.8
MCM-41 (22 Å)	4.2	673	4.2 ± 1.4
MCM-41 (27 Å)	2.9	673	4.0 ± 1.5
MCM-41 (31 Å)	3.5	673	4.9 ± 1.9
Activated carbon	5.2	673	5.0 ± 1.6

by DP (Table 4.3). With the preparation methods involving the use of preformed gold colloids (Section 4.3.6) and the impregnation followed by washing with ammonia (Section 4.2.6), this is probably the only method for obtaining small gold particles on silica. DFT calculations have shown that the negatively charged oxygen atoms of the planar $Me_2Au(acac)$ may interact with surface defect sites and/or surface hydroxyls without steric hindrance.[97] The advantages and drawbacks of this method are the same as stated at the end of Section 4.3.3.

4.3.5. Cation exchange and adsorption

As far as we know, only the bis-ethylenediamine (or 1,2-diaminoethane) Au^{III} cation, $[Au(en)_2]^{3+}$, has been used to prepare gold catalysts by cation exchange of zeolites, or cation adsorption on oxide supports; synthesis of the complex with chloride counter-ions is easy.[98]

Cation exchange is an efficient way of introducing highly dispersed metal particles into zeolites. With HY zeolite containing 4 wt.% gold, thermal treatment (He, 423 K) gave gold particles 1–4 nm in size, located within the zeolite framework, with a few larger particles on the outside.[99,100] Other methods for introducing gold into zeolites are described in Section 4.5.1.

The $[Au(en)_2]^{3+}$ cation has also been used to deposit gold onto activated carbon fibres by ion exchange of protons in acidic (phenolic) surface groups;[101] reduction in hydrogen gave 2–5 nm particles (see also Section 4.5.4).

This cation has also been used for depositing gold onto oxide supports by cation adsorption; the PZC of the oxide must be lower than the pH of the precursor solution to make the surface negatively charged. Also the preparation has to be performed at room temperature, as the cation decomposes above 333 K. Au/TiO_2 made at 298 K and pH 9 gave gold 2.5 nm particles after calcination at 573 K,[37] but the gold content (1 wt.%) was lower than the expected value of 3 wt.% calculated on the basis of all surface hydroxyls being exchanged, three of them interacting with the trivalent cation.

4.3.6. Deposition of colloidal gold onto supports

Methods for preparing colloidal gold and also bimetallic colloids containing gold have been described in Section 3.2.3. The gold particles may be immobilised on supports by dipping the supports in the colloidal suspension, followed by washing and drying. In a successful preparation, the particles once supported should not be significantly larger than they were in the sol; achieving this depends on striking a delicate balance between various parameters such as the nature and concentration of the stabiliser, the stabiliser/gold ratio and of course the nature of the support (Table 4.8). For instance, polyvinylalcohol (PVA), which is an appropriate stabiliser for depositing gold on carbon, is not suitable for silica or alumina.[102]

Au/TiO$_2$ has been prepared by depositing colloidal gold in di-*iso*propylether onto titania; calcination at 573 K removed all the organics.[103] Gold particles ($\sim$5 nm) made in this way were found to sinter much more easily than those made by DP. Gold particles $\sim$2 nm in size made by reducing HAuCl$_4$ with a basic solution of THPC, (tetrakis (hydroxymethyl)phosphonium chloride) have been deposited on titania and zirconia at pH 2, which is below their PZC:[104,105] part of the adsorbed ligands were destroyed on exposure to air; calcination at 673 K removed all the carbon, and gave particles 4–6 nm in size. In other work,[102] this type of sol could be immobilised on carbon or alumina irrespective of its pH.

The method is also applicable to making Au/SiO$_2$, by using a sol derived from an ethanolic solution of Au(PPh$_3$)$_3$Cl reduced by sodium borohydride.[106] 1 wt.% Au/SiO$_2$ contained particles 2.7 nm in size, increased only slightly by calcination at 673 K to eliminate the phosphine ligands.

Using preformed particles is advantageous because particle size is independently controllable, the size distribution is narrow and the gold already reduced. The mean size can be varied by well-established methods, and thermal post-treatment can be avoided for liquid-phase reactions if the stabiliser is appropriately chosen, and does not cover the particles completely.

Table 4.8: Examples of catalysts prepared from gold sol deposited on various supports (1 wt.% Au).[102]

Support	Stabiliser[a]	$\bar{d}_{Au}$ in the sol (nm)	$\bar{d}_{Au}$ on support (nm)
Carbon	PVA	n.r.	4
Carbon	PVP	6.0	6.9
SiO$_2$	PVP	2.2	2.6
Carbon	THPC	2.9	4.3
Al$_2$O$_3$	THPC	3.5	3.8

[a]PVA: polyvinylalcohol; PVP: polyvinylpyrrolidone; THPC: tetrakis(hydroxymethyl)phosphonium chloride.

For gas-phase reactions, however, elimination of the stabiliser is recommended, but this may lead to sintering of the particles.

4.3.7. Deposition of dendrimer-stabilised gold particles

Dendrimers can be used as templates for synthesis of monodisperse gold particles, which can then be deposited onto oxide supports; metal particles are encapsulated within dendrimers, and the size of the particles can be tuned by varying the metal-to-dendrimer ratio prior to reduction.

Two different approaches have been attempted for preparing Au/TiO_2 catalysts (1 wt.%)[107] after the synthesis of Au_{55} particles encapsulated by amine-terminated fourth-generation dendrimers (G4-NH$_2$): (i) they were deposited onto commercial titania via impregnation to incipient wetness; after drying and calcination in air at 773 K to remove the dendrimers, the gold particles were much larger than initially (7.2 instead of 1.9 nm); (ii) they were incorporated into an amorphous titania network prepared by sol–gel through a controlled hydrolysis of $Ti(O^iPr)_4$ alkoxide. In this case, the dendrimer had a dual templating role, as it defined the size and monodispersity of both the gold particles and the pore structure in the titania framework. After preparation, drying and calcination at 773 K, the gold particles were only 2.7 nm, i.e. they had only increased by 40%.

Another type of dendrimer, that has been used to encapsulate Au_{32} particles is hydroxyl-terminated fifth-generation dendrimers (G5-PAMAM). The particles were adsorbed on silica (0.3 wt.% Au),[108] and after washing and removal of the dendrimer under oxygen at 573 K, followed by reduction under hydrogen at 573 K, the supported gold particles were even larger than above (14.5 nm), probably because silica is not a support that easily leads to small particles.

4.4. Less Conventional Methods

4.4.1. Sol–gel method

The sol–gel method involves a single step, as does coprecipitation; for preparing oxide-supported gold catalysts, it usually involves the hydrolysis of a metal alkoxide in a water–alcohol solution of $HAuCl_4$, the gold being reduced thermally or by UV radiation. It was developed for preparing films of a gold-containing oxide supported on wafers, for studies of their optical properties, but the gold particles are somewhat too large ($\sim$20 nm)[109–112] for use in catalysis; nevertheless it has been used for preparing gold catalysts, and Au/Al$_2$O$_3$ catalysts made in this way from various gold precursors (gold acetate (Au(OAc)$_3$), hydrogen tetranitratoaurate (HAu(NO$_3$)$_4$), HAuCl$_4$) and aluminum tri-*sec*-butoxide, all contained particles about 30 nm in size.[113]

Since the use of alcohol in the sol–gel process can lead to aggregation of the colloidal metal, Au/SiO$_2$ catalysts have been prepared without alcoholic solution, using tetramethyl orthosilicate (TMOS) as a water-soluble silicic precursor and colloidal gold by reducing aqueous HAuCl$_4$ with magnesium citrate.[114] However, the size of the gold particles was not reported, but reduction by citrate ion does not usually produce small particles.

An alternative approach was to prepare the support by the sol–gel method in the presence of gold colloids ($\sim$2 nm) protected by alkanethiolate to avoid aggregation.[115] After calcination at 700 K to remove the ligands and crystallize the titania support, the gold particle size was $\sim$6 nm (6 wt.%). The sol–gel method does not therefore seem to be suitable for preparing small gold particles in an oxide matrix. Moreover, even if the oxide matrix turns out to be porous, some of the gold is probably embedded in the support, and therefore not accessible to reactants in a catalytic reaction. The method does however find more convincing application where the oxide support has a mesoporous structure (see Section 4.5.2).

4.4.2. Photochemical deposition

Metal cations with appropriate redox potentials can be reduced by photoelectrons created by band gap illumination of semiconductors, such as the oxides of zinc, tungsten and titanium; the mechanism of photoreduction has been described.[116] Au/TiO$_2$ was prepared using aqueous colloidal titania containing *iso*propanol to which HAuCl$_4$ has been added: UV irradiation reduced 98% of the gold, to give 4 wt.% Au/TiO$_2$ having particles mainly larger than 6 nm.[116] The method has also been used to prepare Au/TiO$_2$ using water–methanol as solvent,[73] and to deposit gold on anatase nanofibres in the presence of organic capping reagents[117] (PVA, PVP and PEG, see Section 3.2.3), but in both these cases the gold particles were quite large (>5 nm). Supports not possessing a suitable bandgap (Fe$_2$O$_3$ and SnO$_2$) are ineffective.[117] The main disadvantage of this method of preparation is that its efficiency is very sensitive to both bulk and surface structural features. Such problem has been overcome recently by reducing the size of titania particles from 50 to 200 nm (commercial titania) to 6–15 nm (CVD);[118] with such a type of support and in contrast with larger titania particles, small gold particles were obtained; an increase in gold concentration (10^{-5}–10^{-3} M) or irradiation times resulted in no detectable changes in gold particle size (<1 nm with a gold loading that could reach ~11 wt.%).

The advantages of this method are that the extent of gold deposition is almost quantitative, and thermal treatment is unnecessary because gold is already reduced by the UV irradiation. The gold particles can be very small, but the method has not been used very much, although it has been applied to preparing bimetallic catalysts (see Section 4.6.2.4).

4.4.3. Sonochemical techniques

Gold particles can be deposited on the surface of supports by ultrasound, which induces chemical changes due to cavitation phenomena caused by the formation, growth and implosive collapse of bubbles in a liquid;[119] in the case of water, these lead to the homolytic dissociation into H· and ·OH radicals, the former being able to reduce Au^{III} species in solution. Gold has been deposited on mesoporous silica in this way,[120] the support being first immersed for three weeks in a water–*iso*propanol solution of $HAuCl_4$ to equilibrate the concentrations of the reagents throughout the pores: the support turned purple during sonication. Radicals formed from the *iso*propanol may also have participated in the reduction, which was faster in the pores than in solution, and gave smaller particles (5.2 nm) than in solution (20–30 nm) in the absence of support. Gold has also been deposited on silica microspheres (100–1000 nm) in this way, but ammonia had to be added to activate the silica surface by forming reactive silanol groups[121] (see Section 4.1.3 safety note). About 7 wt.% Au/SiO_2 containing 4 nm particles was obtained after sonication for 45 min. Sonication has also been used to assist the deposition of a gold colloid onto carbon nanotubes[122,123] (see Section 4.5.4). This method although not much used yet, seems to be promising.

4.4.4. Spray techniques

Using a solution-spray technique,[124] an aqueous solution of $HAuCl_4$ and titanium tetrachloride was atomised by an ultrasonic device to produce a mist without separation of the components: this was then calcined, and the fine particles collected on a glass filter at the outlet. Samples of 1 wt.% Au/TiO_2 contained 4 nm particles when the spray reaction temperature was

673 K. Alternatively, one may start with an aqueous ultrasonically dispersed suspension of titania containing $HAuCl_4$, and proceed in the same way,[125] but at a spray temperature of 673 K the gold particle size was larger (20 nm). All samples contained Au^0 and Au_2O_3; the presence of Au_2O_3 is surprising after such a temperature of treatment (Section 4.1.1).

4.4.5. Low-energy cluster or atom beam deposition

Methods for deposition of gold clusters or atoms usually on single crystals or flat surfaces, can also be used for powder supports. For this purpose, the powders are stirred in a rotating device located in front of the cluster or atom beam.

One method consists in the vaporisation of metal atoms from a rod with a laser beam to create a plasma;[126] cluster nucleation and growth occur when a continuous flow of inert gas is introduced into the vacuum chamber to cool the plasma. Finally, a beam of clusters of well controlled size (2–3 nm) is obtained, but only the neutral clusters are deposited on the substrates; after deposition, the samples can be transferred in the air. Hence, gold particles with the same average size (2.6–2.9 nm) and the same size distribution could be deposited on various powder supports (Table 4.9). The drawback of the equipment as it has been designed, is that the gold loading is low (∼0.1 wt.% Au) even after a few hours of laser vaporisation; however, these samples can be evaluated in dynamic conditions of reaction as for the other catalysts. This method is suitable for studying support effects since the gold particle sizes are identical, and also to study bimetallic particles because the composition of each particle is the same as that of the original rod (see Section 4.6.4.4).

The other method consists of the sputtering of a high-purity gold target with argon ions, followed by the subsequent deposition of the sputtered gold atoms on the surface of the moving powder support.[127] The

Table 4.9: Characteristics of Au/oxide catalysts prepared by laser vaporisation.[126]

Support	BET surface area (m^2g^{-1})	[Au] (wt.%)	$\bar{d}_{Au}$ (nm)
Al_2O_3	221	0.08	2.6
ZrO_2	79	0.05	2.9
TiO_2	60	0.02	2.9

equipment allows the deposition of much larger gold loading ($\sim$1.4 wt.%) in a rather shorter time of sputtering (195 min) than by laser vaporisation. The gold particles were small (2.3 nm) on alumina, but is not certain that with this method, the gold particle size could be the same on other supports. The 3 M Company have scaled up the size of preparations using this vacuum sputtering technique on agitated batches of powdered support.[127]

4.4.6. Direct oxidation of bulk alloy

Au/ZrO_2 containing a high loading of metal has been prepared by oxidation of $Au_{50}Zr_{50}$ alloy freshly prepared by arc-melting wires of the two metals under argon.[128] It was roughly ground and then allowed to oxidize in air at 298 K for 1 month; this led to an intimate mixture of gold and zirconia particles, each about 7 nm in size, with a surface area of $82 \, m^2 \, g^{-1}$. The gold particles formed 2D filaments some tens of nm in length. No heating was needed before use as an efficient catalyst for oxidation of carbon monoxide and for preferential oxidation of carbon monoxide in the presence of excess of hydrogen (PROX) (see Chapter 7). Catalysts were also obtained by activation of rapidly quenched metal alloys such as Au_5FeZn_{14} and Au_5AgZn_{14} under CO oxidation conditions at 553 K.[129] The products of oxidative decomposition gave the corresponding zirconia-supported gold, promoted by the second element.

4.4.7. Solvated metal atom dispersion or impregnation (SMAD or SMAI)

This method of preparation of supported metal catalyst requires a closed reactor to perform the preparation in the absence of water, so both the organic solvent and the oxide support must be carefully dehydrated. The method is based on the following principle: the metal is evaporated and co-condensed with the organic to 77 K on the walls of the reactor. Under dynamic vacuum, the co-condensate is then warmed up to 195 K, and melted. The oxide support is impregnated with the solvated metal atom (cluster) at the same temperature, After a given time of contact, the slurry is warmed up to ambient temperature, and the solvent is eliminated, after which the sample can be dried.

This method has been used to prepare Au/TiO_2 (3 wt.% Au) with toluene as solvent.[130] The bis(toluene)Au^0 complex formed at 77 K was unstable and decomposed at 173 K into Au^0 and toluene. After drying at room temperature, the sample had particles 1.8 nm average size. After elimination of the carbonaceous residues by a thermal treatment under vacuum at ~573 K, the gold particles were still small (2.2 nm). Au/CeO_2 (3 wt.% Au) was also prepared by this method[55] with acetone as solvent. After drying at room temperature and calcination at 673 K, the gold particles were 2.5 nm in size.

The advantage of this technique is that Au^0 is directly deposited onto the support, the particles are small, the sample is free of chloride, and it is easy to control the gold loading. The disadvantage is that the preparation must be performed in the absence of water and requires a specific reactor and an evaporator.

4.5.   Preparation of Gold Catalysts on Specific Supports

4.5.1.   Gold in zeolites

Apart from the use of cation exchange with $[Au(en)_2]^{3+}$ [99,100] (Section 4.3.5), or CVD of $Me_2Au(acac)$[94,95] (Section 4.3.4), there are several other methods for introducing gold into the pores of zeolites when the objective is to stabilise the Au^I state after reduction with carbon monoxide. However, with few exceptions, they seem much less satisfactory for forming small metallic particles, as they are usually larger than the pore size, and there is scarcely any evidence for particles being really located inside the pores.

Gold has been introduced into various zeolites (NaY, Na-mordenite, Na-ZSM5 and ZSM5) by mechanically mixing $AuCl_3$ with the dehydrated zeolite, followed by moderate heating under vacuum to disperse the salt throughout the zeolite cavities, either as a monolayer or as an amorphous phase;[131-133] the temperature must not be so high (~340 K) that the salt decomposes and the resultant gold particles aggregate; and microwave radiation can be used to assist the $AuCl_3$ diffusion.[133] Gold has been placed in a CaA zeolite by subliming $AuCl_3$ in vacuum in a closed ampoule after dehydrating the zeolite; after heating in vacuum to 543 K, small gold clusters were found within the zeolite cavities.[134] The method has been used to make Au/ZSM5;[135] it was thought that the $AuCl_3$ reacted with acidic protons to form HCl and $[AuCl_2]^+$-OZ^- entities mainly located at cation exchange sites; these were easily reduced to Au^0, but very large (30–40 nm) particles were formed after calcination at 453 K.

One can find methods using $KAu(CN)_2$ or $HAuCl_4$ precursors, which clearly give anionic species in aqueous solution; amazingly in spite of the fact that zeolites are not expected to show anion exchange, it seems possible to introduce anionic species into the pores of zeolites after surface modification. For example, NaA zeolite first exchanged with potassium acetate reacted with aqueous $KAu(CN)_2$ to yield Au^I species in the zeolite, with loadings claimed to be between 0.2 and 42 wt.%.[136] This has been done on various zeolites (Y, β and mordenite) after surface acidity modification with sodium nitrate at pH 6 and drying.[137,138] Then, the zeolite was suspended in a solution of $HAuCl_4$ adjusted to pH 6 by sodium hydroxide. After drying at 333 K, the best results, i.e. smallest gold particles (1 nm, 1.7 wt.% Au), were obtained for Y zeolite with high Al/Si ratio, as well as Au^{III} species mainly located in the supercages. It is sometimes possible to introduce anionic species into the pores of zeolites without surface modification. This has been done with Y zeolites;[139] at moderate pH (5 or 6) gold was deposited and small gold particles were obtained. In both cases, the method resembles DP (Section 4.2.3), but the mechanism of deposition of the gold species present at pH 6 onto the zeolite is unclear, and the samples have not been calcined.

4.5.2. Gold in ordered mesoporous silica

The preparation of gold particles in mesoporous materials covers a wide field, relating not only to catalysis, but also to materials science and domains of physics, including optical and electrical properties. Many methods of preparation have been used, but one wonders why some people have laboured to develop very sophisticated methods while others have been content with a much simpler method, apparently giving the same result, namely, the desired small gold particles.

The simplest method consists of soaking the support (MCM-41) in a solution of $HAuCl_4$ for 15 days;[140] the gold (0.9 wt.%) was found reduced into metal (5 nm) after drying at 353 K, but the investigators authors claim that reduction is mediated by the pore walls. Immersion in an *iso*propanol solution of $HAuCl_4$ for 3 weeks followed by sonication has also been described.[120] Treating another mesoporous support (HMM-2) with a solution of $HAuCl_4$ adjusted to pH 11.7 for 24 h led after hydrogen reduction at 473 K to 3.2 nm particles.[141] These results show that it is easier to obtain small gold particles in mesoporous silica than on amorphous silica with HAuCl$_4$ precursor (see Section 4.2). Materials made in this way show very low activity in oxidation of carbon monoxide[142−144] as is also the case for silica support (see also Section 6.3.3.4).

Gold has also been introduced into MCM-41 of various pore sizes (2.2–3.1 nm) by CVD with Me$_2$Au(acac) followed by calcination at 673 K;[97] particle sizes were 3–4 nm, smaller than on amorphous silica and with a narrower size distribution (Table 4.7). The gold particles, some of which were rod-like, appeared to be well distributed throughout the pore structure, although their sizes were slightly larger than the pore diameters.

More sophisticated methods involve functionalising the support before exposure to HAuCl$_4$ solution. This may be achieved by incorporation of a reagent during the synthesis that produces either propylamine, or propylthiol groups on the surface of MCM-41[145] and SBA-15.[146]

Post-synthesis treatment of SBA-15 by silanation with [NR′R$_3$]$^+$Cl$^-$ (R′ = silane) gives a cationic function, which has allowed adsorption of Au(OH)$_n$Cl$_{4-n}^-$ anions at various values of pH.[147,148] These methods give usefully small particle sizes (<2.5 nm) and sometimes quite high metal loading (16 wt.%), but they exhibit poor catalytic properties in oxidation of carbon monoxide[148] (Section 6.3.3.4).

Gold may also be incorporated during the sol–gel synthesis of the mesoporous support, e.g. by introducing HAuCl$_4$ during the synthesis of MCM-41.[149] The addition of a bifunctional aminosilane ligand in the synthesis mixture favours the simultaneous formation of the MCM-41 and gold incorporation.[150] Indeed, the ligand complexes Au^{III} *via* the amine functional groups, and covalently bonds to the porous silica matrix *via* the siloxane groups during the sol–gel synthesis. This approach can be generalised to prepare small gold particles in other mesoporous materials under neutral or basic conditions,[151,152] but not under acidic conditions where protonation of the amino-group would occur.

Gold colloid may also be introduced during the gelling step of mesoporous silica synthesis.[153] Calcination at 823 K gave 2–8 nm particles at loadings of up to 10 wt.%, homogeneously distributed. The gold colloids may be coated with the surfactant used in synthesising the mesoporous material. Particles smaller than the pore size have been successfully incorporated into materials of the MCM-41 type,[142,154] into SBA-15[155] and into MCM-48 [154] in this way.

It has to be stressed that the objective of these methods was not always to produce active catalysts, so the fate of the gold particles after calcination to decompose the stabiliser has not always been established, or their behaviour in catalytic reactions studied.

The difficulty with this type of support and with zeolites as well, is to determine whether the metal particles are located in the pores, and if they remain there after thermal treatment. TEM of ultramicrotome cuts is one method for investigating this; another is to use bright-field electron tomography.[156] One can also report a detailed characterisation of gold particles by high-resolution TEM/STEM and SEM.[157] After introduction of HAuCl$_4$ into various mesoporous silicas after post-synthesis amine functionalisation, highly dispersed gold particles (<2 nm) were obtained within the pores (pore sizes ranging from 2.2 to 6.5 nm and different pore structures (2D-hexagonal (MCM-41), 3D-hexagonal (SBA-12, HMM-2), and cubic (SBA-11)). After reduction in H$_2$ at 473 K, sintering of gold particles was found to depend on pore size, pore wall thickness and pore connectivity. In each of the samples, the average particle size exceeded the diameter of the silica pores, destroying the pore walls in the vicinity of the gold particles. When the mesoporous silicas have a 2D hexagonal structure, the majority of the gold particles remained within the pores, whereas when they have a 3D interconnected pore structure, the gold particles migrated outside the pores.

4.5.3. Gold in titanosilicate

Gold in mesoporous titanosilicate catalysts have been developed for their promising application in the vapour phase epoxidation of propene (Section 8.2.1). Gold can be introduced by DP using NaOH into the pores of Ti–MCM41 (Ti introduced during synthesis or post synthesis)[158,159] to give up to 1 wt.% gold, and particles smaller than about 3 nm. Ti–MCM48 has also been employed, also leading to small gold particles.[160] Both supports have been made hydrophobic by coating with methyl groups;[161] the DP method worked with Ti–MCM41, but with Ti–MCM48 it was necessary to use Me$_2$Au(acac), which reacted with both titanium and silicon sites. Very small gold particles (<1 nm) have been deposited by DP into mesoporous SBA-15 coated by an atomic layer of titania *via* sol–gel.[162]

Gold in titanosilicate TS-1 has been prepared by DP, but the low site density is a key problem with this support; for instance, at a pH of 9–10, only 1–3% of the gold in the solution was deposited.[163] In order to improve the yield of deposition and avoid the huge loss of gold in solution, TS-1 support was modified by ammonium nitrate to increase fourfold the yield of deposited gold, probably through the formation of a gold ammine complex.[164] However, large gold particles were obtained (30 nm for 5 wt.% Au).

4.5.4. Gold on carbon supports

The DP is not a suitable method for obtaining small gold particles on activated carbon;[68] this is attributed to the reducing nature of carbon, which would limit the deposition of gold in a dispersed state. Carbon is, however, a very promising support for gold in liquid-phase oxidation of organic compounds (Section 8.3), and alternative methods have therefore been sought. CVD with $Me_2Au(acac)$ is one such method[97] (Table 4.7), but most work has been performed with gold colloids. Their immobilisation on carbon has been developed using PVA and PVP as stabilisers (Table 4.8), particles 6–8 nm in size being obtained;[68,102,165] for use in liquid-phase reactions, removal of the stabiliser is unnecessary as long as some gold surface remains exposed. Other stabilisers have proved effective,[166] giving gold particles in the desirable size range; dextrin was one of the best, with gold particles of 2.8–3.6 nm. Colloids prepared in basic solution have also been tried,[167] and gold has been deposited on carbon from a water/organic microemulsion.[168]

Considerable work has been done with carbon nanotubes (CNT). About 7 nm particles have been directly obtained on single-walled CNT by merely immersing them in $HAuCl_4$ solution for 3 min when spontaneous redox reactions take place between metal ions and the nanotubes.[169] The CNT can also first be activated by depositing Pd^{2+} 'seeds' (electroless plating); then optimising the parameters of the plating process led to a high concentration of 3–4 nm particles on the external surface of the CNT.[170] Refluxing acid-treated CNT in $HAuCl_4$ solution containing THPC (tetrakis(hydroxymethyl)phosphonium chloride) as reducing agent, and mild sonication, has given 2 nm gold particles;[122] sonication of CNT suspended in a gold colloid has also been used.[123] Other still more sophisticated methods involve CNT functionalisation with aliphatic bifunctional thiols through a direct solvent-free procedure, followed by addition of $HAuCl_4$ and citric acid to reduce the gold; the smallest gold particles (1.7 nm) could be obtained with 1,6-hexanedithiol.[171] Others involve assembly of monolayer-capped gold particles on CNT before thermal treatment at 573 K to remove the capping thiolate shell to give 5 nm particles.[172]

4.6. Preparation of Supported Bimetallic Catalysts Containing Gold

4.6.1. Introduction

Supported bimetallic catalysts of which gold is one component have found a number of applications, in some cases being distinctly superior to a simple gold catalyst. The principles of their use were introduced in Sections 1.6, and 2.6, something was said about the properties of bimetallic phases. Some discussion of their mode of action is presented in Section 8.1.2, since it is in selective oxidation that their beneficial character first emerges. Sections 8.3–8.5 provide further examples of their use. A number of methods are available for making bimetallic phases in highly dispersed but unsupported forms (Section 3.2.4); in the main they require only the use of procedures that work for single metals, and they normally produce materials having the desired properties without difficulty.

The preparation of a successful supported bimetallic catalyst is quite a difficult proposition. The main problem is to ensure that the two components reside in the same particle in the finished catalyst, and to know that it is so. The main physical techniques to characterise bimetallic particles are hydrogen chemisorption, XRD, TEM, EDX, XPS, XAFS,[197]Au Mössbauer (Section 3.3) and CO chemisorption coupled by IR spectroscopy (Section 5.3). The characterisation of bimetallic catalysts is not always thoroughly done, and there is the further complication of structural changes (particularly of the surface) during use. *In situ* or post-operative characterisation would reveal them, but it is rarely done.

The now classical methods used for the preparation of supported gold catalysts are hardly capable of giving particles that are both small and bimetallic, when the precursors in solution do not interact strongly with each other. During the subsequent thermal treatment performed to get metal particles, the metals must have enough mobility to migrate on the support, interact with each other, and form bimetallic particles. However, phase separation can be a common problem, especially when the metal ratio falls in the miscibility gap (Section 2.6), or if the intended composition is not thermodynamically stable.

The methods used for preparing gold-containing bimetallic catalysts may be divided into three main classes: (i) the methods that do not involve interaction between the two metal precursors in solution (Section 4.6.2); (ii) those that involve sequences of surface reactions designed to create these interactions (Section 4.6.3); (iii) those which use bimetallic precursors (Section 4.6.4). A literature search on this topic does not produce a very large number of papers, but two reviews can be recommended.[173,174]

4.6.2.   Methods without interaction between precursors in solution

4.6.2.1.   *Co-impregnation*

The simplest and long-practised method is co-impregnation of a support by a solution containing precursors of the two metals, but if chloride salts are used, there is the risk of uncontrolled sintering during subsequent thermal treatment (Section 4.2.1). Platinum–gold on silica catalysts made by co-impregnation to incipient wetness with the chloride precursors ($HAuCl_4$ and H_2PtCl_6) indeed led after reduction to some large particles ($>20\,nm$) and to severe phase separation, irrespective of the platinum:gold ratio.[175–177] All of these preparations were made with ratios that fall within the miscibility gap (Section 2.6), so formation of bimetallic phases is unfavourable. There was a tendency for gold to be present in the larger particles, but at higher gold concentrations (e.g. $Pt_1Au_{0.7}$) both metals were present in all sizes, and smaller particles ($<10\,nm$) had the Pt_3Au composition.[177]

   $PdAu/TiO_2$[178] and $PdAu/Fe_2O_3$[179] ($Pd:Au = 2$) have been prepared by co-impregnation of the chloride precursors. After calcination at $673\,K$, both samples exhibit metal particles with a bimodal distribution of size ($4–10\,nm$ and $30–70\,nm$). These particles, at least the largest ones, are core-shell type alloy with a Pd-rich shell.

   Impregnating pre-reduced $5\,wt.\%$ Ni/SiO_2 with a solution of $[Au(NH_3)_4](NO_3)_3$ ($0.4\,wt.\%$ Au) gave after drying small nickel particles, some partly covered by gold, and large ($10–20\,nm$) gold particles.[180]

4.6.2.2.   *Co-adsorption of cations*

Palladium–gold forms a continuous range of solid solutions (Section 2.6) and supported bimetallic particles are easily made. XAFS measurements confirmed that co-adsorption of $[Au(en)_2]^{3+}$ (*en* = ethylenediamine) and $[Pd(NH_3)_4]^{2+}$ cations at natural pH onto silica gave small bimetallic particles ($<5\,nm$) after reduction at $623\,K$, irrespective of the Pd:Au ratio.[181] The same result was obtained by controlling the pH at 10 ($Pd:Au = 1$), but at pH 7 ($Pd:Au = 1.3$), reduction at $573\,K$ gave particles with a gold-rich core decorated by palladium.[182] Small bimetallic particles of platinum–gold and palladium–gold (<3 nm) inside the supercages of Y zeolites have been prepared by simultaneous or sequential exchange with the appropriate *en* complex ions (M:Au = 1.5 and 4), followed by various thermal treatments.[183]

4.6.2.3. *Co-deposition–precipitation*

This method has not been used very much to prepare bimetallic particles. It is always performed at fixed pH as in Section 4.2.3, i.e. the same 'recipe' was applied without attempt to understand the underlying chemistry. Platinum–gold and palladium–gold (Pt or Pd:Au = 5:95) have been deposited this way on titania-silica supports.[66] After calcination at 673 K, the particles are small (<5 nm), and no separate platinum or palladium metal particles are found. With Pd-Au/CeO$_2$ (Pd:Au = 2 to 100) calcined at 673 K, however, only large gold particles (>8 nm) and small palladium particles (<3 nm) were found at gold loadings above 1 wt.%.[184]

4.6.2.4. *Photoreduction*

Bimetallic PdAu/Nb$_2$O$_5$ (Pd:Au = 6) catalysts have been prepared by sequential photoreduction of KAu(CN)$_2$ and Pd(NO$_3$)$_2$ using photons of energy greater than the niobia band gap (3.4 eV);[185] reduction at 673 K gave particles of 7.6 nm mean size, including some that were monometallic. AuPt/TiO$_2$ catalysts have also been made by irradiation of platinum acetylacetonate (Pt(acac)$_2$) in the presence of Au/TiO$_2$ made by DP; metal particles grew in size but not in number as the Pt:Au ratio increases from 1 to 2, so that they must have a gold core and a platinum shell.[186]

4.6.3. Redox methods for preparing bimetallic catalysts

Bimetallic catalysts can be prepared by a direct redox method when a cationic complex of a metal of higher electrochemical potential is reduced by another metal of lower electrochemical potential that has been deposited and reduced first[187] (see Table 4.10); PdAu/SiO$_2$[188] and PdAu/C[189] have been made in this way, gold being deposited after the palladium. Small amounts of the metals were found in filtrates, and XRD and temperature-programmed decomposition of palladium hydride indicated a substantial

Table 4.10: Standard redox potentials of noble metal complexes.

Redox reaction	E^0 (V)
$[PdCl_4]^{2-} + 2e^- \Leftrightarrow Pd + 4Cl^-$	0.62
$[PtCl_4]^{2-} + 2e^- \Leftrightarrow Pt + 4Cl^-$	0.73
$[PtCl_6]^{2-} + 2e^- \Leftrightarrow Pt + 6Cl^-$	0.74
$[AuCl_4]^{2-} + 2e^- \Leftrightarrow Au + 4Cl^-$	1.00

degree of mixing of the components, with gold enriching the surfaces. Particle sizes were not reported.

Hydrogen adsorbed on one metal can be used to reduce the precursor of a second metal, which becomes deposited on it.[187] PtAu/SiO$_2$,[190] PtAu/Al$_2$O$_3$[191] and PtAu/C[192] have been prepared using hydrogen adsorbed on the supported platinum. Complete deposition of the gold was achieved, but with PtAu/Al$_2$O$_3$ some sintering of the platinum occurred during the deposition of the gold. Application of this method to PdAu/TiO$_2$ also gave bimetallic particles, whereas impregnation of Au/TiO$_2$ made by DP with a solution of palladium nitrate gave separate particles of the two metals.[193]

These methods are attractive, but are intricate and time-consuming since re-oxidation of the first metal after reduction must be prevented.

4.6.4. Use of a bimetallic precursor

Supported bimetallic catalysts can be made by adsorption of a bimetallic precursor such as molecular cluster compounds, colloidal particles or dendrimer-stabilised particles. In several cases, 'homogeneous' bimetallic particles have been found where the compositions lie within the miscibility gap of the bulk alloy (e.g. with PtAu particles). This suggests that when the particles are small enough and do not possess metallic properties, the normal rules do not apply.

4.6.4.1. *Adsorption of bimetallic molecular clusters*

Bimetallic molecular cluster compounds containing gold will adsorb intact on supports, and the ligands may then be removed by heating, but only a few such compounds are available. Simple carbonyl complexes do not exist, but while several bimetallic clusters stabilised by phosphine ligands are known,[194] they have not been used for making catalysts. Only the complex $Pt_2Au_4(C\equiv C^tBu)_8$ has been employed;[175] after adsorption from hexane solution, and oxidation to 523 K to remove the ligands, a Pt_2Au_4/SiO_2 catalyst was obtained, with small ($\sim$2.5 nm) and uniform bimetallic particles.

4.6.4.2. *Deposition of bimetallic colloids*

Methods for preparing bimetallic colloids containing gold have been described in Section 3.2.3. Their composition may be easily tuned since solutions of the mixed chloride precursors are normally used, and are reduced in a variety of ways after addition of a stabiliser. As already mentioned in Section 4.3.6, for gas-phase catalytic reactions, after depositing the colloid on an oxide support the stabilisers must be removed, but for liquid-phase reactions they may be retained providing access to the metal is still possible.

A number of supported bimetallic colloids have been prepared, and their properties described; they include $PdAu/TiO_2$,[195] $PdAu/SiO_2$,[196,197] $PdAu/C$,[198] $PtAu/C$[199] and $AgAu/MCM$-41.[200] Calcination is usually performed to remove the stabiliser (573–673 K), followed by reduction in hydrogen to obtain the metallic state (473–673 K), this sometimes transforms a core-shell structure into a random bimetallic phase.[197] Particle sizes are often satisfactorily small, but it must be remembered that for selective oxidations a small particle size is not always preferred (Section 8.5). Moreover, it does not automatically follow that all particles will have exactly the same composition, arising from a fault in the colloid preparation stage. The composition of the particles naturally depends on the way in which they are synthesized. Thus, depositing palladium onto colloidal gold or forming a colloid from the mixed salt solution gave small Pd_4Au_6 particles, the excess palladium forming a separate phase, while depositing gold onto a palladium colloid gave Pd_2Au_8 plus some palladium particles.[198] Carbon was used as a support in this instance, and it has also found use as a support for a platinum–gold colloid ($Pt_{18}Au_{82}$);[199] homogeneous particles of this composition were obtained after an oxidation-reduction cycle. $PdAu/SiO_2$ has been made by constructing the support in the presence of the colloid and polyethyleneglycol stearate by base-hydrolysing tetraethoxysilane.[197] Similarly $AgAu/MCM$-41 was formed by using the colloidal dispersion in the sol–gel synthesis of the zeolite;[200] a quaternary ammonium surfactant served both as stabiliser for the colloid and template for the zeolite.

4.6.4.3. *Deposition of dendrimer-stabilised bimetallic particles*

The use of dendrimer-encapsulated bimetallic particles is an elegant way for preparing supported catalysts, because their composition may be tuned, and their size is generally retained after removing the dendrimer. The process for converting them into a supported catalyst closely resembles that used with colloidal particles; after adsorption on a support,[108] or incorporation into a sol–gel synthesis of a support,[201] the dendrimer is removed from the complex by calcination (573–773 K) followed by reduction (573–773 K). In this way $PtAu/SiO_2$ has been made from $Pt_{16}Au_{16}$ (cluster of 32 atoms) particles in G5-PAMAM dendrimer (see Section 4.3.7), the particle size (2–6 nm) being notably smaller than for gold by itself.[108] Similarly, $PtAu/TiO_2$ has been prepared from $Pt_{27.5}Au_{27.5}$ particles (cluster of 55 atoms) in G4-NH_2 dendrimer, via a sol–gel synthesis of the support by hydrolysis of $Ti(O^iPr)_4$.[201]

4.6.4.4. *Low-energy cluster deposition*

The method of low-energy cluster deposition using laser vaporisation described in Section 4.4.5 is also applicable to forming bimetallic particles of controlled size on oxide supports. Thus, for example gold–titania particles (2.3 nm) having the same composition as the $Au_{25}Ti_{75}$ rod from which they were formed, were deposited on alumina, but ambient air-oxidation led to $Au–TiO_2/Al_2O_3$ (J.L. Rousset, personal communication). Nickel–gold particles have also been deposited on graphite in this way,[202] although their surface was enriched in gold. The originality of this technique lies in the fact that the composition of the particles is precisely determined by that of the rod, and the average particle size is controlled as well, between 2.3 and 2.6 nm. However, it cannot be considered as a routine technique, since it requires a UHV chamber and laser, and the metal loading on powder supports is only low (~0.1 wt.%) after several hours of evaporation.

4.7. Influence of the Thermal Treatment on Gold Particle Size

Gold particle size depends not only on the method of preparation, but also on the conditions used for the subsequent treatment performed to obtain the metal. Indeed, as mentioned in the introduction to this chapter, for most methods gold is in the +3 oxidation state after drying (the so-called 'as prepared' sample), and heating to reduce it to Au^0 can be performed under any gas because of the instability of Au_2O_3; air is most often used. Such treatment does not necessarily ensure that all the gold will be as Au^0; some ionic species may remain. This is an important issue for explaining the catalytic properties of gold (see Chapter 6), and is not easy to establish qualitatively and still less quantitatively. Decomposition of the precursor and reduction to Au^0 begins as low as 373 K, and the extent of reduction may moreover depend on many parameters, such as the nature of the support.[203,204]

4.7.1. Nature of the gas

Heating 'as prepared' DP Au/TiO_2 in hydrogen to 393 K gave complete reduction to Au^0, and smaller particles (2.1 nm) than when heated in air to 423–473 K (3.3 nm).[62,204,205] This may be because (i) reduction starts at a lower temperature, leading to a stronger interaction between particle and support, and less sintering[62] and (ii) hydrogen removes residual chloride ion as HCl,[10] whereas chloride is still present on the support after calcination in air even at 873 K, which induces sintering.[50] Heating in carbon monoxide[62] or in argon[206] is also preferable to heating in air to obtain small gold particles. Effects of treating Au/TiO_2 prepared from $[Au_6(PPh_3)_6](BF_4)_2$ (Section 4.3.2) in hydrogen and/or air have also been reported.[87]

4.7.2. Gas flow rate and sample weight

Gold particle size strongly depends on the flow rate of air and the sample weight used for calcination: the higher the heating rate (0.1–4 Kmin^{-1}), the larger the gold particles (3.4–4.9 nm) in Au/Fe_2O_3 samples prepared by impregnation with a gold phosphine complex.[12] With an 'as prepared' Au/TiO_2 catalyst made by DP, the average gold particle size was halved (4.3–2.2 nm) when flow rate was increased from 50 to 100 ml min^{-1}, and even more (5.7–2.7 nm) when sample weight was lowered from 450 to 24 mg.[206] Shorter contact time lowers particle size because of more efficient elimination of water and/or of remaining chloride. The height of the bed used is also likely to be a relevant parameter.

4.7.3. Effect of the temperature of thermal treatment

The size of gold particles increases with the final temperature of pre-treatment,[5,62,206,207] and more in air than in hydrogen, for the reasons given in Section 4.7.1.[206] The effect may be small below 573 K, however, and the useful point has been made that the surface of titania (and perhaps other oxides) starts dehydrating below this temperature, and this may affect the ease of migration and coalescence of small particles. Indeed diffusion of adatoms and clusters has been observed by TEM with Au/TiO_2 between 573 and 673 K, particles being trapped and growing at the interface between support particles.[5] In other investigations, however, no migration was observed,[51,203,206,208] and particles were only found on flat areas of titania surfaces on calcination at 673 K. Temperature does not have the same drastic effect as change of contact time: particle size may be mainly determined by the temperature at which reduction of the precursor starts (423–473 K in air).[206] Once fully reduced, the particle size of gold does not increase drastically with temperature. It appears that mobility of the precursor, especially if chloride is still retained, is largely instrumental in creating larger particles; its reduction at as low temperature as possible and short contact time with gas are recommended.

4.8. Effect of Conditions of Storage on Size of Gold Particles

Gold catalysts are not easy to handle. To start with, aqueous solutions containing Au^{III} can be photoreduced by natural light, and hence it is recommended that preparations be performed in the dark.[49,52,206,207] Second, supported Au^{III} precursors and Au^0 particles are also very sensitive to ambient air and light: only a few groups are aware of this problem[52,83,206,207,209,210] reporting that supported Au^{III} could be reduced for instance, during drying on a suction filter in air[209] or once dried, by exposure to air in the dark.[207] The colour then changes from yellow or grey to purple (presence of Au^0). This is especially true when gold is supported on titania; gold on a support with no band gap, such as alumina, is much less sensitive to the light. Metallic gold particles in samples stored in air are also reported to grow; thus ∼2 nm particles in Au/TiO_2 made by DP grew to more than 5 nm in size within a few weeks,[206] while small particles in Au/MCM-41 mesopores migrated to the external surface and grew in size, leading to a binodal distribution with maxima at ∼2–5 and 12 nm.[152]

The literature recommends that 'as-prepared' samples be stored in a refrigerator below 273 K,[207] that calcined catalysts should also be kept cold,[207,210] and that after drying, samples should be stored in a vacuum desiccator in the dark, reduction being performed immediately before use.[83,206]

Since, very few people seem to be aware of the process of ageing in air, it is probable that some of the published results regarding the oxidation states of gold in dried samples are inaccurate, as well as the catalytic results, especially in oxidation of carbon monoxide (Chapter 6). Furthermore, Au^{III} can be reduced under the influence of radiation, especially when applied in high vacuum; this may occur with electrons as in TEM[204,208] and with photons as in XAFS (J.T. Miller, personal communication) and XPS,[204] so that special care is needed to characterise unreduced samples. Ease of reduction may also depend on the kind of support, and materials made from an Au^I precursor may also suffer reduction.

4.9. Conclusion

By way of conclusion, one can claim that small gold particles can be made on almost any type of support providing the appropriate preparation method is chosen. With certain of these methods, it is also possible to control *a priori* the gold loading, and to avoid the loss of gold in solution. The methods that do not involve chloride or where chloride ion can be eliminated are definitively desirable because of its detrimental effect on gold particle size and on catalytic activity, especially in oxidation of carbon monoxide. For some methods, success directly depends on the point of zero charge of the oxide support. Again, it must be stressed that it is important to pay attention to the conditions of preparation and storage of the samples, so as to avoid uncontrolled reduction and gold particle sintering.

The World Gold Council (www.gold.org) can provide several gold reference catalysts: $1.5\,wt.\%$ Au/TiO_2 prepared by deposition-precipitation (with NaOH), $4.5\,wt.\%$ Au/Fe_2O_3 prepared by coprecipitation, $0.3\,wt.\%$ Au/Fe_2O_3 on alumina beads prepared by deposition-precipitation, and $0.8\,wt.\%$ Au/C prepared from gold sol. Recipients of these catalysts are being encouraged to share their results both in the characterization and activity spheres, thus providing valuable comparisons of results between the groups using them.

References

1. G.C. Bond, *Heterogeneous Catalysis: Principles and Applications*, Clarendon Press, Oxford (1987).
2. L.A. Hermans and J.W. Geus, *Stud. Surf. Sci. Catal.* **4** (1979) 113.
3. J.A. van Dillen, J.W. Geus, L.A. Hermans and J. van der Meijden Proc. 6th Intern. Congr. Catal., London, 1976, G.C. Bond, P.B. Wells, F.C. Tompkins (eds.), The Chemical Society, London (1977) p. 677.

4. P. Burattin, M. Che and C. Louis, *J. Phys. Chem. B* **102** (1998) 2722.
5. T. Akita, P. Lu, S. Ichikawa, K. Tanaka and M. Haruta, *Surf. Interf. Anal.* **31** (2001) 73.
6. M. Haruta, *Chem. Record* **3** (2003) 75.
7. M. Mavrikakis, P. Stoltze and J.K. Norskov, *Catal. Lett.* **64** (2000) 101.
8. M. Valden, X. Lai and D.W. Goodman, *Science* **281** (1998) 1647.
9. G.C. Bond and D.T. Thompson, *Catal. Rev.-Sci. Eng.* **41** (1999) 319.
10. M. Haruta, *Cattech* **6** (2002) 102.
11. M. Haruta, *Catal. Today* **36** (1997) 153.
12. A.I. Kozlov, A.P. Kozlova, K. Asakura, Y. Matsui, T. Kogure, T. Shido and Y. Iwasawa, *J. Catal.* **196** (2000) 56.
13. A.I. Kozlov, A.P. Kozlova, H. Liu and Y. Iwasawa, *Appl. Catal. A* **182** (1999) 9.
14. M. Haruta, *Catal. Surveys Japan* (1997) 61.
15. M. Haruta, H. Kageyama, N. Kamijo, T. Kobayashi and F. Delannay, *Stud. Surf. Sci. Catal.* **44** (1988) 33.
16. G.C. Bond, *Metal-Catalysed Reactions of Hydrocarbons*, Springer, New York (2005).
17. Y.A. Nechayev and G.V. Zvonareva, *Geokhimiya* **6** (1983) 919.
18. J.A. Peck and G.E. Brown, *Geochim. Cosmochim. Acta* **55** (1991) 671.
19. F. Farges, J.A. Sharps and G.E. Brown, *Geochim. Cosmochim. Acta* **57** (1993) 1243.
20. P.J. Murphy and M.S. LaGrange, *Geochim. Cosmochim. Acta* **62** (1998) 3515.
21. S. Ivanova, V. Pitchon, C. Petit, H. Herschbach, A.V. Dorsselaer and E. Leize, *Appl. Catal. A* **298** (2006) 203.
22. N. Bjerrum, *Bull. Soc. Chim. Belg.* **57** (1948) 432.
23. H. Chateau, M.C. Gadet and J. Pouradier, *J. Chim. Phys.* **63** (1966) 18.
24. C.F. Baes Jr and R.R. Mesmer, *The Hydrolysis of Cations*, John Wiley & Sons, New York (1986).
25. S.-J. Lee and A. Gavriilidis, *J. Catal.* **206** (2002) 305.
26. S. Ivanova, C. Petit and V. Pitchon, *Appl. Catal. A* **267** (2004) 191.
27. F. Moreau, G.C. Bond and A.O. Taylor, *Chem. Commun.* (2004) 1642.
28. V.I. Belevantsev, G.R. Kolonin and S.K.R. Ryakhovskaya, *Russ. J. Inorg. Chem.* **17** (1972) 1303.
29. P.L. Murphy, G. Stevens and M.S. LaGrange, *Geochim. Cosmochim. Acta* **64** (2000) 479.
30. J.H. Yang, J.D. Henao, C. Costello, M.C. Kung, H.H. Kung, J.T. Miller, A.J. Kropf, J.-G. Kim, J.R. Regalbuto, M.T. Bore, H.N. Pham, A.K. Datye, J.D. Laeger and K. Kharas, *Appl. Catal. A: Gen.* **291** (2005) 73.
31. J.W. Mellor, *A Comprehensive Treatise on Inorganic and Theoretical Chemistry*, Longmans, Green and Co, London, **3** (1941) 583.
32. J.A. Cusumano, *Nature* **247** (1974) 456.
33. J.M. Fisher, *Gold Bull.* **36** (2003) 155.
34. L. Delannoy, N.E. Hassan, N.Nguyen Le To and C. Louis, Submitted to *J. Phys. Chem. B*.
35. Q. Fu, S. Kudriavtseva, H. Saltsburg and M. Flytzani-Stephanopoulos, *Chem. Eng. J.* **93** (2003) 41.
36. Q. Xu, K.C.C. Kharas and A.K. Datye, *Catal. Lett.* **85** (2003) 229.
37. R. Zanella, L. Delannoy and C. Louis, *Appl. Catal. A: Gen.* **291** (2005) 62.
38. R.J. Puddephatt, *The chemistry of gold*, Elsevier Scientific Publishing Company, Amsterdam (1978).
39. W.R. Mason and H.B. Gray, *J. Am. Chem. Soc.* **90** (1968) 5721.

40. E. Weitz, *Leibig's Ann. Chem.* **410** (1915) 117.
41. P.A. Sermon, G.C. Bond and P.B. Wells, *J. Chem. Soc. Faraday Trans. I* **75** (1979) 385.
42. K. Blick, T.D. Mitrelias, J.S.J. Hargreaves, G.J. Hutchings, R.W. Joyner, C.J. Kiely and F.E. Wagner, *Catal. Lett.* **75** (1998) 385.
43. N.W. Cant and W.K. Hall, *J. Phys. Chem.* **75** (1971) 2914.
44. S. Galvagno and G. Parravano, *J. Catal.* **55** (1978) 178.
45. W. Vogel, D.A.H. Cunningham, K. Tanaka and M. Haruta, *Catal. Lett.* **40** (1996) 175.
46. S.D. Lin, M. Bollinger and M.A. Vannice, *Catal. Lett.* **17** (1993) 245.
47. J.Y. Lee and J. Schwank, *J. Catal.* **102** (1986) 207.
48. H.H. Kung, M.C. Kung and C.K. Costello, *J. Catal.* **216** (2003) 425.
49. H.S. Oh, J.H. Yang, C.K. Costello, Y.M. Wang, S.R. Bare, H.H. Kung and M.C. Kung, *J. Catal.* **210** (2002) 375.
50. *Report of the Osaka National Research Institute* **393** (1999).
51. M. Haruta, S. Tsubota, T. Kobayashi, H. Kageyama, M.J. Genet and B. Delmon, *J. Catal.* **144** (1993) 175.
52. R. Zanella, S. Giorgio, C.R. Henry and C. Louis, *J. Phys. Chem. B* **106** (2002) 7634.
53. M. Haruta, T. Kobayashi, H. Sano and N. Yamada, *Chem. Lett.* **2** (1987) 405.
54. M. Haruta, N. Yamada, T. Kobayashi and S. Iijima, *J. Catal.* **115** (1989) 301.
55. A.M. Venezia, G. Pantaleo, A. Longo, G.D. Carlo, M.P. Casaletto, F.L. Liotta and G. Deganello, *J. Phys. Chem. B* **109** (2005) 2821.
56. M. Haruta, *Stud. Surf. Sci. Catal.* **110** (1997) 123.
57. R.M. Finch, N.A. Hodge, G.J. Hutchings, A. Meagher, Q.A. Pankhurst, M.R.H. Siddiqui, F.E. Wagner and R. Whyman, *Phys. Chem. Chem. Phys.* **1** (1999) 485.
58. A.M. Visco, A. Donato, C. Milone and S. Galvagno, *Reac. Kinet. Catal. Lett.* **61** (1997) 219.
59. R.D. Walters, J.J. Weimer and J.E. Smith, *Catal. Lett.* **30** (1995) 181.
60. S.D. Gardner, G.B. Hoflund, M.R. Davidson, H.A. Laitinen, D.R. Schryer and B.T. Upchurch, *Langmuir* **7** (1991) 2140.
61. S. Tsubota, M. Haruta, T. Kobayashi, A. Ueda and Y. Nakahara, *Stud. Surf. Sci. Catal.* **72** (1991) 695.
62. S. Tsubota, D.A.H. Cunningham, Y. Band o and M. Haruta, *Stud. Surf. Sci. Catal.* **91** (1995) 227.
63. C.-K. Chang, Y.-J. Chen and C.-T. Yeh, *Appl. Catal. A: Gen.* **174** (1998) 13.
64. F. Moreau and G.C. Bond, *Appl. Catal. A: Gen.* **302** (2006) 110.
65. A. Wolf and F. Schüth, *Appl. Catal. A: Gen.* **226** (2002) 1.
66. A. Zwijnenburg, M. Saleh, M. Makkee and J.A. Moulijn, *Catal. Today* **72** (2002) 59.
67. F. Moreau, G.C. Bond and A.O. Taylor, *J. Catal.* **231** (2005) 105.
68. L. Prati and G. Martra, *Gold Bull.* **32** (1999) 96101.
69. Y.A. Nechayev and N.V. Nikolenko, *Geochem. Intern.* **11** (1985) 1656.
70. M.L. Machesky, W.O. And rade and A.W. Rose, *Geochim. Cosmochim. Acta* **5** (1991) 769.
71. R.J.H. Grisel, P.J. Kooyman and B.E. Nieuwenhuys, *J. Catal.* **191** (2000) 430.
72. G.K. Bethke and H.H. Kung, *Appl. Catal. A* **194–195** (2000) 43.
73. G.R. Bamwenda, S. Tsubota, T. Nakamura and M. Haruta, *Catal. Lett.* **44** (1997) 83.

74. D.A.H. Cunningham, W. Vogel, H. Kageyama, S. Tsubota and M. Haruta, *J. Catal.* **177** (1998) 1.

75. M. Okumura, S. Nakamura, S. Tsubota, T. Nakamura, M. Azuma and M. Haruta, *Catal. Lett.* **51** (1998) 53.

76. C.K. Costello, M.C. Kung, H.-S. Oh and K.H. Kung, *Appl. Catal. A: Gen.* **232** (2002) 159.

77. M.A.P. Dekkers, M.J. Lippits and B.E. Nieuwenhuys, *Catal. Lett.* **56** (1998) 195.

78. M.A.P. Dekkers, M.J. Lippits and B.E. Nieuwenhuys, *Catal. Today* **54** (1999) 381.

79. R.J.H. Grisel and B.E. Nieuwenhuys, *Catal. Today* **64** (2001) 69.

80. M. Khoudiakov, M.-C. Gupta and S. Deevi, *Appl. Catal. A: Gen.* **291** (2005) 151.

81. N.S. Patil, B.S. Uphade, P. Jana, S.K. Bharagava and V.R. Choudhary, *J. Catal.* **223** (2004) 236.

82. S. Ivanova, V. Pitchon, Y. Zimmermann and C. Petit, *Appl. Catal. A: Gen.* **298** (2006) 57.

83. W.-C. Li, M. Comotti and F. Schüth, *J. Catal.* **237** (2006) 190.

84. W.N. Delgass, M. Boudart and G. Parravano, *J. Phys. Chem.* **72** (1968) 3563.

85. Y. Yuan, K. Asakura, H. Wan, K. Tsai and Y. Iwasawa, *Catal. Lett.* **42** (1996) 15.

86. Y. Yuan, A.P. Kozlova, K. Asakura, H. Wan, K. Tsai and Y. Iwasawa, *J. Catal.* **170** (1997) 191.

87. T.V. Choudhary, C. Sivadinarayana, C.C. Chusuei, A.K. Datye, J.P.F. Jr and D.W. Goodman, *J. Catal.* **207** (2002) 247.

88. Y. Yuan, K. Asakura, H. Wan, K. Tsai and Y. Iwasawa, *Catal. Today* **44** (1998) 333.

89. T.J. Mathieson, A.G. Langdon, N.B. Milestone and B.K. Nicholson, *Chem. Commun.* (1998) 371.

90. M. Okumura and M. Haruta, *Chem. Lett.* (2000) 396.

91. J. Guzman, B.G. Anderson, C.P. Vinod, K. Ramesh, J.W. Niemantsverdriet and B.C. Gates, *Langmuir* **21** (2005) 3675.

92. J.C. Fierro-Gonzalez and B.C. Gates, *J. Phys. Chem. B* **109** (2005) 7275.

93. J. Guzman and B.C. Gates, *Langmuir* **19** (2003) 3897.

94. J.C. Fierro-Gonzalez and B.C. Gates, *J. Phys. Chem. B* **108** (2004) 16999.

95. J.C. Fierro-Gonzalez, B.G. Anderson, K. Ramesh, C.P. Vinod, J.W. Niemantsverdriet and B.C. Gates, *Catal. Lett.* **101** (2005) 265.

96. P. Serp, P. Kalck and R. Fenner, *Chem. Rev.* **102** (2002) 3085.

97. M. Okumura, S. Tsubota and M. Haruta, *J. Molec. Catal. A: Chem.* **199** (2003) 73.

98. B.P. Block and J.J.C. Bailar, *J. Am. Chem. Soc.* **73** (1951) 4722.

99. D. Guillemot, V.Y. Borovskov, V.B. Kazansky, M. Polisset-Thfoin and J. Fraissard, *J. Chem. Soc., Faraday Trans.* **93** (1997) 3587.

100. D. Guillemot, M. Polisset-Thfoin and J. Fraissard, *Catal. Lett.* **41** (1996) 143.

101. D.A. Bulushev, I. Yuranov, E.I. Suvorova, P.A. Buffat and L. Kiwi-Minsker, *J. Catal.* **224** (2004) 8.

102. F. Porta, L. Prati, M. Rossi, S. Coluccia and G. Martra, *Catal. Today* **61** (2000) 165.

103. S. Tsubota, T. Nakamura, K. Tanaka and M. Haruta, *Catal. Lett.* **56** (1998) 131.

104. J.-D. Grunwaldt, C. Kierner, C. Wogerbauer and A. Baiker, *J. Catal.* **181** (1999) 223.

105. J.-D. Grunwaldt, M. Maciejewski, O.S. Becker, P. Fabrizioli and A. Baiker, *J. Catal.* **186** (1999) 458.

106. G. Martra, L. Prati, C. Manfredotti, S. Biella, M. Rossi and S. Coluccia, *J. Phys. Chem. B* **107** (2003) 5453.

107. R.W.J. Scott, O.M. Wilson and R.M. Crooks, *Chem. Mater.* **16** (2004) 5682.

108. H. Lang, S. Maldonado, K.J. Stevenson and B.D. Chandler, *J. Am. Chem. Soc.* **126** (2004) 12949.
109. F.B. Li and X.Z. Li, *Appl. Catal. A* **228** (2002) 15.
110. H. Kozuka and S. Sakka, *Chem. Mater.* **5** (1993) 222.
111. F. Akbarian, B.S. Dunn and J.I. Zink, *J. Raman Spect.* **27** (1996) 775.
112. P. Innocenzi, G. Brusatin, A. MArtucci and K. Urabe, *Thin Solid Films* **279** (1996) 23.
113. E. Seker and E. Gulari, *Appl. Catal. A: Gen.* **232** (2002) 203.
114. D.D. Smith, L.A. Snow, L. Sibille and E. Ignont, *J. Non-Cryst. Solids* **285** (2001) 256.
115. J.J. Pietron, R.M. Stroud and D.R. Rolison, *Nano Lett.* **2** (2002) 545.
116. A. Fernandez, A. Caballero, A.R. Gonzalez-Elipe, J.-H. Herrmann, H. Dexpert and F. Villain, *J. Phys. Chem.* **99** (1995) 3303.
117. D. Li, J.T. McCann, M. Gratt and Y. Xia, *Chem. Phys. Lett.* **394** (2004) 387.
118. S.C. Chan and M.A. Barteau, *Langmuir* **21** (2005) 5588.
119. K.S. Suslick and D.A. Hammerton, *IEEE Trans. Sonics Ultrason* **143** (1986) SU.
120. W. Chen, W. Cai, L. Zhang, G. Wang and L. Zhang, *J. Coll. Inter. Sci.* **238** (2001) 291.
121. V.G. Pol, A. Gedanken and J. Calderon-Moreno, *Chem. Mater.* **15** (2003) 1111.
122. B.C. Satishkumar, E.M. Vogt, A. Govindaraj and C.N.R. Rao, *J. Phys. D: Appl. Phys.* **29** (1996) 3173.
123. A. Fasi, I. Palinko, J.W. Seo, Z. Konya, K. Hernadi and I. Klricsi, *Chem. Phys. Lett.* **372** (2003) 848.
124. T. Uematsu, L. Fan, T. Maruyama, N. Ichikuni and S. Shimazu, *J. Molec. Catal. A: Chem.* **182–183** (2002) 209.
125. L. Fan, N. Ichikuni, S. Shimazu and T. Uematsu, *Appl. Catal. A: Gen.* **246** (2003) 87.
126. S. Arii, F. Mortin, A.J. Renouprez and J.L. Rousset, *J. Am. Chem. Soc.* **126** (2004) 1199.
127. G.M. Veith, A.R. Lupini, S.J. Pennycook, G.W. Ownby and N.J. Dudney, *J. Catal.* **231** (2005) 151; L.A. Brey, T.A. Wood, G.M. Bucellato, M.E. Jones, C.S. Chamberlain and A.R. Siedle, WO Patent 2005/030382 A2.
128. M. Lomello-Tafin, A.A. Chaou, F. Morfin, V. Caps and J.-L. Rousset, *Chem. Commun.* (2005) 388.
129. A. Baiker, M. Maciejewski, S. Tagliaferri and P. Hug, *J. Catal.* **151** (1995) 407.
130. S.-H. Wu, X.-C. Zheng, S.-R. Wang, D.-Z. Han, W.-P. Huang and S.-M. Zhang, *Catal. Lett.* **96** (2004) 49.
131. S. Qiu, W. Pang, W. Xu and R. Xu, *Stud. Surf. Sci. Catal.* **84** (1994) 1059.
132. M.M. Mokhtar, T.M. Salama and M. Ichikawa, *J. Coll. Inter. Sci.* **224** (2000) 366.
133. T.M. Salama, T. Shido, H. Managawa and M. Ichikawa, *J. Catal.* **152** (1995) 322.
134. K. Kuge and G. Calzaferri, *Micro. Meso. Mater.* **66** (2003) 15.
135. Z.-X. Gao, Q. Sun, X. Wang and W.M.H. Sachtler, *Catal. Lett.* **72** (2001) 1.
136. S.M. Kanan, C.P. Tripp, R.N. Austin and H.H. Patterson, *J. Phys. Chem. B* **105** (2001) 9441.
137. J.-N. Lin and B.-Z. Wan, *Appl. Catal. B: Env.* **41** (2003) 83.
138. J.-H. Chen, J.-N. Lin, Y.-M. Kang, W.-Y. Yu, C.-N. Kuo and B.-Z. Wan, *Appl. Catal. A: Gen.* **291** (2005) 162.
139. K. Okumura, K. Yoshino, K. Kato and M. Niwa, *J. Phys. Chem. B* **109** (2005) 12380.
140. C. Kan, W. Cai, C. Li, G. Fu and L. Zhang, *J. Appl. Phys.* **96** (2004) 5727.

141. A. Fukuoka, H. Araki, J.-I. Kimura, Y. Sakamoto, T. Higuchi, N. Sugimoto, S. Inagaki and M. Ichikawa, *J. Mater. Chem.* **14** (2004) 752.
142. Y.-S. Chi, H.-P. Lin, C.-N. Lin, C.-Y. Mou and B.-Z. Wan, *Stud. Surf. Sci. Catal.* **141** (2202) 329.
143. Y.-S. Chi, H.-P. Lin and C.-Y. Mou, *Appl. Catal. A: Gen.* **284** (2005) 199.
144. Z.-P. Liu, S.J. Jenkins and D.A. King, *Phys. Rev. Lett.* **94** (2005) 196102.
145. A. Ghosh, C.R. Patra, P. Mukherjee, M. Sastry and R. Kumar, *Micro. Meso. Mater.* **58** (2003) 210.
146. Y. Guari, C. Thieuleux, A. Mehdi, C. Reyé, R.J.P. Corriu, S. Gomez-Gallardo, K. Philippot and B. Chaudret, *Chem. Mater.* **15** (2003) 2017.
147. K.-J. Chao, M.-H. Cheng, Y.-F. Ho and P.-H. Liu, *Catal. Today* **97** (2004) 49.
148. C.-M. Yang, M. Kalwei, F. Schuth and K.-J. Chao, *Appl. Catal. A: Gen.* **254** (2003) 289.
149. G. Lü, R. Zhao, G. Qian, Y. Qi, X. Wang and J. Suo, *Catal. Lett.* **97** (2004) 115.
150. H. Zhu, B. Lee, S. Dai and S.H. Overbury, *Langmuir* **19** (2003) 3974.
151. B. Lee, H. Zhu, Z. Zhang, S.H. Overbury and S. Dai, *Micro. Meso. Mater.* **70** (2004) 71.
152. S.H. Overbury, L. Ortiz-Soto, H. Zhu, B. Lee, M.D. Amiridis and S. Dai, *Catal. Lett.* **95** (2004) 99.
153. S. Cheng, Y. Wei, Q. Feng, K.-Y. Qiu, J.-B. Pang, S.A. Jansen, R. Yin and K. Ong, *Chem. Mater.* **15** (2003) 1560.
154. Z. Konya, V.F. Funtes, I. Kiricsi, J. Zhu, J.W. Ager, M.K. Ko, H. Frei, A.P. Alivisatos and G.A. Somorjai, *Chem. Mater.* **15** (2003) 1242.
155. J. Zhu, Z. Konya, V.F. Funtes, I. Kiricsi, C.X. Miao, J.W. Ager, A.P. Alivisatos and G.A. Somorjai, *Langmuir* **19** (2003) 4396.
156. A.H. Janssen, C.-M. Yang, Y. Wang, F. Schüth, A.J. Koster and K.P.d. Jong, *J. Phys. Chem. B* **107** (2003) 10552.
157. M.T. Bore, H.N. Pham, E.E. Switzer, T.L. Ward, A. Fukuoka and A.K. Datye, *J. Phys. Chem. B* **109** (2005) 2873.
158. B.S. Uphade, M. Okumura, S. Tsubota and M. Haruta, *Appl. Catal. A: Gen.* **190** (2000) 43.
159. A.K. Sinha, S. Seelan, T. Akita, S. Tsubota and M. Haruta, *Appl. Catal. A: Gen.* **240** (2003) 243.
160. B.S. Uphade, T. Akita, T. Nakamura and M. Haruta, *J. Catal.* **209** (2002) 331.
161. C. Qi, T. Akita, M. Okumura, K. Kuraoka and M. Haruta, *Appl. Catal. A: Gen.* **253** (2003) 75.
162. W. Yan, B. Chen, S.M. Mahurin, E.W. Hagaman, S. Dai and S.H. Overbury, *J. Phys. Chem. B* **108** (2004) 2793.
163. N. Yap, R.P. Andres and W.N. Delgass, *J. Catal.* **226** (2004) 156.
164. L. Cumaranatunge and W.N. Delgass, *J. Catal.* **232** (2005) 38.
165. F. Porta and L. Prati, *Recent Res. Devel. Vacuum Sci. Tech.* **4** (2003) 99.
166. F. Porta and M. Rossi, *J. Molec. Catal. A: Chem.* **204–205** (2003) 553.
167. S. Biella, F. Porta, L. Prati and M. Rossi, *Catal. Lett.* **90** (2003) 23.
168. F. Porta, L. Prati, M. Rossi and G. Scari, *Coll. Surf. A* **211** (2002) 43.
169. H.C. Choi, M. Shim, S. Bangsaruntip and H. Dai, *J. Am. Chem. Soc.* **124** (2002) 9058.
170. X. Ma, N. Lun and S. Wen, *Diamond Rel. Mater.* **14** (2005) 68.
171. R. Zanella, E.V. Basiuk, P. Santiago, V.A. Basiuk, E. Mireles, I. Puente-Lee and J.M. Saniger, *J. Phys. Chem. B* **109** (2005) 16290.

172. L. Han, W. Wu, F.L. Kirk, J. Luo, M.M. Maye, N.N. Kariuki, Y. Lin, C. Wang and C.-J. Zhong, *Langmuir* **20** (2004) 6019.
173. O.S. Alexeev and B.C. Gates, *Ind. Eng. Chem. Res.* **42** (2003) 1571.
174. L. Guczi and A. Sarkany, in *Specialist Periodical Reports: Catalysis*, J.J. Spirey and S.K. Agarwal, (eds.), Roy. Soc. Chem., London, Vol. 11, 1994, p. 318.
175. B.D. Chandler, A.B. Schabel, C.F. Blanford and L.H. Pignolet, *J. Catal.* **187** (1999) 367–384.
176. A. Vàzquez-Zavala, J. Garcia-Gòmez and A. Gòmez-Cortes, *Appl. Surf. Sci.* **167** (2000) 177.
177. A. Sachdev and J. Schwank, *J. Catal.* **120** (1989) 353.
178. J.K. Edwards, B.E. Solsona, P. Landon, A.F. Carley, A. Herzing, C.J. Kiely and G.J. Hutchings, *J. Catal.* **236** (2005) 69.
179. J.K. Edwards, B. Solsona, P. Landon, A.F. Carley, A. Herzing, M. Watanabe, C.J. Kiely and G.J. Hutchings, *J. Mater. Chem.* **15** (2005) 4595.
180. A.M. Molenbroek, J.K. Nørskov and B.S. Clausen, *J. Phys. Chem. B* **105** (2001) 5450.
181. Y.L. Lam and M. Boudart, *J. Catal.* **50** (1977) 530.
182. S.N. Reifsnyder and H.H. Lamb, *J. Phys. Chem. B* **103** (1999) 321.
183. G. Riahi, D. Guillemot, M. Polisset-Thfoin, A.A. Khodadadi and J. Fraissard, *Catal. Today* **72** (2002) 115.
184. M.P. Kapoor, Y. Ichihashi, T. Nakamori and Y. Matsumura, *J. Molec. Catal. A: Chem.* **213** (2004) 251.
185. R. Brayner, D.d.S. Cunhab and F. Bozon-Verduraz, *Catal. Today* **78** (2003) 419.
186. H. Tada, F. Suzuki, S. Ito, T. Akita, K. Tanaka, T. Kawahara and H. Kobayashi, *J. Phys. Chem. B* **106** (2002) 8714.
187. J. Barbier, in *Handbook of Heterogeneous Catalysis*, G Ertl, H Knözinger, J. Weitkamp, (eds.), Wiley-VCH, Weinheim, 1997, Vol. 1, p. 257.
188. M. Bonarowska, J. Pielaszek, W. Juszczyk and Z. Karpinski, *J. Catal.* **195** (2000) 304.
189. M. Bonarowska, J. Pielaszek, V.A. Semikolenov and Z. Karpinski, *J. Catal.* **209** (2002) 528–538.
190. J. Barbier, P. Marécot, G.D. Angel, P. Bosch, J.P. Boitiaux, B. Didillon, J.M. Dominguez, I. Schifter and G. Espinosa, *Appl. Catal. A: Gen.* **116** (1999) 179.
191. G. Espinosa, G.D. Angel, J. Barbier, P. Bosch, V. Lara and D. Acosta, *J. Molec. Catal. A: Chem.* **164** (2000) 253.
192. P.D. Angel, J.M. Dominguez, G.D. Angel, J.A. Montoya, J. Capilla, E. Lamy-Pitara and J. Barbier, *Topics Catal.* **18** (2002) 183.
193. T.V. Choudhary, C. Sivadinarayana, A.K. Datye, D. Kumar and D.W. Goodman, *Catal. Lett.* **86** (2003) 1 .
194. L.H. Pignolet and D.A. Krogstad in *Gold — Progress in Chemistry, Biochemistry and Technology*, H. Schmidbaur, (ed.), John Wiley and Sons, Chichester, 1999 p. 429.
195. L. Guczi, A. Becka, A. Horváth, Z. Koppány, G. Stefler, K. Frey, I. Sajó, O. Geszti, D. Bazin and J. Lynch, *J. Molec. Catal. A: Chem.* **204–205** (2003) 545.
196. A.M. Venezia, V.L. Parola, G. Deganello, B. Pawelec and J.L.G. Fierro, *J. Catal.* **215** (2003) 317.
197. T. Nakagawa, H. Nitani, S. Tanabe, K. Okitsu, S. Seino, Y. Mizukoshi and T.A. Yamamoto, *Ultrasonics Sonochem.* **12** (2005) 249.
198. C.L. Bianchi, P. Canton, N. Dimitratos, F. Porta and L. Prati, *Catal. Today* **102–103** (2005) 203.

199. J. Luo, M.M. Maye, V. Petkov, N.N. Kariuki, L. Wang, P. Njoki, D. Mott, Y. Lin and C.-J. Zhong, *Chem. Mater.* **17** (2005) 3086.

200. A.-Q. Wang, J.-H. Liu, S.D. Lin, H.-P. Lin and C.-Y. Mou, *J. Catal.* **233** (2005) 186.

201. R.W.J. Scott, C. Sivadinarayana, O.M. Wilson, Z. Yan, D.W. Goodman and R.M. Crooks, *J. Am. Chem. Soc.* **127** (2005) 1380.

202. J.L. Rousset, *personal commun.*

203. J.L. Rousset, F.J.C.S. Aires, B.R. Sekhar, P. Mélinon, B. Prevel and M. Pellarin, *J. Phys. Chem. B* **104** (2000) 5430.

204. A. Zwijnenburg, A. Goossens, W.G. Sloof, M.W.J. Crajé, A.M. van der Kraan, L. Jos de Jongh, M. Makkee and J.A. Moulijn, *J. Phys. Chem. B* **106** (2002) 9853.

205. R. Zanella, S. Giorgio, C.-H. Shin, C.R. Henry and C. Louis, *J. Catal.* **222** (2004) 357.

206. R. Zanella, C. Louis, S. Giorgio and R. Touroude, *J. Catal.* **223** (2004) 328.

207. R. Zanella and C. Louis, *Catal. Today* **107–108** (2005) 768.

208. M. Daté, Y. Ichihashi, T. Yamashita, A. Chiorino, F. Boccuzzi and M. Haruta, *Catal. Today* **72** (2002) 89.

209. D. Cunningham, S. Tsubota, N. Kamijo and M. Haruta, *Res. Chem. Interm.* **19** (1993) 1.

210. C.K. Costello, J.H. Yang, H.Y. Law, Y. Wang, J.N. Lin, L.D. Marks, M.D. Kung and H.H. Kung, *Appl. Catal. A: Gen.* **243** (2003) 15.

211. B. Schumacher, V. Plzak, K. Kinne and R.J. Behm, *Catal. Lett.* **2003** (2003) 109.

212. J.T. Miller, *personal commun.*

CHAPTER 5

Chemisorption of Simple Molecules on Gold

5.1. Introduction: Chemisorption and Catalysis

The chemisorption of reactant molecules is usually considered to be the necessary first step in the process of catalysis. It not only fixes them to the surface with sufficient, but not excessive, strength so as to facilitate their reaction, but does it in a manner that prepares them to react in the desired way. The importance of this preliminary step (see Section 1.2) has encouraged a great deal of research by many techniques over many years, so that a good deal of information is now available, and much basic understanding of the process has been obtained. Unfortunately, for our immediate purposes, most of this work has been conducted in the metals of Groups 8–10, because they are all very active in chemisorption and catalysis; but the metals of Group 11 are as a whole much less reactive, a fact which has long been attributed to their having filled d-shells. Copper is however highly effective in catalysing methanol synthesis:

$$CO + 2H_2 \rightarrow CH_3OH \tag{5.1}$$

and the water-gas shift reaction

$$CO + H_2O \rightarrow H_2 + CO_2 \tag{5.2}$$

showing that it can at least chemisorb these reactants, and silver works well for oxidising ethene to oxirane (ethylene oxide). Until some 15 years ago, what was known about the capability of gold in chemisorption could be written on the back of a postage stamp: all that has now changed, and the realisation of its catalytic potential has attracted many scientists to look more carefully at how it acts in chemisorption, and theoreticians have been tempted to try to explain why it should behave as it does.

Many, indeed perhaps most, of the important reactions that gold is currently known to catalyse involve molecular oxygen. At first sight this seems strange and unexpected, because gold is the acme of nobility, and its refusal to corrode under any circumstance implies that it cannot react with oxygen. Its ability to chemisorb oxygen is therefore immediately called into question, and it will be necessary to enquire in some depth into how it can manage to catalyse oxidations so successfully. We may prepare the ground for this discussion by listing three possibilities: (1) perhaps very small particles of gold can chemisorb oxygen; (2) perhaps oxygen reacts from the fluid phase with the other chemisorbed molecule by a kind of Eley–Rideal mechanism; (3) in the case of supported gold, the oxygen may first be chemisorbed on the support, close to a gold particle. Some aspects of these possibilities will be examined below.

This chapter is organised in the following way. Section 5.2 treats the ways in which oxygen interacts with gold, dealing first briefly, and only briefly, with gaseous and matrix-isolated clusters, and then with massive surfaces (powders, film and single crystals) and finally with supported gold catalysts. The extreme nobility for which gold is so well known implies that its affinity for oxygen must be very limited, and so it is, but there are circumstances in which reaction or chemisorption does occur, and analysis of these situations poses some interesting theoretical questions. We then examine the reaction of carbon monoxide with gaseous clusters, and then its chemisorption (Section 5.3), adopting the same progression of massive to microscopic, and endeavouring to present the extensive body of work in an orderly and systematic way. Much less information is available on other molecules, but some work has been performed with nitric oxide (Section 5.4) and a very little with hydrogen (Section 5.5). Infrared spectroscopy has been by far the most widely used technique with carbon monoxide and nitric oxide, because adsorbed amounts are normally very small and this is the only method having sufficient sensitivity. Other methods will be mentioned where relevant. Use of IR spectroscopy to study surfaces on which reaction is actually occurring is deferred until that reaction is considered in its entirety.

5.2. Interaction of Oxygen with Gold

5.2.1. Interaction with gold clusters[1,2]

The term *cluster* is, following common practice, reserved for small assemblies of atoms formed in the vapour phase or in matrix isolation, to distinguish them from *particles*, which are formed by chemical or physical means, often with the collaboration of a support. The preparation of gaseous clusters has been briefly described in Section 3.2.2; they can be used either positively or negatively charged or as neutral species. While it would be stretching the definition of chemisorption past its elastic limit to include reactions of simple molecules with gaseous atoms or clusters,[3] nevertheless study of these processes may provide some glimmer of understanding as to how a small gold particle responds to their presence on its surface, since they may in some measure simulate the behaviour of atoms in low co-ordination number sites.

Before proceeding further it may be useful to review briefly the properties of the oxygen molecule in its various manifestations.[4–7] It is unique among gaseous diatomic species in being paramagnetic while having an even number of electrons, the two most weakly bonded electrons occupying degenerate π orbitals and having parallel spins: this gives a triplet ground state $^3\Sigma_g$. If these two electrons remain unpaired in separate orbitals but adopt opposite spins, we have a singlet excited state $^1\Delta_g$; a singlet state $^1\Sigma_g^+$ also results if they enter a single π^* orbital with (of necessity) opposed spins. These excited states are very important in gas phase oxidations, but their role in catalysed oxidations is unclear. The ionisation potential of the oxygen molecule is $1314\,\mathrm{kJ\,mol^{-1}}$ and its electronegativity (3.5) is exceeded only by that of fluorine (4.0); the bond dissociation energy is $493\,\mathrm{kJ\,mol^{-1}}$, and its electron affinity is $222\,\mathrm{kJ\,mol^{-1}}$. It is an extremely effective oxidising agent, and it is little short of miraculous that life has evolved and is sustained in so aggressive an atmosphere, containing as it does about 20% oxygen: or perhaps it is a miracle.

There have been a number of experimental investigations of the formation of complexes between gold clusters and molecular oxygen.[1,3,8–10] Fortunately, there is a substantial measure of agreement between them, and their conclusions are readily summarised. Uncharged clusters of any size do not react; with cationic clusters only Au_{10}^+ has shown reactivity, but anionic clusters Au_n^- where n is an even number not more than 20 all form complexes, the only exception being Au_{16}^- which does not. There is a very distinct 'odd–even' alternation, the clusters having an odd number of atoms showing a much lower reactivity. Clusters have an effective temperature close to ambient,[3,8,11] and only one oxygen molecule becomes attached to each cluster irrespective of the cluster's size.[9] It is clear therefore that complex formation requires the cluster to have an odd number of electrons, and that the one having the highest energy (i.e. in the HOMO) is passed to the oxygen molecule, which thereby acquires a negative charge as O_2^-. Figure 5.1 shows how the main characteristics of the oxygen molecule and its various ionic forms depend on the number of electrons bonding the two

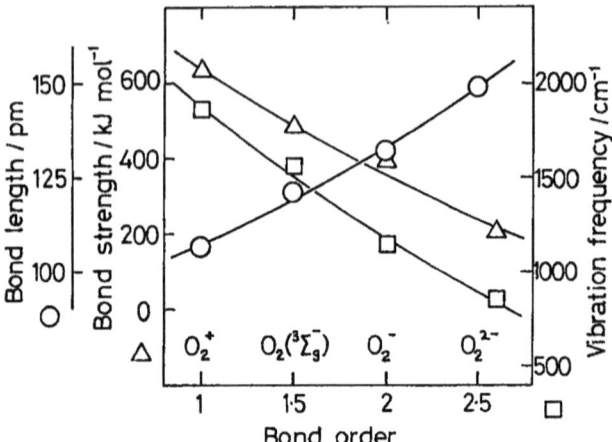

Figure 5.1: Bond length, vibration frequency and bond energy of the oxy-gen molecule ($^3\Sigma_g^-$) and its various ionic forms (oxygenyl, O_2^+, superoxide, O_2^-, peroxide, and O_2^{2-}) as a function of bond order.

atoms.[6,7] The acceptance of an electron into the molecule's lowest unoc-cupied orbital (LUMO), which is the $2p\pi^*$ antibonding orbital, gives the superoxide ion O_2^-, in which the O–O bond is lengthened and weakened. Addition of a further electron gives the very strongly oxidising peroxide ion O_2^{2-} in which the O–O bond is further weakened, and conversely removal of an electron from the $2p\pi^*$ orbital to give the oxygenyl ion O_2^+ shortens and strengthens the O–O bond. A few salts containing this ion are known, but a very electronegative molecule is needed to effect the ionisation, as for example in the complex $[O_2]^+[PtF_6]^-$. Massive gold is not sufficiently electronegative to do this.

The structure of the complexes formed when gold clusters react with oxygen is not always certain, and in particular it is not certain whether the oxygen molecule is dissociated or not.[3] Clearly if it is to dissociate, the transfer of charge from the cluster to the molecule must be the first step, but what must happen thereafter if dissociation into atoms is to occur is far from obvious. Some light is cast on this problem by the observation of the vibrational fine structure in the ultraviolet photoelectron spectrum (UPS) of the Au_2^-–O_2 and Au_4^-–O_2 complexes.[8,9] Subsidiary peaks at 179 and 152 meV compare with those found elsewhere for the O_2^- ion at 80–120 meV and for the O_2^{2-} ion at 135–150 meV (1 meV = 8.06 cm^{-1}); this suggests that in these complexes at least the molecule retains its integrity but is greatly weakened by electron transfer. The valence band structure is completely changed by the acquisition of the oxygen molecule, and not only the HOMOs but also other of the cluster's orbitals having higher binding energies take part in the bonding. These are 'final state' measurements, relating to neutral complexes that have the structure of the anion. Further enlightenment concerning likely structures has to come from the use of theoretical methods (see below).

Oxygen molecules react with Au_n^-–OH clusters only when n is even.[12] It is worth noting that an Au–O_2 complex has been detected in matrix isolation chemistry. When gold atoms and oxygen were co-condensed with a rare gas at extremely low temperature, a green complex was formed in which the oxygen was bonded sideways on.

The concept of an anionic cluster's reaction with oxygen being viewed as a radical–radical reaction, and hence correlating with the cluster's electron affinity, provides a simple and tidy explanation for the fact that neutral and cationic clusters, and anionic clusters having an odd number of atoms, are all either unreactive or of low reactivity, since all of them have even electron counts.[1,3,8,9] Lower electron affinity allows there to be a higher free energy of formation of the Au_n^-–O_2 bond, but by the time the cluster size has risen to 20 the electron affinity has fallen to the point where forming this bond is no longer favoured. This change in reactivity can be expressed in other ways. Thus for example as cluster size grows, the electron on the Au_n^- cluster becomes progressively more delocalised, and efficient radical-radical reaction requires there to be a high charge concentration such as appears in small clusters. We have already discussed in Section 3.4 the evidence for the disappearance of band structure in small particles; this also applies to clusters, those with $n < 20$ having no d-band, although when n exceeds 70 the valence band region is almost the same as that of massive metal. The typical lack of reactivity towards oxygen therefore also correlates with the appearance of overlapping electron energy levels and metallic character. The trend is also shown by the gradual increase in the gap between the top of the valence band and the Fermi level E_F.

There have been a number of theoretical studies of small gold clusters and their reactivity with oxygen,[1,8,13–21] but because of the different DFT methodologies (see Appendix at the end of this chapter) used it is scarcely surprising that the answers are qualitatively different. Most of the calculations confirm the experimental observations, and add some gloss to them; the most recent calculations agree extraordinarily well with experimental results where comparison is possible.[3] It has for example been found that, although the odd–even alternation of n in Au_n^- clusters is reproduced, the odd n values should have some albeit much lower reactivity, which indeed they do; and neutral clusters are also predicted to be marginally active.[3,18] One focus of these studies is the state of the complexed molecule, viz. is it dissociated or not? One study[16] suggested that dissociation occurred only when $n > 3$, while another thought both might happen, the atomic state being the more stable.[18] In the molecular form, the oxygen molecule was predicted to be bonded end-on,[3,18] with an Au–O–O angle of about $119°$.[17,18] The vibration frequency found with the Au_1^-–O_2 complex at 98 meV was thought to be that for the Au–O bond, and DFT calculations gave O–Au–O as the most stable structure. With neutral Au_n^0 (n even) clusters, calculations gave slight charge transfer *from* oxygen *to* gold, consistent with very weak bonding and low reactivity.[3]

A further point of focus, which is also relevant to the problem of chemisorption on massive gold and small gold particles (Section 5.2.2), is the importance of the coordination number (CN) of the atom at which the reaction takes place.[14,15] It has been shown that in the Au_n^- clusters the charge concentrates at atoms of lower CN, so in line with thinking on radical–radical reactions these are the favoured places; the energy gap between E_F and the top of the valence band decreases as CN falls,[14,15] and although mere chemists may have some difficulty in conceiving the band structure of a single atom, theoreticians happily work with the *local density of states* (LDOS), which informs on the manner of electrons likely to be found at that point. With a surface constructed of small square gold pyramids on an Au(100) base, DFT calculations showed that oxygen molecules were adsorbed linearly on the apical atom with a heat of adsorption of 74 kJ mol^{-1}; with sideways bonding it was only[22] 53 kJ mol^{-1}. If however the pyramids were composed of platinum atoms with only an apical gold atom, the strength of bonding on this was unchanged. This may provide some hint as to how bimetallic particles of gold plus a Group 10 metal may operate.

The relevance to small particles and indeed massive surfaces now becomes clear, because the preponderance of low CN atoms increases as particle size goes down, and this may turn out to be the most important factor in determining reactivity[14] — more important than quantum size effects (i.e. the metal $\rightarrow$ nonmetal transition), surface mobility or any of the other properties that are characteristic of very small assemblies of atoms (see Section 3.4). *It becomes possible to imagine that activity in catalytic oxidations is solely due* *to gold particles that are sufficiently small to chemisorb oxygen as O_2^-, and that this is the key species in securing oxidation.*

5.2.2. Chemisorption of oxygen on gold surfaces[1,23–25]

There have been many attempts over the years to establish under what conditions if any molecular oxygen is chemisorbed by gold. The consensus is that under ambient conditions of temperature and pressure it does not take place on massive gold, so that much of the reported work has had the intention of delimiting the range of conditions under which it does not. Early work has been reviewed in some detail,[2] and the current position may be summarised as follows.

On the Au(111) surface, the molecule adsorbed exothermically but only above 773 K;[26] there was no adsorption at 298 K up to a pressure of 10 Torr,[27] nor on polycrystalline foil below 573 K,[27] nor below 1000 K at 6×10^{-3} Pa.[28] On the highly stepped Au(110)(1 × 2) surface at 1400 Torr, there was no interaction between 300 and 500 K,[29] but on Au(111) adsorption occurred at low pressure (5×10^{-6} mbar) above 873 K and at 1 bar above 773 K to give a ($\sqrt{3} \times \sqrt{3}$)R30° overlayer of atoms.[30] Adsorption of molecular oxygen was also seen at high temperature on gold powder[31] (maximum at 473 K), on film[32,33] (slow at 678 K), and on wire.[34] The validity of these early reports, which date mainly between 1970 and 1985, is clouded by the suspicion that adsorption may have been triggered by the presence of impurities such as calcium,[35] silicon[23] or silver,[36] forming under oxidising conditions a small degree of surface coverage by their oxides, which could then act as active centres for the dissociation of molecules into atoms; these could then diffuse away to cover much of the surface. This has recently been confirmed by observing that barium atoms deposited on Au(100) catalyse oxygen dissociation and the formation of a monolayer over the gold surface by spillover at low temperature;[37] presumably the barium atoms have first oxidise to ions. Auger electron spectroscopy (AES) may lack enough sensitivity to spot the impurities initially, although after oxidation they have been detected by high-resolution electron-energy loss spectroscopy (HREELS).[38] *We may conclude that, impurities notwithstanding, chemisorption of oxygen molecules on massive gold does not take place under normal conditions, but that it does occur (possibly aided by impurities) if the temperature is high enough.*[39]

There is one other circumstance in which the chemisorption of oxygen molecules on gold has been observed: this is by field-ion microscopy (FIM) using a very fine gold tip.[40] Exposure of the tip to oxygen at 300–450 K and 100–1000 mbar, followed by neon imaging at 77 K, provided clear evidence for oxygen atoms decorating the edges of (100) and (111) planes, with a suggestion of some patches of Au_2O_3. This occurred in the absence of an electric field, but was enhanced by a field of 12–15 V nm^{-1}. It would appear that the high concentration of gold atoms of low CN allows adsorption to occur under these conditions, the field enhancement effect arising from polarisation of the oxygen molecule as it approaches the surface. This imparts additional translational energy, and the incoming molecules effective temperature may well exceed 1000 K, but it remains in its ground state. After capture in the near-surface region, it will perform multiple large-amplitude jumps before sticking, so the pressure close to the tip is increased. The enhancement may also be due in part to some reorganisation of the gold's valence electrons.

There is however ample evidence to show that if oxygen is supplied *as atoms* these may be strongly held, and a good deal is known about their properties. Atoms may be formed in several ways: the energy required to break the O–O bond may be provided by a heated filament,[29,41] by an electric or microwave discharge,[23] or by attacking the physisorbed molecule with electrons or UV radiation. Atoms have also been deposited on a model $Au/TiO_2(110)$ at 77 K by a supersonic RF-generated plasma-jet source.[42] Perhaps the best method is to use ozone.[25,29] Decomposed by UV radiation under UHV conditions, extensive oxidation of the $Au(110)(2 \times 1)$ surface occurred, giving adsorbed atoms plus several monolayers of auric oxide Au_2O_3, which disappeared on heating to 423–473 K. This oxide is known to decompose at about 473 K.[43] The layer of oxygen atoms produced by decomposing ozone at 300 K over Au(111) has been studied by a variety of physical techniques.[25] No ordered LEED pattern appeared at any coverage. The change in work function $\Delta\Phi$ was +0.8 eV at one monolayer, showing that the Au–O bond was polarised negative outwards; the O1s binding energy had a normal value of 530.1 eV. Everything had desorbed at 600 K, there being a single TPD peak at 520 K at low coverage, rising to 550 K at high coverage. The first-order desorption had an activation energy of 126 kJ mol^{-1}, and the Au–O bond dissociation energy was estimated at 235 kJ mol^{-1}. Very similar values have been found with $Au(110)(1 \times 2)$.[29] With islands of controlled thickness formed on $TiO_2(110)$, the desorption temperature was higher (741 K) on thin islands than on thick (545 K), showing that oxygen atoms are stabler on small particles.[41]

Transition of oxygen molecules on Au(110) from the physisorbed to chemisorbed state has been achieved by UV radiation at 28 K.[1] This probably excites them from the triplet to the singlet $^1\Delta_g$ state in which all electrons are paired; although the bond length is almost unchanged, the dissociation energy (396 kJ mol^{-1}) is much decreased, and it appears that in this state it is more easily chemisorbed, probably without dissociation. Bombardment of this surface by O$^+$ ions led on TPD to recognition of a number of bound states.[1]

Attempts to deposit oxygen atoms by the decomposition of nitrous oxide (N$_2$O) have not been successful.[44]

Before diving into the depths of density-functional theory the better to understand the interaction of oxygen with gold, we may pause to see how what has been discussed can be portrayed by a simple two-dimensional potential energy diagram (Figure 5.2).[24] Curve A represents a molecule of oxygen approaching the surface, and initially feeling a weak attraction due to physical adsorption; when there are no sympathetic orbitals emerging from the gold surface with which its own orbitals can interact, repulsion sets in, and at low temperatures this is the terminus. At higher temperatures, wishing to proceed further, it encounters a formidably large'potential energy barrier before it can meet curve B, which describes the energy barrier of the dissociated molecule. This implies an activation energy E_{ads} that cannot normally be surmounted. If, however circumstances are such that electron transfer to the molecule is possible — as with Au$_n^-$ clusters, for example — another curve (C) is accessible, and the physisorbed molecule can change into an adsorbed O$_2^-$ with zero or small activation energy. With most clusters, this represents the final stage; as we have seen, however, dissociation to give two adsorbed atoms can take place with Au$_1^-$. The precise location of curve C will depend on the availability of electrons at the point of encounter: the activation energy to be overcome in passing from A to C will for example increase with cluster size due to the increase in (E_d-E_F), where E_d is the energy of the top of the valence band; for the same reason with particles and extended surfaces it will *rise* as the CN of the gold atom goes up.

There are several possible reasons why massive gold should chemisorb oxygen at high temperature, apart from effects due to impurities: (1) electronic excitation of electrons from the 5d to the 6s level may occur,

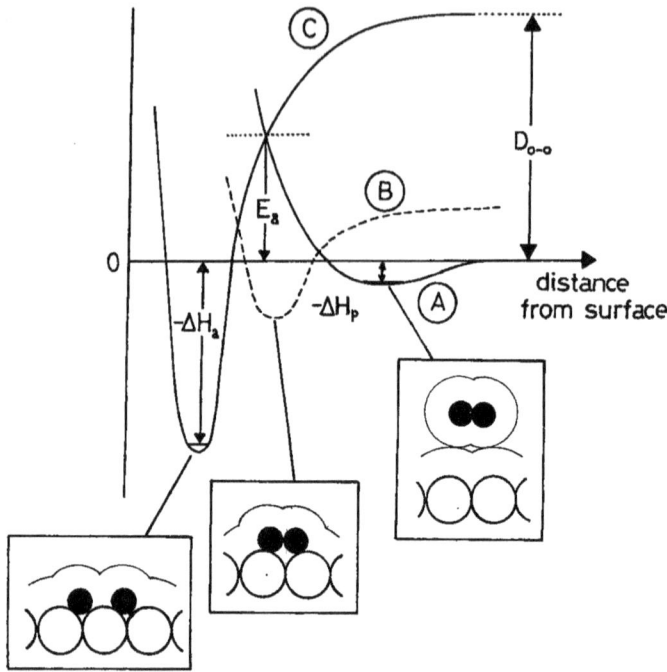

Figure 5.2: Potential energy curves representing the interaction of molecular oxygen with a gold cluster or a gold surface. (A) Gold surface *or* anionic cluster + oxygen molecule. (B) Gold surface *or* cluster + O_2^-. (C) Gold surface *or* cluster + two oxygen atoms. D_{O-O} is the dissociation energy of the oxygen molecule; ΔH_P is the heat of physical adsorption; ΔH_a that of atomic oxygen; and E_a the activation energy for dissociative chemisorption of the molecule.

making the formation of O_2^- easier; (2) the greater translational energy of the arising molecules may be enough to overcome the E_{ads} barrier; (3) the surface gold atoms will vibrate with greater amplitude, their energy contributing that needed to reach the transition state.

We may turn now to consider the results provided by the several sets of DFT calculations (see Appendix) that relate to massive gold surfaces.[13–15,20,21,24,25,45] These treat in particular the likelihood of molecular adsorption, the activation energy for dissociative chemisorption and the strength of adsorption (binding energy) of oxygen atoms on Au(111), Au(211) (a stepped surface), Au(221) (a kinked surface), as well as the first two of these where the lattice has been expanded by 10%. The reason for this is that there is evidence of lattice expansion in small particles formed in model systems. Some of the numerical results are shown in Table 5.1. They show that at best the molecule can only be very weakly held on these surfaces, and would not remain at room temperature. Calculated activation energies (E_a) are very high, and except perhaps for the strained Au(211) are enough to discourage reaction, except at high temperatures. It was estimated using reasonable assumptions that the sticking coefficient S_0 of the oxygen molecule on Au(111) would be about 10^{-21}! For the oxygen atom however the position is more optimistic: remembering that, for its chemisorption from a molecule, the bond energy has to exceed half the molecule's dissociation energy $(493\,\text{kJ mol}^{-1})$, we see that stable states are allowed on both the strained surfaces, and perhaps also on the stepped and kinked surfaces. All of this work emphasises the overriding importance of the surface atom's CN: as with the gaseous clusters, the lower the CN the greater the ease with which electrons can be mobilised to move to the oxygen molecule,[14,15] and hence, as Figure 5.2 showed, the more likely is dissociative chemisorption to ensue. The Au–O bonds are, however, estimated as being at best quite weak, as one might expect qualitatively when two electronegative atoms compete for the same electrons.

Some further refinement can be added to this picture. Mulliken charge analysis has shown[15] that gold atoms at kinks and steps have a smaller

Table 5.1: Results of DFT calculations for the reaction of oxygen with various gold surfaces (energies in kJ mol^{-1}).

	Au(111)	Au(111) str[a]	Au(211)	Au(211) str[a]	Au(221)	Reference
O_2 binding energy	0	8	14	25	—	24
O_2 binding energy	19	—	—	—	—	25
$E(O_2)$ adsorption[c]	—	132	108	61	—	24
$E(O_2)$ adsorption[c]	150	—	—	—	—	25
$E(O_2)$ adsorption[c]	215	—	90	—	112	15
O_a binding energy	245	303	267	296	—	24
O_a binding energy	268[a]	—	281	—	298	15
O_a binding energy	225	—	—	—	—	25

[a]Lattice stretched by 10%. [b]At 11% of a monolayer; falls to $238\,\text{kJ mol}^{-1}$ at 25% monolayer. [c]Activation energy for adsorption.

number of d-electrons ($\sim$9.7) than the free atom (10), and they become more positively charged as CN decreases, electron depletion mainly being in the sp-orbitals. It was concluded that the bonding of the atom has both covalent and ionic character, with the latter predominating, the polarisation being Au^+–O^- as expected from electronegativities. The ionic contribution wins because the electrons having energy near to E_F are mainly sp, their proportion falling with increase in CN, while the d-band centre moves towards E_F. With clusters, as their size goes down, metallic character also disappears due to the decrease in conduction (mainly sp) electrons. The bonding of oxygen atoms to small clusters is therefore weak.

A somewhat different MO description of the process for dissociative chemisorption of oxygen has also been presented.[24]

Attempts have also been made to look for the adsorption of oxygen on supported gold catalysts,[23] but until recently these have not been notably successful. This may be because it was not appreciated that the process was activated, and could therefore only be observed at elevated temperatures. However, as long ago as 1979 it was noted that Au/SiO_2 ($d_{Au} \sim 5\,nm$) took up molecular oxygen at 473 K ($\sim$200 μmol g_{Au}^{-1}), the uptake passing through a maximum at 623 K.[46] From the observed uptakes with several Au/SiO_2 and Au/MgO samples, a mean particle size was derived, assuming an Au:O ratio of 2:1. Table 5.2 shows that values thus obtained agree quite well with those found by TEM and WAXS (wide-angle X-ray scattering, see Section 3.3.1). More recently, Au/Al_2O_3 catalysts prepared by DP have shown strong adsorption of oxygen between 448 and 523 K, the maximum uptake being at 473 K.[47] This oxygen did not desorb at 673 K, but was reduced by hydrogen to regenerate a clean surface. Using the 'strong' component (i.e. that not easily removed by evacuation), which was the larger part, dispersions of 32–60% were established, again using the Au:O ratio of 2:1. Consistent sizes were also obtained by titrating the adsorbed oxygen with hydrogen (see also Table 5.2), but there seemed to be several types of oxygen present, and that found by titration was not a linear function of that observed volumetrically. Low oxygen coverages have also been measured calorimetrically with Au/SiO_2 at 400 K; uptakes declined rapidly with dose number, but no water was formed: presumably it was adsorbed as hydroxyl groups that resulted.[48] Heats of reaction were supported by DFT calculations, which have been performed a number of times.

It is unlikely that adsorption on the support is occurring with these materials, although this may prove to be a complication if the support is

Table 5.2: Comparison of particle sizes (nm) estimated by oxygen chemisorption and hydrogen titration of adsorbed oxygen with those obtained by physical techniques.

Support	[Au] (%)	O$_2$ chemisorption	H$_2$ titration	TEM	WAXS	Reference
Al$_2$O$_3$	0.70	2.0	2.5	1–2	—	47, 134
Al$_2$O$_3$	0.94	2.1	2.4	0.8–2	—	47, 134
Al$_2$O$_3$	2.14	3.6	2.8	1.2–2.5	—	47, 134
Al$_2$O$_3$	0.77	~70	~90	60–80	—	47, 134
SiO$_2$	3.11	13	—	—	9.0	46
SiO$_2$	1.30	5.9	—	8.5	—	46
SiO$_2$	1.24	6.1	—	6.0	4.0	46
MgO	2.0	4.2	—	—	—	46
MgO	3.46	8.7	—	8.3	—	46

reducible. Impurities in the metal are also improbable, and it seems most likely that activated chemisorption can take place above room temperature on particles that are sufficiently small, where every surface atom has a low coordination number and may (if theory is to be believed) have some *d*-band vacancy in their LDOS. Indeed if this was not so, it is hard to see how supported gold catalysts could be so active in the oxidation of organic molecules. However, non-dissociative chemisorption of the oxygen molecule as O$_2^-$ implies that the bonding is ionic, viz. Au$^+$–O$_2^-$, and the separation of charges means the surface of gold is in effect oxidised.

5.2.3. Finale

It may be helpful to try to summarise what we have learnt from work on the interaction of oxygen with gold that may be relevant to catalysed oxidations. Molecular oxygen may become involved in one of four ways: either (i) by becoming adsorbed on the support in an activated form adjacent to a gold particle to which the other reactant is attached; or (ii) by directly reacting with that reactant by a kind of Rideal–Eley mechanism, without prior chemisorption, for which there is little evidence; or (iii) by extracting charge from gold atoms, most probably on very small gold particles, having a high concentration of low CN surface atoms, with the formation of an ionic bond such as Au$^+$–O$_2^-$; or (iv) by dissociative chemisorption into atoms on the same kind of surface site. The existence of a slightly expanded lattice through epitaxial contact with the support might assist these processes. The first possibility is often mentioned and there is much evidence in its favour (see Chapter 6), but the extreme dependence of rate on particle size (at least for carbon monoxide oxidation[45]) is strong support for the last two. It is sometimes suggested, again with supporting evidence, that the form of gold active in carbon monoxide oxidation is a cationic species such as AuO(OH). What is quite certain is that Au^{III} or even Au^{I} species will be incapable of bonding the oxygen molecule as O_2^- and hence of dissociating it. If these are the only species present, its participation in an oxidation must then of necessity proceed by one of the other mechanisms.

5.3. Chemisorption of Carbon Monoxide on Gold

5.3.1. Reaction of carbon monoxide with gaseous gold clusters[1]

Complex formation between carbon monoxide and gold clusters has been less studied than the process involving oxygen; the reaction occurs best with anionic clusters having two, three or some larger number of atoms,[49,50] the monatomic Au^- ion being unreactive. A recent study[17] has been made by photoelectron spectroscopy of anionic clusters of the form $Au_x(CO)_y^-$ ($x = 2$–5, $y = 0$–7); as the value of y increased, there was a significant red shift in the PE spectrum, but past a critical number, which equated to the number of apical gold atoms, there was little further change. This shift was ascribed as 'chemisorption' mutating into 'physisorption'. This provides further evidence for the importance of atoms having low coordination number in binding carbon monoxide molecules, the red shift betokening a lowering of the electron bonding energy caused by a flow of charge to the metal. Water assisted the formation of $Au_2(CO)^-$, $Au_3(CO)^-$ and $Au_4(CO)_2^-$ clusters in a high-pressure fast-flow reactor, perhaps by removing excess energy resulting from the reaction.[51]

Interesting results have been obtained with mixtures of carbon monoxide and oxygen.[1,52] The latter reacts at least ten times faster than the former, thus preventing the formation of gold carbonyls, but complexes of the type $Au_x(CO)(O_2)^-$ ($x = 2$ or 3) and $Au_3(CO)(O_2)_2^-$ have been detected.[52] *It therefore appears likely that the adsorption of the two molecules is a cooperative process, with σ electron donation from carbon monoxide to gold facilitating oxygen chemisorption as O_2^- at a neighbouring site.*[1,8] This could be an important factor, additional to the size factor in oxygen chemisorption noted above (Section 5.2.1), explaining the high activity of gold in oxidising carbon monoxide.

Evidence has also been obtained for the reaction of carbon monoxide with oxygen in experiments with clusters.[1,53,54] Using fast-flow reactor mass-spectrometry, the complex Au_nO^- ($n = 1$ or 3) was found to react *via* an intermediate $Au-O-C\equiv O$ species, and with temperature-dependent r.f. ion-trap mass-spectrometry the Au_2^- cluster was seen to react via $Au_2CO_3^-$ or the peroxoformate[54] (Scheme 5.1). Complexes containing a single gold atom and an oxygen atom AuO^- and AuO_3^- have also been reported to react with carbon monoxide.[55] Earlier work has been reviewed;[1] under certain conditions $Au_6(CO)_y^-$ clusters reacted very rapidly with oxygen to give $Au_6(CO)_{y-1}O^-$, with the production of carbon dioxide by implication, and in other work the complex $Au_2(CO)(O_2)^-$ was thought to react as follows.

$$Au_2(CO)(O_2)^- + CO \rightarrow Au_2^- + 2CO_2. \qquad (5.3)$$

In matrix isolation work using a solid inert gas matrix, a peroxoformate intermediate has been directly observed;[56] gold atoms formed a complex with oxygen, and this reacted with carbon monoxide to form a gold carbonyl peroxoformate (Scheme 5.1), which at 30–40 K released carbon dioxide in two stages, $Au(CO)(O)$ being an intermediate. Experiments of this kind are indicative of the way in which the catalytic process may operate, at least on the unsupported metal,[57] but there is less than full agreement on the mechanism by which clusters act. *Quot homines, tot sententiae.*

Many experimental studies are accompanied by DFT calculations; several papers report results of such calculations without the benefit of supporting experimental evidence.[10,58,59]

Mono- and di-carbonyls have been formed at the surface of a gold field-emitter tip at room temperature in the presence of an electric field;[60] they could be field-desorbed to give gaseous cations carrying one or two charges. The driving force for these reactions was thought to be the local positive

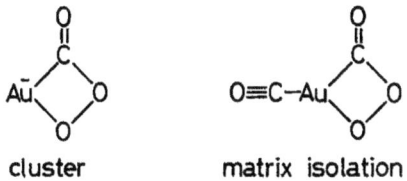

cluster   matrix isolation

Scheme 5.1: Structures of peroxoformate complexes.

field-induced charge gathering at the atom to which the molecules were attached.

5.3.2. Chemisorption of carbon monoxide on massive gold surfaces[1,23]

Unlike the situation with oxygen, there is no doubt that carbon monoxide can be chemisorbed by film and single crystals of gold, and the adsorbed state can be studied by a number of physical techniques, especially temperature-programmed desorption[61–63] (TPD; or thermal desorption spectroscopy, TDS) and infrared spectroscopy.[23,63–65] Early work, performed on metal films, has been summarised; the main conclusion seems to be that chemisorption occurs preferentially or perhaps only at defect sites, at which there are atoms of unusually low coordination number. Desorption from condensed and physically adsorbed states occurred in four steps below 67 K, and from chemisorbed states at 145 K on Au(110)(1 × 2),[61] at 110 and 170 K on film[62] and at similar values on Au(332),[63] a kinked surface. The reconstructed 'missing-row' Au(110)(1 × 2) surface is formed from the normal Au(110) (Figure 5.3) as shown in Figure 5.4; it exposes a series of short plateaux of (111) structure, and has been used in several

Figure 5.3: The (110) surface of gold.

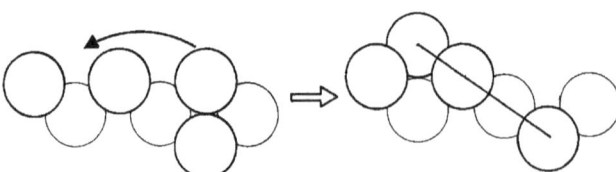

Figure 5.4: Rearrangement of the (110) surface to give the 'missing row' reconstruction.

studies.[61,64,66] Just about everything detectable by TPD after massive gold is exposed to carbon monoxide therefore occurs below ambient temperature.

Carbon monoxide has proved to be a most attractive molecule to study in the adsorbed state by infrared spectroscopy.[7,67] Where the incident radiation cannot penetrate the sample, the spectrum is obtained by examining what is specularly reflected, this technique being termed *reflection–absorption IR spectroscopy* (RAIRS). With powder samples the reflected radiation is diffuse and not specular, and the technique is then *diffuse reflectance Fourier transform spectroscopy* (DRIFTS): signals are generally weak, because not much of the reflected radiation can be collected. The molecule is strongly held by many metals, and is indeed a notorious catalyst poison, but its main advantage lies in the ease and accuracy with which the C≡O vibration frequency can be measured. It is usually quoted to the nearest wave number,[68] occasionally to half a wave number, and significance is attached to differences of 1 or $2\,cm^{-1}$, although the instrumental resolution is not always so high. It is very sensitive to the exact location of the molecule on the surface and to the chemical nature of its adsorption site. M–CO vibrations absorb at much lower frequencies and are not easily accessible. The fundamental for the free molecule is at $2143\,cm^{-1}$, although in the gas-phase spectrum one only sees the P and R branches on either side of the fundamental.

There are close analogies between the structures adopted in chemisorption and those found in carbonyl complexes. There are two common forms: (i) the *linear*, bonded through a formally single M–CO bond, and (ii) the *bridged*, in which the carbon atom, rehybridised to sp^2, makes σ bonds to two metal atoms (Figure 5.5). The latter is rarely reported to occur on gold surfaces, perhaps because of a reluctance to provide the necessary electrons. In addition especially on the central transition metals, the molecule sometimes dissociates, and occasionally it is thought to chemisorb sideways on (see also Figure 5.5).[1,61] Its interaction with metal surfaces is therefore a mine of information, but the situation becomes more complicated with supported metal catalysts, as we shall see shortly (Section 5.34). The nature of its bonding to surfaces and the light cast by DFT calculations will be considered in the next section.

Early work on a very thin gold film revealed[65] a single band at $2110\,cm^{-1}$; bands close to this have been observed on many occasions with a variety of adsorbents, and are universally assigned to a carbon monoxide molecule linearly adsorbed on top of a single gold atom. This is one thing on which everybody agrees.[64] The frequency red-shifts (i.e. the wavelength

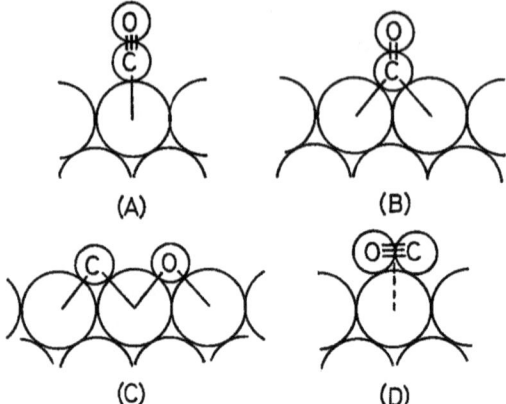

Figure 5.5: Ways of chemisorbing carbon monoxide on a gold surface: (A) linear, (B) bridged, (C) dissociated and (D) sideways on.

decreases) as coverage is increased; this also is often observed[63,66] and will be discussed later (Section 5.3.3). On Au(110) below 125 K, low-temperature core and valence photoemission spectra have been recorded, and the $5\sigma/1\pi$- and 4σ-derived levels and associated shake-up satellites have been identified[69] by auto-ionisation of the excited core levels. By comparison with copper and silver surfaces, gold seems to resemble copper more than silver, the strengths of adsorption being Cu > Au > Ag. The proximity of the top of the d-band to the Fermi energy (E_d-E_F) appears to be the determining factor.

Recognition of the easier chemisorption of carbon monoxide on atoms of low CN as indicated by its preference for 'rough' films formed at low temperature led to a study of the stepped Au(332) surface,[63] although even here desorption was complete when temperature was raised to ambient. The use of isotopically labelled molecules showed that the IR band at $2120\,\mathrm{cm}^{-1}$ contained four components, although TPD revealed only two. The band at $2120\,\mathrm{cm}^{-1}$ red shifted, and the surface potential increased linearly, with increasing coverage.

It is also generally agreed that gold surfaces saturate at low pressures of carbon monoxide (e.g. $10\,\mathrm{Torr}^{64}$) corresponding to somewhat low coverages. Perhaps for this reason, ordered LEED structures are never observed. Heats of adsorption, measured isosterically[61] or by the change in work function,[70] fall quickly with increasing coverage, from initial values[61,63,66,70,74] of close to 60 to about $20\,\mathrm{kJ\,mol^{-1}}$ at saturation.[63,74] The fall has been attributed to the mutual repulsion of dipoles, suggesting that although coverage may be low the adsorbed molecules are crowded together on sites of a special kind. The change in work function ($-0.95\,\mathrm{eV}$ at saturation on $\mathrm{Au(110)(1\times 2)}$[64] points to a polarised Au–CO bond with negative end inwards, i.e. the bonding involves a net transfer of charge *from* the molecule *to* the metal.

Two studies have been reported on the use of carbon monoxide chemisorption at higher than normal pressures and ambient or higher temperatures:[64,71] one[71] covered an astronomic pressure range ($\times 10^{13}$) from UHV to 2 bar. Application of surface X-ray diffraction[71] or STM[64] showed lattice expansion and substantial reorganisation of the surface.

An interesting observation has been made which throws some possible light on the mode of action of bimetallic systems formed from gold and a metal of Group 10 (e.g. Pd). Such systems sometimes exhibit catalytic performance that is superior to either component acting alone (see, e.g. Sections 8.3–8.5). The first and second monolayers of gold atoms deposited on a Pd(110) surface adsorbed carbon monoxide quite strongly, due to some kind of unspecified electronic interaction: desorption maxima in TPD occurred at 225 and 195 K, respectively.[72] This is in line with the DFT calculations concerning oxygen adsorption on a gold atom sitting on top of a platinum pyramid mentioned above.

We may conclude that the chemisorption of carbon monoxide on gold surfaces is generally somewhat weak; it cannot for example lift the reconstruction of the $\mathrm{Au(110)(1\times 2)}$ surface to restore the normal (1×1) phase.[61] It prefers atoms of low CN and it desorbs below room temperature. Nevertheless, under the appropriate conditions considerable information can be obtained about its adsorbed state.

5.3.3. The bonding of carbon monoxide to gold and results of DFT calculations

A simple view of the mode of bonding of the carbon monoxide molecule to the surfaces of the transition metals and in their carbonyl complexes is well known and is thoroughly described in the literature.[4] These metals have unfilled d-orbitals (or holes in their d-band), and the molecule in the linear form is held by a 'push-pull' bond in which charge is transferred from the 5σ orbital of the molecule into the metal's d-band, while there is back-donation of charge from the top of the d-band into the molecule's vacant

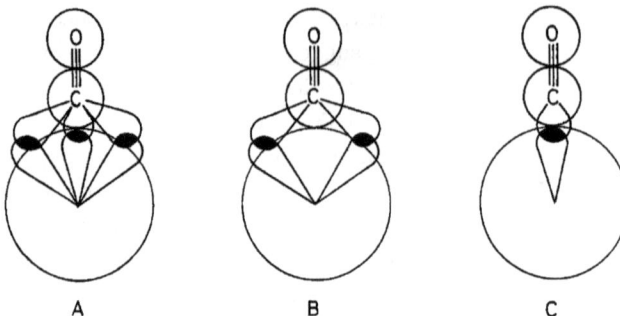

Figure 5.6: Modes of chemisorption of carbon monoxide. (A) Combination of donation from the 5σ orbital to the metal's vacant d-orbital + back-donation from a filled d-orbital to the $2\pi^*$ antibonding orbital. (B) Donation from the 5σ orbital to the metal's vacant d-orbital only. (C) Back-donation from the filled d-orbital to the $2\pi^*$ orbital only.

$2\pi^*$ antibonding orbital (Figure 5.6(A)). This makes for electroneutrality and quite a strong bond. This model may in fact be oversimplified, and other orbitals may make some contribution,[73] but it is adequate for most purposes. The marked weakening in bonding on passing from Groups 10 and 11 might therefore be due to the filling of the d-band, which then falls totally below the Fermi level E_F, and hence to the absence of any vacant d-orbitals; this would leave only the back-donation from the $5d$ level (or the $6s$ level) into the molecule's antibonding $2\pi^*$ orbital (Figure 5.6(B)). This is not entirely unreasonable, because the relativistic effect (Section 2.2) raises the energy of the d-electrons to the point where they can participate in bonding, and indeed this type of bonding is sometimes advocated.[61,64] It could explain why carbon monoxide is adsorbed more strongly by gold than by silver. However, it requires transfer of charge *away* from the metal, which is contrary to the observed work function change.[61] The alternative option is to consider donation from the molecule's 5σ orbital into either the $6s$ level or more probably into residual d-level vacancies that, as we have seen, are postulated by DFT calculations[15] to occur at low CN atoms on steps (Figure 5.6(C)). On the other hand XANES measurements of the white line on the L$_3$ edge (Section 3.4.4) indicate that small particles, rich in coordinatively unsaturated atoms, have a *higher* concentration of d-electrons than massive metal. All DFT[14,15,45] calculations agree however that chemisorption must be stronger there than on atomically flat surfaces, most likely for this reason, and several publications support this bonding mode,[49,68,75] usually citing its original proposal.[63] Some authors however opt to have the best of both worlds, believing that both types contribute, as in Figure 5.6(A).[66] By far the greater number of papers decline (perhaps wisely) to mention this problem at all, and in some theoretical papers its solution is buried at an impenetrable depth. Donation into a small amount of d-band vacancy would account for preferred chemisorption at steps and defects, and since they are somewhat few in number the process would appear less favoured than on metals of Groups 8–10. With the small gaseous clusters the situation is even less clear, but the greater reactivity of Au_n^- clusters with odd electron counts (as with oxygen) suggests that the back-donation mode is important.

There have been several attempts to explain the red shift that occurs as surface coverage by chemisorbed carbon monoxide is increased,[49,63] not all of them consistent in their approach. If indeed the principal bonding mode involves inward electron movement, increasing coverage will deplete the availability of d-band vacancies, so less charge will be removed from the antibonding orbital, and the C≡O bond will get weaker. A much more detailed explanation has been offered[69] to account for the red shifts found with all Group 11 metals, in terms of competing 'chemical' and 'vibrational' effects between adjacent molecules; but this concept has not been adopted generally.

For the application of DFT calculations to the oxidation of carbon monoxide,[49,76] see Chapter 6.

5.3.4. Chemisorption of carbon monoxide on supported gold catalysts[23,77–79]

Chemisorption on supported metals is necessarily more complex than on massive metals; there are a greater diversity of types of gold atom at the surface of particles, because in addition to those found on stepped or kinked surfaces, some are close to the support surface and may be influenced by it, and there is the strong possibility of adsorption on the support as well. At this stage we are not concerned, except incidentally, with the co-adsorption of carbon monoxide and oxygen, since this is intimately connected with their reaction, and this is discussed later (Chapter 6). Figure 5.7 shows diagrammatically the various locations for carbon monoxide molecules on Au/TiO_2 for which there is some spectroscopic evidence.

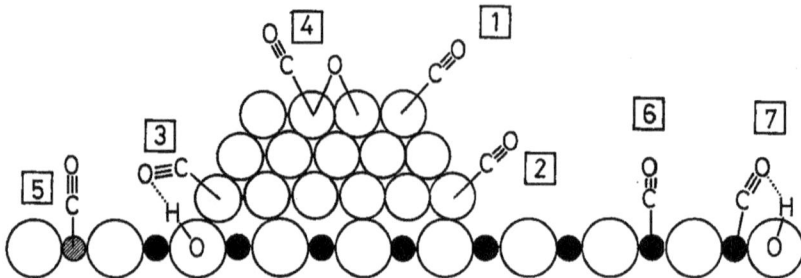

Figure 5.7: Sites for chemisorbing carbon monoxide on a supported gold catalyst. (1) On a gold atom of low CN (corner or step). (2) On a gold atom close to the support and adjacent to an oxide ion. (3) The same but adjacent to a support hydroxyl group. (4) On a gold atom close to an oxygen atom. (5) On an oxidised gold species (Au^{3+}, Au^{1+} or Au_2O_3). (6) On a support cation. (7) The same but hydrogen-bonded to a support hydroxyl group.

'Model' catalysts are made by depositing atoms or particles onto the flat surface of a supporting oxide (Section 3.2.2), so no higher oxidation state is involved at any stage, and the attraction between the metal and the underlying oxide is purely physical in nature. The same may also be true of supported catalysts made from a gold salt (normally $HAuCl_4$) or from a complex containing gold in the zero oxidation state, but with metals of Groups 8 to 10 the use of such methods sometimes leads to a layer of unreduced metal cations between the metal particle and the support; this provides a kind of *chemical glue*[80] that keeps the particle in place, and is visible by XAFS. There is no direct evidence for this in the case of gold, although such a layer has been postulated to play a role in the oxidation of carbon monoxide[44,75,77,81] (Section 6.6.2). We must therefore watch out for differences between results obtained on 'model' and practical catalysts, in case they are significant. Moreover, not all techniques are equally applicable to both; for example, TPD is notoriously difficult to perform on supported metal catalysts, because of the likelihood of multiple re-adsorption–desorption events, although it works well with 'model' systems.[1,74,82]

Thermal desorption studies have indicated that, as with massive gold, adsorbed carbon monoxide desorbs from small gold particles below room temperature. Thus, with 'model' Au/Al_2O_3 there were two desorption peaks, both associated with the metal, because the molecule did not adsorb on fully hydroxylated Al_2O_3, although it might have on bare Al^{3+} ions acting as Lewis acid centres.[1] Sintering studies strongly indicated that chemisorption was much stronger on gold atoms of low coordination number: morphology was changed and the surface made smoother just by annealing, but with Au/Fe_2O_3 for example sintering at 500 K gave particles so large they performed as the massive metal.[74] On this material, thermal desorption continued to 300 K, so, remembering that the oxidation of carbon monoxide can be catalysed down to at least 200 K, species detected in this range may well be relevant to the catalysis. The chemical nature of the support seems to have little effect on the forms and stability of the adsorbed molecule;[1] particle morphology appears to be much more important.

Volumetric measurements at room temperature showed that by far the greater part of the adsorption of carbon monoxide on Au/TiO_2 occurred on the support; it followed the Langmuir equation and most of it was removable by pumping.[23,83] About one-third of the titania surface was able to retain it, but there was little uptake on Au/SiO_2. Use of the double-isotherm method with Au/MgO showed that adsorption onto the metal was complete at about 1 atm, but on various samples the coverage never rose above 18%. On 'model' $Au/MgO(100)$ the maximum coverage attained using a pulsed molecular beam at room temperature was $< 10\%$.[54]

Extensive work has been carried out on characterising adsorbed carbon monoxide by *infrared spectroscopy* both at ambient and low temperatures;[1,23,77,78,84,85] compilations of observed frequencies and their assignments are given in the cited references. There is very considerable agreement about the interpretation of what is seen, but there are also areas of disagreement. There are four spectral ranges of interest: (i) 1000–1800 cm^{-1}, often described as 'the carbonate–carboxylate region', (ii) around 2110 cm^{-1}, where the molecule is chemisorbed on metallic gold, (iii) 2120–2200 cm^{-1}, where it may be adsorbed on support cations or on gold cations and (iv) 2800–3000 cm^{-1}, where C–H vibrations in adsorbed intermediates are located. These categories will be discussed in further detail, linking (i) and (iv). Other regions such as 1700–2100 cm^{-1} and around 3630 cm^{-1} sometimes show bands, the latter due to support hydroxyls.

We focus first on the region close to 2110 cm^{-1}. Table 5.3 shows a selection of the results available, most of which have been obtained at 90 K and only a few at 298 K. They are quite remarkable in their uniformity:

Table 5.3: Chemisorption of carbon monoxide on supported gold catalysts: IR bands assigned to linear CO on Au^0 particles (1) and associated bands (2).

Support	ν (cm^{-1}) (1)[a]	ν (cm^{-1}) (2)[b]	$\Delta\nu$ (cm^{-1}) ($\uparrow$P)[c]	$\Delta\nu$ (cm^{-1}) (O_2)[d]	Reference
CeO_2	2100	—	—	—	91
CeO_2	2103	2124sh	—	—	92
TiO_2	2119	—	—	—	135
TiO_2	2112	2128sh	—	—	136
TiO_2	2115	—	—	—	82
TiO_2[e]	2109	2009, 2054	—	—	86
TiO_2	2107	—	-8	—	137
TiO_2	2112	—	—	—	138
TiO_2	2112	—	-4	—	75
TiO_2	2110	2087	—	—	87
TiO_2	2105	—	-8	—	97
TiO_2	2105	—	-5	$+11$	88
TiO_2[g]	2088	—	—	—	139
TiO_2	2114	—	-7	—	140
$Ti(OH)_4$[h]	2119	—	—	—	141
Fe_2O_3	2107	—	—	$+7$	96
Fe_2O_3	2116	—	—	—	95
α-Fe_2O_3	2110.5	2075	—	—	68
Fe_2O_3[f]	2110	—	-6	$+2$	90
FeO[g]	2122	—	-8	—	74
Fe_2O_3	2107	—	—	$+7$	96
$Fe(OH)_3$[h]	2136	2170	—	—	141
ZnO	2116	—	-10	—	88
ZrO_2	2100	2091	-12	—	87
ZrO_2	2119–2129	—	—	—	136
SiO_2	2110	—	-10	—	23, 90
SiO_2	2129	—	-23	—	23, 79

[a]Band always assigned to linear CO on Au^0. [b]Other associated band. [c]Change of ν on increasing CO pressure. [d]Change of ν on introducing O_2. [e]Room temperature, 2 bar CO pressure. [f]353 K. [g]'Model' catalyst. [h]Freshly precipitated hydroxide.

on all the oxide supports used, a sharp band is seen close to this value, and is always assigned to the linear form of the molecule adsorbed on Au^0. Sometimes there is a shoulder at $\sim$2125 cm^{-1}, and sometimes a band at $\sim$2080 cm^{-1}; bands at lower frequency seen at 2 bar pressure suffered marked changes during evacuation, but firm assignments could not be made.[86] Brave efforts have been made to identify different locations for the molecule on the gold particle.[87,88] The major band peaking at about $2110\,cm^{-1}$ often tails towards the low frequency side, and has been deconvoluted into two Lorentzian bands, the one at higher frequency being assigned to carbon monoxide at steps or edges on the Au^0 particles, while the other has been attributed to the molecule on atoms at the periphery of the particle, perhaps under the influence of the adjacent support. It is thought that it may supply negative charge to the particle because of the small energy difference between the relevant energy levels,[87] made even lower if, as has been suggested,[89] the Fermi level of the gold is lowered by contact with titania; such charge would however probably be confined to atoms close to the support.[14] However, it is impossible to detect any consistent or significant variation in the frequency of the major band with the nature of the support, which implies the absence of any extensive electronic effect across the support-metal interface; a red shift has however been seen as particle size was increased with the model Au/Fe_2O_3 system.[74] The lower frequency band ($\sim 2080\,cm^{-1}$) disappeared the more quickly on evacuation, showing the species responsible to be the more weakly adsorbed. The distinction between the two forms has however been questioned, because possible effects of the particle size distribution had not been taken into account.[90] Bands sometimes observed between 2100 and $2075\,cm^{-1}$ have been ascribed to chemisorbed molecules on particles carrying some degree of negative charge.

Greater ambiguity arises when the spectral region above $2120\,cm^{-1}$ is considered, because bands that lie mainly between about 2150 and $2180\,cm^{-1}$ have been variously attributed to carbon monoxide attached to support cations or to oxidised gold species, which may be isolated Au^+ or Au^{3+} cations or small particles of Au_2O_3 or $AuO(OH)$. Table 5.4 illustrates the difficulty of trying to arrive at definite conclusions; the following notes strive to put some order in the chaos. (1) With silica and alumina, bands at $2175\,cm^{-1}$ are unlikely to be caused by adsorption on support cations; on reducible supports however those assigned to Au^{x+} are mainly between 2150 and $2165\,cm^{-1}$. (2) Comparison between catalyst and the support by itself ought to help identification, and has indeed been tried; but the presence of the gold can sometimes induce changes to the support (e.g. with ceria[91]) which may upset the conclusions. (3) Bands usually assigned to molecules adsorbed on support cations tend to appear at slightly higher frequencies ($2170–2190\,cm^{-1}$), and these species are more easily removed

Table 5.4: Chemisorption of carbon monoxide on supported gold catalysts.

Support/cation	ν (cm^{-1})	Reference
(a) IR bands attributed to CO–support cation		
Ce^{4+}	2151, 2170	91
Ce^{3+}	2157, 2140	91
Ce^{4+}	2186, 2170	93
Ce^{3+}	2133	92
Ti^{4+}	2180	82
Ti^{4+}	2178	86
Ti^{4+}	2163, 2177, 2155[a]	87
Ti^{4+}	2071	97
Ti^{4+}	2186, 2202[a]	88
Ti^{4+}	2188	135
Zr^{4+}	2181, 2169	87
(b) IR bands attributed to CO–Au^{x+}		
TiO$_2$	2151	138
TiO$_2$	2154	88
Fe$_2$O$_3$	2159	95
Fe$_2$O$_3$	2165	68
Fe$_2$O$_3$	2160	74
ZnO	2133	88
SiO$_2$	2175	23
Al$_2$O$_3$	2175	79
CeO$_2$	2124–2133	92

[a]Due to cations in different sites.

by evacuation than those on gold cations; separate bands are sometimes thought due to cations in different coordinations.[88] (4) Molecules attached to Au^I are stabler than those on Au^0. (5) Assignment of bands in this region to molecules on Au^{III} is unlikely to be valid, because this oxidation state is easily reduced by the adsorbate, even at 90 K (see Section 6.7) and (6) On ceria itself, the surface of which is reducible by carbon monoxide, separate bands due to Ce^{4+} and Ce^{3+} ions have been recognised, and have been compared with those seen with Au/CeO$_2$: the available results[91–93] are, however, by no means in agreement.

A major objective of the work employing infrared spectroscopy is the identification of the species involved in the reactions that gold is adept at catalysing ((selective) oxidation of carbon monoxide, water-gas shift, etc.), especially whether the gold site is an Au^{x+} species or an Au^0 particle. Quite clearly much, indeed most, of the adsorbed carbon monoxide is held by the support, although it is unsure whether it is attached to O^{2-} ions[94] or as is generally supposed to the cation (Ti^{4+} and Fe^{3+}, etc.). It may migrate back from the support to the gold during evacuation.[87] The general consensus is that in its reactive state it resides on the Au^0 particles, preferring low CN atoms, but covering typically only ~20% of the surface, and that oxygen adsorbs only on the smallest (1–2 nm in size[91]), this possibly accounting for the marked size sensitivity of its oxidation.[14] Larger particles adsorb only carbon monoxide, so oxygen has to be activated differently, perhaps utilising surface anion vacancies created by partial reduction of the support. This effect was particularly notable with the easily reduced ceria; Au/CeO_2 adsorbed much more carbon monoxide than ceria itself, and coordinatively unsaturated (CUS) Ce^{3+} sites appeared, adjacent to gold particles, suggesting that spillover had taken place. Stress is also placed on the importance of those particles close to reduced cations[91] (Ti^{3+}, Ce^{3+} and Fe^{2+}) and acquiring a negative charge, which, as with gaseous clusters, helps to activate the oxygen molecule by forming the O_2^- ion. On ferric oxide, Au^{x+} species were seen to be reduced by pulses of carbon monoxide, by weakening of the band at $2159\,cm^{-1}$;[95] they showed greater activity in the catalytic oxidation, and their removal led to loss of activity. Bands due to $CO–Au^{x+}$ were not seen when Au/Fe_2O_3 was thoroughly reduced or in the presence of hydrogen.[90] This provides a possible way of distinguishing between adsorption on Au^{x+} and on support cations, whose state of oxidation might not be altered.

The application of infrared spectroscopy has been extended to examine the effects of co-adsorption of oxygen with carbon monoxide and of adding oxygen to already adsorbed carbon monoxide,[82,88,90,91,94,96] as well as changes produced by the presence of water[68,75,88] and of carbon dioxide.[68,94] As noted in Table 5.4, the addition of oxygen gives a *blue* shift to the C≡O vibration frequency, and this is attributed to the effect of oxygen atoms being adsorbed adjacently, presumably on Au^0 particles, as sketched in Figure 5.7(4). This effect may also be caused by a lowering of the surface coverage due to competitive adsorption by carbon dioxide.[97] At the same time, with Au/CeO_2 the band at $2060\,cm^{-1}$ arising from $CO–Au^-/Ce^{3+}$ decreased in intensity.[91] The further consideration of the application of spectroscopy to the unravelling of reaction mechanisms is deferred to the next and subsequent chapters.

We must now comment on the plethora of bands that are often observed in the 1000–$1800\,\text{cm}^{-1}$, but more particularly between 1200 and $1700\,\text{cm}^{-1}$. This is the 'carboxylate-carbonate' region,[98] and the species giving rise to them are referred to as being 'carbonate-like'. In the absence of oxygen they may be formed by reaction of carbon monoxide with support hydroxyl groups HO–**S**, as for example:

$$\text{CO} + \text{HO–}\mathbf{S} \rightarrow \text{HCOO–}\mathbf{S}. \qquad (5.4)$$

A preliminary heat treatment will remove many if not all hydroxyl groups, and bands in the 1200–$1700\,\text{cm}^{-1}$ then do not appear; conversely when water is present they are seen more strongly, and with a reducible support such as zinc oxide the reaction

$$\text{CO} + \text{H}_2\text{O–}\mathbf{S} \rightarrow \text{HCO}_2^-\text{–}\mathbf{S} + \text{H}^+ \qquad (5.5)$$

can take place, and the proton effects a partial reduction of the Zn^{2+} ions, forming positive holes that lower the IR transparency dramatically.[88] If the support is easily reducible, carbon monoxide reacts with oxide ions as follows.

$$\text{CO} + 2\text{O}^{2-} \rightarrow \text{CO}_3^{2-} + 2\text{e}^-, \qquad (5.6)$$

this process being preferred even when molecular oxygen is present. The electrons released will reduce adjacent support cations. The HCO_2^- ion can be formulated either as bicarbonate or formate, the latter being recognised by C–H stretching vibrations in the 2800–$3000\,\text{cm}^{-1}$ region, and at much higher frequencies the support hydroxyl vibrations at around $3630\,\text{cm}^{-1}$ can be used to follow reactions in which they are involved. Three studies[75,86,91] discuss the analysis of the 1200–$1700\,\text{cm}^{-1}$ region in depth, but it would be tedious and of little value to rehearse these in detail. Table 5.5 however, gives an indication of how one of these studies[75] assigned bands observed in this region, but the structural implication of some of the terms employed is not always clear. A general feature of the analyses is the recognition that the carbonate ion CO_3^{2-} makes regular appearances, and it is frequently blamed for activity loss during oxidation of carbon monoxide. Quite what it does, however, is rarely explained: perhaps it occupies the anion vacancies created by partial reduction of the support, and on which the oxygen molecule is waiting to adsorb.

Table 5.5: Comparison of low frequency bands observed in chemisorbing CO onto Au/TiO$_2$ with literature.

ν (cm^{-1}) (observed)	ν (cm^{-1}) (literature)	Allocation
1363	1330–1370	Mono- and bidentate CO$_3^{2-}$
1381		and HCOO$^-$
(1445)^a	1440	Monodentate or free CO$_3^{2-}$
1552	1560	Bidentate CO$_3^{2-}$
	1585	Bidentate CO$_3^{2-}$
1672	1690	Bidentate CO$_3^{2-}$
	1725	Bridging CO$_3^{2-}$

^{a}A weak band appearing after oxidative pretreatment.

5.4. Chemisorption of the Oxides of Nitrogen on Gold Surfaces

The chemisorption and reactivity of the oxides of nitrogen on metal surfaces are of great environmental interest because of their connection to the reaction with carbon monoxide, leading to innocuous products, e.g.

$$NO + CO \rightarrow CO_2 + 1/2N_2. \tag{5.7}$$

This is a vital component of the system of reactions involved in treating the exhaust from internal combustion engines (see Chapter 11). The methodology of surface science has been extensively applied to this problem, mainly using platinum and rhodium surfaces,[99] as these are high on the list of components of choice for practical use,[100] but it is only recently that gold has come to be seen as having a possible role to play.

On gold single crystal surfaces, nitric oxide (NO) appears to be somewhat more strongly chemisorbed than carbon monoxide. This is shown by its ability to lift the hexagonal reconstruction of the Au(100) surface (see Section 2.5.2), which it did quickly at 170 K and 10^{-6} mbar pressure.[99] It desorbed over a broad range of temperature (170–230 K) in three or four poorly resolved steps. On Au(111) however neither nitric oxide nor nitrous oxide (N$_2$O) remained adsorbed above 95 K. On the rougher stepped Au(310), nitric oxide decomposed at a temperature as low as 80 K to give nitrous oxide (N$_2$O) and adsorbed oxygen atoms. The molecule has one electron in its $2\pi^*$ antibonding orbital, so it is able to dimerise, and this was believed to precede its dissociation. Chemisorption of the monomer presumably occurs as for carbon monoxide, by donation of this electron either into a partially vacant d-band on the steps, or into the $6s$ level. The occurrence of both dimerisation and dissociation of nitric oxide has been confirmed by its behaviour on field-emitter tips:[101] using pulsed field-desorption mass-spectrometry (PFDMS) and dynamic flow at $300\,K$, the ions NO^+, N_2O^+ and $(NO)_2^+$ were formed at the stepped surface region between the (111) and (001) planes.

When nitric oxide was first introduced to Au/TiO_2 at room temperature, examination by DRIFTS showed a single weak band at $1796\,cm^{-1}$, blue-shifting with time to $1810\,cm^{-1}$ due to $Au-NO^-$; the spectrum was however dominated by a broad band centred on $1657\,cm^{-1}$ which may have been the $\nu_{asym}(NO_2)$ band and/or the bending $\delta(HOH)$ band.[86] This diminished with time and was replaced by bands at 1820 and $1740\,cm^{-1}$ due to $Au(NO)_2$; bands due to N_2O (2239 and $2202\,cm^{-1}$) and chelating NO_2^- were also recognised. It would appear that nitric oxide was reacting in much the same way as on massive gold surfaces, with the additional possibility of combining with oxygen drawn from the support to form the dioxide. Very detailed and extensive studies of the reactions of nitric oxide on Au/TiO_2 making careful use of DRIFTS have subsequently been reported;[102–104] these include lengthy tables recording published values of IR frequencies shown by species derived from this molecule.

Nitrogen dioxide (NO_2) adsorbed reversibly on Au(111), multilayer desorption taking place at $150\,K$ and monolayer desorption at $220\,K$.[105] It was held vertically, being bonded to the surface by its two oxygen atoms.

5.5. The Chemisorption of Hydrogen and of Other Molecules on Gold

5.5.1. Hydrogen

Molecular hydrogen does not chemisorb by dissociation on massive gold at or below room temperature,[55,106] nor is there any *direct* evidence that it does so at any higher temperature. However, it catalysed *para*-hydrogen[107] and hydrogen–deuterium equilibration[108,109] under conditions where thermal excitation of electrons from the $5d$ to the $6s$ band may create conditions that allow dissociation to occur: the former reaction does not necessarily require dissociation, but the latter does. Reactions involving hydrogen are considered further in Chapter 9. Molecular hydrogen was adsorbed by rough unsintered gold film at 78 K, desorbing at 125 K,[110,111] but this form is unlikely to be of importance in catalysis. Both hydrogen and deuterium were however dissociated at 100 K on thin gold film grown epitaxially on Ir(111),[112,113] but this was apparently not due to an electronic interaction, and indeed nuclear activation analysis using the reaction $^1H(^{15}N,\alpha\gamma)^{12}C$ showed that the effect was independent of gold film thickness. It did not occur with Au/Pt(111), so its cause remains something of a mystery; it was suggested that a narrowing of the $6s^1$ band might be responsible.

Hydrogen atoms formed by radiative dissociation chemisorbed on Au(110)(1 × 2) at 150 K, but they desorbed at 216 K. They also dissolved exothermically $(9\,kJ\,mol^{-1})$ in gold film, the amount dissolved, detected by TPD, being only 4.4×10^{-3} atoms per gold atom at 273 K; E_{des} was[114,115] $57\,kJ\,mol^{-1}$. Hydrogen is atomised by gold filament at very high temperature $(>1200\,K)$,[116,117] and at normal temperature gold surfaces efficiently catalyse hydrogen atom recombination.[108] Theoretical studies of the gold-hydrogen system do little except confirm experimental observations;[106,118−120] the activation energy for dissociation was calculated to be $45\,kJ\,mol^{-1}$.

The situation regarding the dissociative chemisorption on small gold particles has been much less clear. There have been a number of intimations of it happening from the observation of enhanced activity in hydrogenation when very small particles are employed,[106,121,122] and the occurrence of exchange support hydroxyl groups with deuterium[122] clearly implies the spillover of deuterium atoms. Indeed the fact that this occurred with Au/SiO$_2$ even at room temperature provided one of the earliest examples of the spillover phenomenon.[123] Many supported gold catalysts can oxidise hydrogen at quite low temperatures;[48,124] this and other hydrogenations are discussed in later chapters (see Chapters 7 and 9).

Only recently however has definitive evidence for dissociative chemisorption of hydrogen on supported gold particles been obtained.[125] Au/Al$_2$O$_3$ catalysts having mean particles sizes between 1 and 5 nm have been studied by XAFS/XANES and volumetrically; new features on the near-edges of the L$_2$ and L$_3$ X-ray absorption bands were attributed to chemisorbed hydrogen atoms. Adsorption isotherms measured volumetrically at 298 K approached plateaux at about 5 kPa pressure, and total amounts adsorbed at 100 kPa are shown as H/Au$_{tot}$ ratios in Figure 5.8(A); unlike the behaviour of Pt/SiO$_2$ (also shown in this figure) the H/Au$_{tot}$ tended to increase slightly with temperature, showing that the process was activated. Adsorption was

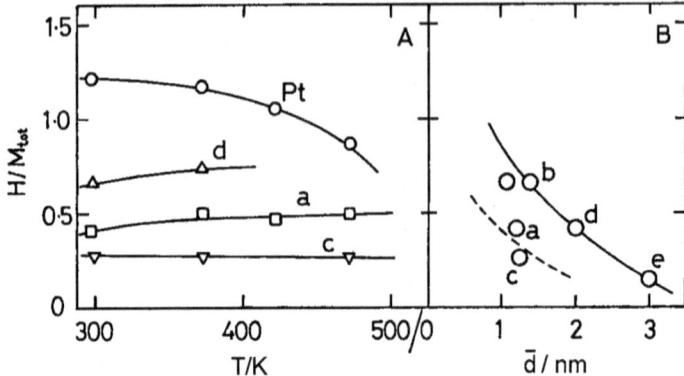

Figure 5.8: Chemisorption of hydrogen on Au/Al_2O_3 at 298 K. (A) H/M_{tot} as a function of temperature (O, Pt/SiO_2; others are Au/Al_2O_3). (B) H/Au_{tot} dependence on mean size ($P_H = 300$ Torr). a: 0.6% Au; b: 0.4% Au; c: 1.2% Au; d: 0.4% Au; e: 1.6% Au.

thought to be limited to atoms of low coordination number, as the H/Au_{tot} ratio increased rapidly with decreasing size (see Figure 5.8(b)), although the degree of scatter suggests that other forces are at work; migration from steps and edges to face sites did not occur (although such sites must be few and far between on such small particles). Some of the adsorbed hydrogen atoms were strongly held, resisting evacuation for 2 h. The Au–Au distance was smaller than the value for the massive state, but was not changed by adsorbed hydrogen: with platinum the Pt–Pt distance is increased thereby because of the electron-withdrawing character of the hydrogen atoms, but the effect does not apply with gold because of its greater electronegativity (Chapter 2). Two of the catalysts effected hydrogen–deuterium equilibration at 298–373 K, although the possible exchange of support hydroxyls was not explored. This careful work therefore confirms long-held suspicion of the ability of very small gold particles, having an abundance of low coordination number atoms, to activate the hydrogen molecule at room temperature. Even so, its adsorption is much weaker than that of carbon monoxide.[123]

Deuterium reacts with small cationic clusters ($n < 16$) and with neutral clusters where n is three and seven; its way of bonding must therefore differ from that of carbon monoxide, but it has not been much discussed, although inevitably it has been explored using DFT methodology.[126]

5.5.2. Other molecules[1]

Chlorine chemisorbs strongly on Au(111). Hydrocarbons such as ethene, propene and ethyne are also chemisorbed, but in the main only below room temperature. DFT calculations have been performed[127] on the adsorption of propene on Au(111) and gaseous clusters, stimulated by gold's ability to oxidise it selectively to methyloxirane (see Section 8.2): likely sites for its adsorption were predicted, but unfortunately there are no experimental results with which to compare them. Organic molecules containing oxygen or some functional group have also been extensively examined,[1,128] but many of them have not been subjected to catalytic processing, so their behaviour need not delay us. Where they have fuller information will be given under the heading of the appropriate reaction.

Isotherms for the reversible adsorption of sulfur deposited from hydrogen + hydrogen sulfide on the three low index planes of a gold single crystal have been measured between 373 and 623 K; heats of adsorption per half mole of S_2 were similar (176–$197 \, kJ \, mol^{-1}$).[129] There appeared to be attractive lateral interactions, but atoms in the close-packed monolayer were out of register with the gold lattice because of the disparity between the sizes. The adsorption of elemental sulfur (S_8) on gold particles supported on a number of oxides has been examined; with Au/TiO_2 the molecule first dissociated to give Au–S bonds, but at higher coverages the S_8 ring structure was maintained.[130] UV radiation ($\lambda_{ex} > 300 \, nm$) of the catalyst suspended in water at room temperature led to partial desorption of the sulfur, but in ethanol it was complete; visible light was ineffective. Adsorbed sulfur was therefore reductively desorbed by band-gap excitation of the support. Hydrogen sulfide was chemisorbed reversibly on Au(100) and (111),[1] and probably also on the (110) surface. Sulfur dioxide was only very weakly held, but it was more strongly chemisorbed on Au/MgO(100) than on Au(111); far more dissociation took place on Au/TiO_2(110). On Au/CeO_2 and on 'model' Au/CeO_2 its adsorption needed anion vacancies in order to proceed.[131] Alkanethiols, alkyl sulfides and disulfides have been extensively studied in the context of self-assembling monolayers.

There is little information of the chemisorption of water on gold surfaces, although its presence has a marked acceleratory effect on the rate of carbon monoxide oxidation over Au/SiO_2;[132] since the support is not expected to be involved in the reaction, it was considered that it might help the adsorption of oxygen on gold panicles. Water occupies oxygen vacancies on titania,[133] and this is expected to affect the gold–titania interfacial energy, and hence particle shape and mobility.

Acknowledgements

Thanks are due for advice and assistance from Dr. Eric Short, Dr. Adrian Taylor, Dr. Dave Willock and Professor Norbert Kruse.

Appendix: Introduction to Density Functional Theory

by Eric L. Short

In 1985, Roberto Car and Michele Parrinello published a short, but seminal, paper entitled "Unified Approach for Molecular Dynamics and Density Functional Theory" (*Phys. Rev. Lett.* **55** (1985) 2471). Up to then, chemists had been carrying out numerous quantum mechanical calculations on molecules but, as the molecules to be studied became larger and/or contained more and more heavy metal atoms, the dimensions of the corresponding matrices which arise in the mathematical theory became enormous and the computing time required to diagonalise them in order to obtain the energy levels and MOs was simply too long and far too expensive even using supercomputers.

However, physicists and material scientists needed to study metals, semiconductors, etc. and it was shown in the above paper that there was, in fact, a way around this bottleneck. Computer packages for solids and semiconductors, such as CASTEP used by P. Hu (*Phys. Rev. Lett.* **91** (2003) 266102), implement many of the concepts in the Car and Parrinello paper.

Now, in any *ab initio* quantum mechanical calculation, a starting geometry is supplied either by giving the x-, y- and z-coordinates of each atom or by supplying the same data in the form of internal coordinates, that is, bond lengths, bond angles and dihedral (twist) angles.

In addition, it is necessary to state the charge on the molecule, its spin multiplicity (singlet, triplet, etc.) and the basis set.

This last requirement essentially tells the computer which orbitals (s, p, d and f) are to be included in the calculation and to what degree of accuracy they are going to be represented, that is, there are minimal basis sets for not so accurate calculations and very high level basis sets for more refined calculations.

The basis sets themselves, which include relativistically corrected pseudopotentials for heavy atoms, are built into the program. The idea behind a pseudopotential is as follows. In the case of an atom like gold, it is very expensive on computer resources to treat all of the 89 electrons and since most of these are not directly involved in the bonding (but, of course, are indirectly involved via the relativistic effects of the core electrons), the valence electrons ($5d$ and $6s$ in the case of gold) are assumed to move in a potential due to the core electrons which has been corrected for relativistic effects. This means that, in the case of a gold atom, one is explicitly dealing with 11 electrons and not 89. It is clear that the saving of computer time is enormous, for example, in the case of Hu's calculations where he is considering systems containing quite a lot of gold atoms.

Once this input data has been supplied, the computer will carry out a calculation using the given input geometry for the molecule and find the eigenfunctions (MOs) and eigenvalues (the energy levels) and, also, calculate any other properties which have been requested (NMR parameters, dipole and quadrupole moments, for example). However, the input geometry does not, in general, correspond to the equilibrium (minimum energy) geometry of the system and methods to find this equilibrium geometry had to be developed. When a system is not at equilibrium, there are net forces acting on the atoms so it is logical to try to calculate these forces and then to calculate a vector of the changes in the coordinates which would be necessary in order to reduce these forces. Once this vector is known and there are standard procedures for obtaining it, the Cartesian coordinates of the individual atoms can be modified to give a new geometry that is lower in energy than the previous one. Another quantum mechanical calculation is then carried out and the whole procedure is repeated until some specified criterion is satisfied (i.e., that the residual forces should not exceed a certain value, e.g. <0.0001 atm). When this criterion is satisfied, equilibrium is deemed to have been achieved and the geometry of the system corresponds to a (the) minimum energy state.

Now, the above procedure can also be very time-consuming when carrying out calculations on extended arrays of atoms as in crystals and other solids since convergence is often very slow and many iterations are required before the optimisation criterion is satisfied. So, these conventional methods of optimisation also became difficult on large systems.

In the Car-Parrinello paper, they proposed a way of circumventing both the difficulty in the calculation of the orbitals and energy levels of large systems and the problem of the energy optimisation of such systems.

They took the standard energy functional in DFT theory but, rather than simply solving this, they sought a way of achieving the solution of the Kohn-Sham DFT equations, geometry relaxation and volume and strain relaxation simultaneously.

This was achieved using 'dynamic simulated annealing', which is a technique in molecular dynamics, and combining it with DFT theory; the resulting dynamical equations being solved simultaneously rather than sequentially.

If the total energy of a system is not at an energy minimum, there will be forces on the atoms but, according to Newton, force is equal to mass times acceleration and there will therefore be accelerations. If atomic positions can be found such that these accelerations are equal to zero then these correspond to the total energy of the system being a minimum, that is, the atoms are at their equilibrium (minimum total energy) positions. The computational procedure yields the molecular orbitals and energy levels as well as the total energy of the system at this point of minimum energy. In Hu's gold paper, he has put an oxygen molecule on various gold crystal surfaces and found the positions of minimum energy using the above procedure. The total energy of the oxygen–gold system is then known, and the adsorption energy can be calculated by subtracting the energy of an oxygen molecule and the energy of the gold alone, calculated by the same method, from the total energy.

The ideas outlined above, which have subsequently led to the development of computer programs such as CASTEP, are clearly very important as they have opened up the possibility of carrying out quite accurate and realistic calculations on solids, particularly those containing heavy elements, and molecules adsorbed on solids, due to the great amount of computer time saved by adopting the Car and Parrinello approach.

References

1. R. Meyer, C. Lemire, Sh.K. Shaikhutdinov and H.-J. Freund, *Gold Bull.* **37** (2004) 72.
2. M.B. Knickelbaum, *Ann Rev. Phys. Chem.* **50** (1999) 79.
3. Q. Sun, P. Jena, Young Dok Kim, M. Fischer and G. Ganteför, *J. Chem. Phys.* **120** (2004) 6510.
4. N.N. Greenwood and A. Earnshaw, *Chemistry of the Elements*, Butterworth-Heinemann, Oxford, 1997, Ch.14.
5. A.G. Sykes and J.A. Weil, *Progr. Inorg. Chem.* **13** (1970) 1.

6. K.F. Purcell and J.C. Kotz, *Inorganic Chemistry*, W.B. Saunders, Philadelphia, 1977.
7. *Infrared and Raman Spectra of Inorganic and Raman Spectra*, K. Nakamoto, (ed.), Wiley: New York, 1986.
8. D. Stolcic, M. Fischer, G. Gauteför, Young Dok Kim, Q. Sun and P. Jena, *J. Am. Chem. Soc.* **125** (2003) 2848.
9. B.E. Salisbury, W.T. Wallace and R.L. Whetten, *Chem. Phys.* **262** (2000) 131.
10. M. Okamura, Y. Kitigawa, M. Haruta and Y. Yamaguchi, *Appl. Catal. A: Gen.* **291** (2005) 37.
11. Young Dok Kim, M. Fischer and G. Ganteför, *Chem. Phys. Lett.* **377** (2003) 170.
12. W.T. Wallace, R.B. Wynwas, R.L. Whetten, R. Mitrič, and V. Bonačić-Koutecký, *J. Am. Chem. Soc.* **125** (2003) 8408.
13. N. Lopez and J.K. Nørskov, *J. Am. Chem. Soc.* **124** (2002) 11262.
14. N. Lopez, T.V.W. Janssens, B.S. Clausen, Y. Xu, M. Mavrikakis, T. Bligaard and J.K. Nørskov, *J. Catal.* **223** (2004) 232; N. Lopez and J.K. Nørskov, *J. Am. Chem. Soc.* **124** (2002) 11262.
15. Z.-P. Liu, P. Hu and A. Alavi, *J. Amer. Chem. Soc.* **124** (2002) 14770.
16. B. Yoon, H. Häkkinen and U. Landmann, *J. Phys. Chem. A* **107** (2003) 4066.
17. E.L. Short, unpublished work.
18. A. Francechetti, S.J. Pennycook and S.T. Pantelides, *Chem. Phys. Lett.* **374** (2003) 474-475.
19. D.H. Wells Jr., W.N. Delgass and K.T. Thomson, *J. Chem. Phys.* **117** (2002) 10597.
20. B. Yoon, H. Häkkinen, U. Landmann, A.S. Wörz, J.-M. Antonietti, S. Abbet, K. Judi and U. Heiz, *Science* **307** (2005) 40.
21. G. Mills, M.S. Gordon and H. Metiu, *J. Chem. Phys.* **118** (2003) 4198.
22. F. Tielens, J. Andrés, M. van Brussel, C. Buess-Hermann and P. Geerlings, *J. Phys. Chem. B* **109** (2005) 7624.
23. G.C. Bond and D.T. Thompson, *Catal. Rev.-Sci. Eng.* **41** (1999) 319.
24. Y. Xu and M. Mavrikakis, *J. Phys. Chem. B* **107** (2003) 9298.
25. N. Saliba, D.H. Parker and B.E. Koel, *Surf. Sci.* **410** (1998) 270.
26. M.A. Chesters and G.A. Somorjai, *Surf. Sci.* **52** (1975) 21.
27. P. Legaré, L. Hilaire, M. Sotto and G. Maire, *Surf. Sci.* **91** (1980) 175.
28. D.D. Eley and P.B. Moore, *Surf. Sci.* **76** (1978) L599.
29. A.G. Sault and R.M. Madix, *Surf. Sci.* **169** (1986) 347.
30. J. Cao, N. Wu, S. Qi, K. Feng and M.S. Zei, *Chin. Phys. Lett.* **6** (1989) 92.
31. W.R. MacDonald and K.E. Hayes, *J. Catal.* **18** (1970) 115.
32. P.C. Richardson and D.R. Rossington, *J. Catal.* **20** (1971) 420.
33. R.R. Ford and J. Pritchard, *Chem. Commun.* (1968) 362.
34. N.V. Kul'kova and L.P. Levchenko, *Kinet. Katal.* **6** (1965) 765.
35. M.E. Schrader, *Surf. Sci.* **78** (1978) L227.
36. Y. Iizuka, A. Kawamoto, K. Akita, M. Daté, S. Tsubota, M. Okumura and M. Haruta, *Catal. Lett.* **97** (2004) 203.
37. A.F. Carley, P.R. Davies, M.W. Roberts and A.M. Shah, *Catal. Lett.* **101** (2005) 1924.
38. J.J. Pireaux, M. Chtaib, J.P. Delrue, P.A. Thiry, M. Liehr and R. Caudano, *Surf. Sci.* **141** (1984) 211.
39. L.G. Carpenter and W.N. Mair, *Trans. Faraday Soc.* **55** (1959) 1924.
40. T. Bär, T. Visart de Bocarmé, B.E. Nieuwenhuys and N. Kruse, *Catal. Lett.* **74** (2001) 127.

41. V.A. Bondzie, S.C. Parker and C.T. Campbell, *Catal. Lett.* **63** (1999) 143.
42. J.D. Stiehl, T.S. Kim, S.M. McClure and C.B. Mullins, *J. Am. Chem. Soc.* **126** (2004) 1606, 13574.
43. E. Raub and W. Plate, *Z. Metallkunde* **48** (1957) 529.
44. C.P. Vinod, H.(J.W.) Niemantsverdriet and B.E. Nieuwenhuys, *Appl. Catal. A: Gen.* **291** (2005) 93.
45. M. Mavrikakis, P. Stolze and J.K. Nørskov, *Catal. Lett.* **64** (2000) 101.
46. T. Fukushima, S. Galvagno and G. Parravano, *J. Catal.* **57** (1979) 177.
47. H. Bernt, I. Pitsch, S. Evert, K. Struve, M.M. Pohl, J. Radnik and A. Martin, *Appl. Catal. A: Gen.* **244** (2003) 169.
48. D.G. Barton and S.G. Podkolzin, *J. Phys. Chem. B* **109** (2005) 2262.
49. C. Lemire, R. Meyer, Sh.K. Shaikhutdinov and H.-J. Freund, *Surf. Sci.* **552** (2004) 27.
50. H.-J. Zhai and L.-S. Wang, *J. Chem. Phys.* **122** (2005) 051101.
51. W.T. Wallace, B.R. Wyrwasza and R.L. Whetten, *Phys. Chem. Chem. Phys.* **7** (2005) 930.
52. J. Hagen, L.D. Socaciu, M. Elyazyfer, U. Heiz, T.M. Bernhardt and L. Wöste, *Phys. Chem. Chem. Phys.* **4** (2002) 1707.
53. M.L. Kimble, A.W. Castleman Jr., R. Mitrić, C. Bürgel and V. Bonačić-Koutecký, *J. Am. Chem. Soc.* **126** (2004) 2526.
54. L.D. Socaciu, J. Hagen, T.M. Brenhardt, L.Wöste, U. Heiz, H. Häkkinen and U. Landmann, *J. Am. Chem. Soc.* **125** (2003) 10437.
55. A.G. Sault, R.J. Madix and C.T. Campbell, *Surf. Sci.* **169** (1986) 347.
56. D. McIntosh and G.A. Ozin, *Inorg. Chem.* **15** (1976) 2869; **16** (1977) 51.
57. G.C. Bond, *Catal. Today,* **72** (2002) 5.
58. N.S. Phala, G. Klatt and E. van Steen, *Chem. Phys. Lett.* **395** (2004) 33.
59. S. Kandoi, A.A. Gokhale, L.C. Grabow, J.A. Dumesic and M. Mavrikakis, *Catal. Lett.* **93** (2004) 93.
60. T.-D. Chau, T.V. de Bocarmé, N. Kruse, R.L.C. Wang and H.J. Kreuzer, *J. Chem. Phys.* **119** (2003) 12605.
61. J.M. Gottfried, K.J. Schmidt, S.L.M. Schroeder and K. Christmann, *Surf. Sci.* **536** (2003) 206.
62. J.J. Stephan and V. Ponec, *J. Catal.* **42** (1976) 1.
63. A. Ruggiero and P. Hollins, *J. Chem. Soc. Faraday Trans.* **92** (1996) 4829.
64. Y. Jugnet, F.J. de Cadet Santos Aires, C. Deraulot, L. Piccolo and J.C. Bertolini, *Surf. Sci.* **521** (2002) L639.
65. A.M. Bradshaw and J. Pritchard, *Proc. Roy. Soc. A* **316** (1970) 169.
66. D.C. Meier, V. Bukhtiyarov and D.W. Goodman, *J. Phys. Chem. B.* **107** (2003) 12668.
67. J.W. Niemantsverdriet, *Spectroscopy in Catalysis,* VCH: Weinheim, 1993.
68. M.M. Schubert, A. Venngopal, M.J. Kahlich, V. Plzak and R.J. Behm, *J. Catal.* **222** (2004) 32.
69. P. Dumas, R.G. Tobin and P.L. Richards, *Surf. Sci.* **171** (1996) 579.
70. G. McElhiney and J. Pritchard, *Surf. Sci.* **60** (1976) 397.
71. K.F. Peters, P. Steadman, H. Isern, A. Alwarez and S. Ferrer, *Surf. Sci.* **467** (2000) 10.
72. P.J. Schmitz, H.C. Kang, W.-Y. Keung and P.A. Thiel, *Surf. Sci.* **248** (1991) 287.
73. P. Hu, D.A. King, M.-H. Lee and M.C. Payne, *Chem. Phys. Lett.* **246** (1995) 73.
74. A. Ruggiero and P. Hollins, *Surf. Sci.* **377** (1977) 583.
75. B. Schumacher, Y. Denkwitz, V. Plzak, M. Kinae and R.J. Behm, *J. Catal.* **224** (2004) 449.

76. A. Shiga and M. Haruta, *Appl. Catal. A: Gen.* **291** (2005) 6.
77. M.C. Kung, C.K. Costello and H.H. Kung, in: *Specialist Periodical Reports: Catalysis*, J.J. Spivey and G.W. Roberts, (eds.), Roy. Soc. Chem., London, **17** (2004) 152.
78. J. Margitfalvi and S. Göbölös, in: *Specialist Periodical Reports: Catalysis*, J.J. Spivey and G.W. Roberts, (eds.), Roy. Soc. Chem., London, **17** (2004) 1.
79. K.I. Hadjiivanov and G.N. Vayssilov, *Adv. Catal.* **47** (2002) 307.
80. G.C. Bond, in: *Handbook of Heterogeneous Catalysis*, G. Ertl, H. Knözinger and J. Weitkamp, (eds.), VCH, Weinheim, 1997, Vol. 2, p. 752.
81. G.C. Bond and D.T. Thompson, *Gold Bull.* **33** (2000) 41.
82. J.-D. Grunwaldt and A. Baiker, *J. Phys. Chem. B.* **103** (1999) 1002.
83. S.D. Lin, M. Bollinger and M.A. Vannice, *Catal. Lett.* **17** (1993) 245.
84. S.H. Overbury, L. Ortiz-Soto, H.-G. Zhu, B. Lee, M.D. Amiridis and S. Dai, *Catal. Lett.* **95** (2004) 99.
85. Y. Iizuka, H. Fujiki, N. Yamauchi, T. Chijiiwa, S. Arai, S. Tsubota and M. Haruta, *Catal. Today* **36** (1997) 115.
86. M.A. Debeila, N.J. Colville, M.S. Scurrell and G.R. Hearne, *Catal. Today* **72** (2002) 79.
87. M. Manzoli, A. Chiorino and F. Boccuzzi, *Surf. Sci.* **532-535** (2003) 377.
88. F. Boccuzzi, A. Chiorino, S. Tsubota and M. Haruta, *J. Phys. Chem.* **100** (1996) 3625.
89. Z.-P. Liu, X.-Q. Gong, J. Kohanoff, C. Sanchez and P. Hu, *Phys. Rev. Lett.* **91** (2003) 266102.
90. M.M. Schubert, M.J. Kahlich, H.A. Gasteiger and R.J. Behm, *J. Power Sources* **84** (1999) 175.
91. T. Tabakova, F. Boccuzzi, M. Manzoli and D. Andreeva, *Appl. Catal. A: Gen.* **252** (2003) 385.
92. S. Carrettin, P. Concepción, A. Corma, J.M. Lopez Nieto and V.F. Puntes, *Angew. Chem. Internat. Edn.* **43** (2004) 2538-2540.
93. H. Idriss, *Platinum Metals Rev.* **48** (2004) 105.
94. N.J. Ossipoff and N.W. Cant, *Topics in Catal.* **8** (1999) 161.
95. S. Minicò, S. Scirè, C. Crisafulli, A.M. Vinco and S. Galvagno, *Catal. Lett.* **47** (1997) 273.
96. A.K. Tripathi, V.S. Kamble and N.M. Gupta, *J. Catal.* **187** (1999) 332.
97. M.A. Bollinger and M.A. Vannice, *Appl. Catal. B: Env.* **8** (1996) 417.
98. G. Stinivas, J. Wright, C.-S. Bai and R. Cook, in: *Proc. 11th Internat. Congr. Catal.*, J.W. Hightower, W.N. Delgass, E. Iglesia and A.T. Bell, (eds.), Elsevier, Amsterdam, 1996, Vol. A, p. 427.
99. E. Rienks, Ph.D. thesis, University of Leiden, 2004, Ch. 4.
100. B.E. Nieuwenhuys, *Adv. Catal.* **44** (1999) 259.
101. T.D. Chau, T. Visart de Bocarmé and N. Kruse, *Catal. Lett.* **98** (2004) 85.
102. M.A. Debeila, N.J. Coville, M.S. Scurrell, G.R. Hearne and M.J. Witcomb, *J. Phys. Chem. B* **108** (2005) 18254.
103. M.A. Debeila, N.J. Coville, M.S. Scurrell and G.R. Hearne, *Appl. Catal. A: Gen.* **291** (2005) 98.
104. M.A. Debeila, N.J. Coville, M.S. Scurrell and G.R. Hearne, *J. Molec. Catal. A: Chem.* **219** (2003) 131.
105. M.E. Bartram and B.E. Koel, *Surf. Sci.* **213** (1989) 137.
106. P. Claus, *Appl. Catal. A: Gen.* **291** (2005) 222.
107. R.J. Mikovsky, M. Boudart and H.S. Taylor, *J. Am. Chem. Soc.* **76** (1954) 3814.

108. G.C. Bond, *Catalysis by Metals,* Academic Press, London, 1962.
109. I. Iida, *Bull. Chem. Soc. Japan,* **52** (1979) 2858.
110. L. Stobinski and R. Dus, *Vacuum* **45** (1994) 299.
111. L. Stobinski and R. Dus, *Czech. J. Phys.* **43** (1993) 1035.
112. M. Okado, M. Nakamura, K. Moritani and T. Kasai, *Surf. Sci.* **523** (2003) 218.
113. M. Okada, S. Ogura, W.A. Diño, M. Wilde, K. Fukutani and T. Kasai, *Appl. Catal. A: Gen.* **291** (2005) 55.
114. L. Stobinski and R. Dus, *Appl. Surf. Sci.* **62** (1992) 62,77.
115. E. Lisowski, L. Stobinski and R. Dus, *Surf. Sci.* **188** (1987) L735.
116. G.K. Boreskov, V.I. Savchenko and V.V. Gorodetskii, *Dokl. Akad. Nauk SSSR* **189** (1969) 537.
117. D. Brennan and P.C. Fletcher, *Trans. Faraday Soc.* **56** (1960) 1662.
118. G. Casalone and G.F. Tandarini, *J. Chem. Soc. Faraday Trans.* **86** (1990) 3793.
119. M.U. Kishyk and I.I. Tret'yakov, *Kinet. Katal.* **15** (1974) 710.
120. H. Stromnes, S. Jusuf, B. Schimmelpfennig, U. Wahlgren and O. Gropen, *J. Molec. Struct.* **567** (1960) 1662.
121. P.A. Sermon, G.C. Bond and P.B. Wells, *J. Chem. Soc. Faraday Trans.* **75** (1979) 385.
122. A. Zanella, C. Louis, S. Giorgio and R. Touroude, *J. Catal.* **223** (2004) 328.
123. D.J.C. Yates, *J. Coll. Interface Sci.* **29** (1969) 194.
124. A.C. Gluhoi, H.S. Vreeberg, J.W. Bakker and B.E. Nieuwenhuys, *Appl. Catal. A: Gen.* **291** (2005) 145.
125. E. Bus, J.T. Miller and J.A. van Bokoven, *J. Phys. Chem. B* **109** (2005) 14581.
126. S.A. Varganov, R.M. Olson, M.S. Gordon, G. Mills and H. Metiu, *J. Chem. Phys.* **121** (2004) 5159.
127. S. Chrétien, M.S. Gordon and H. Metiu, *J. Chem. Phys.* **121** (2004) 3756.
128. T. Kecskés, J. Raskó and J. Kiss, *Appl. Catal. A: Gen.* **273** (2004) 55.
129. M. Kostelitz and J. Oudar, *Surf. Sci.* **27** (1970) 115.
130. H. Tada, T. Soejima, S. Ito and H. Kobayashi, *J. Am. Chem. Soc.* **126** (2004) 15952.
131. J.A. Rodriguez, M. Pérez, J. Evans, G. Lin and J. Hrbek, *J. Chem. Phys.* **122** (2005) 241101.
132. M. Daté, M. Okumura, S. Tsubota and M. Haruta, *Angew. Chem. Int. Ed.* **43** (2004) 2129.
133. R. Schaub, P. Thostrop, N. Lopez, E. Læsgaard, I. Stensgaard and F. Besenbecher, *Phys. Rev. Lett.* **87** (2001) 255104.
134. S.S. Pansare, A. Sirijaruphan and J.G. Goodwin Jr., *J. Catal.* **234** (2005) 151.
135. H. Liu, A. I. Kozlov, A. P. Kozlova, T. Shido, K. Akasura and Y. Iwasawa, *J. Catal.* **185** (1999) 252.
136. J.-D. Grunwaldt, M. Maciejewski, O.S. Becker, P. Fabrizioli and A. Baiker, *J. Catal.* **186** (1999) 458.
137. T.V. Choudhary, C. Sivadinarayana, C.C. Chusei, A.K. Datye, J.P. Fackler Jr. and D.W. Goodman, *J. Catal.* **207** (2002) 247.
138. M.A.P. Dekkers, M.J. Lippits and B.E. Nieuwenhuys, *Catal. Lett.* **56** (1998) 195.
139. M.S. Chen and D.W. Goodman, *Science* **306** (2004) 252.
140. K. Mallick and M.S. Scurrell, *Appl. Catal. A: Gen.* **253** (2003) 527; K. Mallick, M.J. Witcomb and M.S. Scurrell, *Appl. Catal. A: Gen.* **259** (2004) 163.
141. H. Liu, A.I. Kozlov, T. Shido and Y. Iwasawa, *Phys. Chem. Chem. Phys.* **1** (1999) 2851.

CHAPTER 6

Oxidation of Carbon Monoxide

6.1. Introduction

The discovery in 1987 by Masatake Haruta and his associates[1,2] that very small (<5 nm) gold particles supported on suitable oxides are active for carbon monoxide oxidation even below room temperature, was responsible for the astonishing growth of interest in gold as a catalyst that has occurred subsequently. Gold is more active than the noble metals of Groups 8–10;[3] while the latter easily dissociate the oxygen molecule and chemisorb oxygen atoms, and bind carbon monoxide very strongly, leading to low catalytic activity, gold does not adsorb them very strongly (see Chapter 5), and it is therefore nearer the top of the volcano curve for this reaction (see Figure 1.9 in Chapter 1), where it exhibits much greater activity.

The success of Haruta's early work lay in his choice of preparation method and support. Gold particles of the necessary small size were first obtained by coprecipitation (COPPT) and later by deposition–precipitation (DP) (see Sections 4.2.2 and 4.2.3); classical impregnation with $HAuCl_4$ does not work. The choice of support is also critical; transition metal oxides such as ferric oxide and titania work well, whereas the more commonly used supports, such as silica and alumina, do not work well or only less efficiently. This strongly suggests that the support is in some manner involved in the reaction.

There is a truly vast literature on the gold catalysed oxidation of carbon monoxide, the appreciation of which is assisted by a number of reviews.[3–10] One may wonder why this reaction has attracted so great attention: certainly there are a number of possible applications for a catalyst that will perform this reaction at ambient temperature, including carbon dioxide lasers, gas sensors and many environmental applications, such as respirators (see Section 14.2.1). Perhaps of more importance, it is a simple reaction to follow, and the apparatus needed is modest, but also its puzzling mechanism poses a considerable challenge. It has therefore been adopted as a ready means of revealing how the preparation, composition and structure of gold catalysts determine their activity. However, there is some danger in assuming that rates of all other oxidations will follow that of carbon monoxide, and in the synthesis of hydrogen peroxide, for example, the correlation between activity and particle size is inverse (see Chapter 8). Nevertheless the great facility with which carbon monoxide oxidation takes place has proved an attraction.

For such a simple reaction, the mechanism has proved extraordinarily difficult to pin down. The various proposals are reviewed below (Section 6.6); it seems likely that several types of mechanisms are possible, depending on the manner and extent of the support's involvement. The reaction could take place either at the interface between the gold metal particle and the oxide support, or possibly between carbon monoxide adsorbed on the gold particles and di-oxygen activated by the oxide support. Alternatively, the reaction could take place on gold particles only, that are either fully metallic gold particles, or particles that contain ensembles of Au^0 and Au^I-OH or Au^{III} sites. The search for a mechanism has been made all the more difficult because most of the experimental work has not been directed specifically to this task: when orders of reaction and activation are measured, it is not always specified in the papers that mass transport limitation has been avoided. Moreover, the common practice of assessing a catalyst's performance simply by following conversion as a function of temperature has severe limitations (Section 1.4). There is a dearth of reliable kinetic information (orders of reaction, activation energy) and of thorough kinetic modelling, by which possible mechanisms could be tested. In view of this shortcoming, it is scarcely surprising that several different mechanisms have been suggested, some of which arise from a rather limited experimental base.

The goal of this chapter is to consider the difficulties of comparing results from different laboratories, to point out agreement and disagreement in results and their interpretation, to discuss mechanism(s), to raise questions, and to summarise the present position. Because of the great extent of the literature and the speed with which it grows, we may have inadvertently missed some important papers: if we have, we apologise.

6.2. Sensitivity of Gold Catalysts towards Reaction Conditions

Large variations in activity have been reported for catalysts of apparently similar composition.[4,11] Activity depends not only on the obvious variables (method, type of support, gold particle size and preparation method), but also on conditions of pretreatment and storage, and on residual impurities, especially chloride ion. Furthermore, the conditions used for measuring activity also differ considerably (reactant partial pressures, water content, space velocity and temperature), and results are expressed in different ways. Comparison of work from different laboratories is therefore difficult.

6.2.1. Reproducibility of preparative method

The problem of reproducing the preparation of Au/TiO_2 by DP (Section 4.2.3) has been examined using an automated dispenser.[12] One possible difficulty arises from the fact that gold compounds are photosensitive, so that the Au^{III} precursors may be reduced in daylight; this is avoided by performing preparations in the dark,[11,13-16] and premature reduction is also minimised by drying at room temperature or 373 K under vacuum for quite a short time, and storing in the dark away from water vapour, either under vacuum[16,17] or in a refrigerator or a freezer.[13-15] Another variable is the extent to which chloride ion is removed by washing (see Section 4.2), since residual chloride causes agglomeration of precursor species during thermal treatment resulting in large gold particles. It is largely due to the chloride effect that conventional impregnation is generally unsuccessful in giving small particles (Section 4.2.1). Chloride ion also acts as a poison in the oxidation of carbon monoxide,[18-20,21] it is particularly strongly held on alumina and titania, from which it is best removed by hydrolysis of the Au–Cl bond with steam, in water, or in ammonia solution (Section 4.2.6).

Residual sodium from the use of the hydroxide or carbonate for preparations by DP (Section 4.2.3) may also affect catalytic properties, and therefore be a cause of irreproducibility. However, the literature is not agreed on whether sodium acts as a poison[15,22] or as a promoter;[8,20] this may depend on its concentration and on the support used. Nitrate ion may also have a promotional effect depending on the concentration[23] (see Section 4.2.2 for the use of nitrate salts as precursors for preparations by COPPT).

6.2.2. Conditions of activation

The size of gold particles depends on the conditions used for thermal pretreatment (Section 4.7), but there are enormous discrepancies between the various activation procedures. These are most usually performed in air, possibly for convenience; this type of thermal treatment is improperly called 'calcination' in most of the papers, since it leads to the reduction of gold; when it is possible we will use the term 'thermal treatment in air' instead. Use of vacuum or hydrogen may cause catalysts to deactivate quickly,[24] although hydrogen treatment has also been claimed to be beneficial,[25–27] especially when the residual chloride content is high, because it promotes its elimination through HCl formation.[19,28] Moreover, it must be realised that both metallic gold particles and their precursors are unstable. Metallic gold particles may sinter in air.[16] The surface of very small particles (<3 nm) will re-oxidise in ambient air,[29] and Au^{III} species in Au/TiO_2 can be reduced in a reactant mixture of carbon monoxide and oxygen at room temperature[25,26,30,31] or even at 195 K,[27] and thus become active in carbon monoxide oxidation. This is not always the case however, because dried samples of Au/Al_2O_3[30] and of Au/TiO_2,[32] showed no activity, while dried Au/Fe_2O_3 containing 77% Au^{III} was not further reduced under these conditions.[33]

6.2.3. Conditions of reaction and the influence of moisture

Although reactions are always conducted at atmospheric pressure, the conditions used vary greatly (Table 6.1), making comparisons difficult. Carbon monoxide oxidation is very exothermic ($\Delta H^0 = 283\,\text{kJ}\,\text{mol}^{-1}$), so precise temperature control is difficult. The heat of reaction alone raised the temperature of 2.5 wt.% Au/Fe_2O_3 from 293 to 393 K,[34] and various artifices have been adopted to counter this problem: dilution of the catalyst with silicon carbide,[35] or glass powder[36] or use of a water-cooled reactor.[34,37] Although the measured temperature increase may only be slight (~10–15 K), that for the metal particle may be much higher,[38] all the more so because supports often have poor thermal conductivity. Dilution of the catalyst and similar attempted improvements cannot be effective in controlling local temperature.

The amount of water contained in the reactants is usually not controlled, although it may have a drastic influence; reports of its effects on catalytic activity are not consistent (Table 6.2) and general features are hard to discern. Careful studies[44,45] have shown that the effect that water has on catalytic activity depends on the support and on its concentration (Figure 6.1). Activation energies shown by Au/TiO_2 and Au/Al_2O_3

Table 6.1: Examples of experimental conditions used for the oxidation of carbon monoxide.

References	Catalyst weigh(mg)	Gas mixture	Flow rate ($\mathrm{ml\,min^{-1}}$)	GHSV ($\mathrm{h^{-1}\,ml\,g_{cat}^{-1}}$)	Measurement
39	200	1% CO in air	67	20 000	Conv. $= f(T)$
40	50	1% CO in air	17	—	
41, 42	100	1% CO in air[a]	33	20000	Conv. $= f(T)$ TOF at 273 K
20	200	1% CO in air purified[a]	67	20000	Conv. $= f(T)$
35	40	1% CO + 2.5% O$_2$ in He + water vapour 1.5%	200	—	Isotherm at 398 K
11					Conv. $= f(T)$
43	—	3.6% CO + 3.6% O$_2$ in He	—	20000	Conv. $= f(T)$

[a]Purified through a molecular sieve column.

Table 6.2: References to reported effects
of water vapour on catalytic activity.

Support	Positive	Neutral	Negative
FeO_3	47	48, 49	
TiO_2	47, 44, 50	51	52, 53
Al_2O_3	35, 47, 54		
MgO	55		

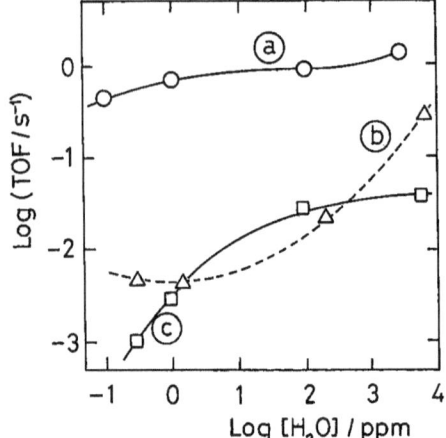

Figure 6.1: Turnover frequency (per surface gold atom) at 273 K over
(a) Au/TiO_2, (b) Au/Al_2O_3 and (c) Au/SiO_2 as a function of moisture
concentration.[45]

were almost independent of water concentration, which clearly affected the
number of active centres but not their structure; for Au/SiO_2, however,
the conversion–temperature curves were changed markedly by the presence
of water. The water-gas shift cannot be responsible for these effects, as it
requires a much higher temperature. The use of isotopic transient analysis
with Au/Al_2O_3 has shown that adding 1800 ppm water increased both the
global rate ($\times 6$), the *intrinsic* TOF ($\times 2$) (see next section), and the CO_x
coverage ($\times 3$).[46] These results suggest that hydroxyl groups at the metal–
support interface are important for efficient reaction. Although the effects
of water addition seem complex, the sensitivity of catalysts to moisture may
well be responsible for some of the reported differences in catalytic activity.

6.2.4. Expression of the results

Many catalysts suffer a more or less rapid initial loss of activity during use, followed by a quasi-stationary state, but the time at which the conversion is reported is not always stated; it may be at time zero or in the steady state. Furthermore, the conditions under which the measurement is made are often not closely specified and the possibility of mass-transport limitation (Section 1.4) is often ignored. Comparison of rates per unit mass of gold, when dispersions or temperatures differ greatly, is not always meaningful. As explained in Section 1.4, it is essential to provide enough information to change a conversion into a rate; thus, while comparison of conversion–temperature curves within a related set of experiments is valid, a wider comparison with the work of others is often impossible. Use of one of the reference catalysts supplied by the World Gold Council as a benchmark is strongly recommended.

Turnover frequencies (TOF) are derived in various ways (Section 1.4). Strictly speaking it should be the rate per unit number of active centres, but this is not generally available for gold catalysts, and so it has been expressed per total number of gold atoms[56] or per number of surface gold atoms as deduced from the average particle size.[11,41,42] Isotopic transient analysis however helpfully evaluates an *intrinsic* TOF (or rate constant) based on the number of reactive intermediates on the catalyst;[46] this comes closest to the proper definition of TOF. Since estimation of the gold surface area by selective oxygen chemisorption has not been generally performed (see Sections 5.2.2 and 5.5.1), reliance has to be placed either on EXAFS, which works well with small particles ($<5\,\mathrm{nm}$),[25,29] or XRD, where peak widths give a mean size but only when they are quite large, or on TEM. For this, the mean size should be calculated from Equation (6.1):

$$d_s = \sum n_i d_i^3 / \sum n_i d_i^2. \tag{6.1}$$

6.2.5. Kinetics of carbon monoxide oxidation

It has been stressed in Section 1.4 that a knowledge of the reaction kinetics, especially of the orders of reaction, is an essential ingredient of any discussion of mechanism. Few measurements of this kind have however been reported and the results are not in good agreement (Table 6.3). Orders in both reactants lie between 0.05 and 0.46, implying moderate to strong chemisorption; the absence of negative orders shows that neither is so

Table 6.3: Kinetic parameters for CO oxidation over supported Au/TiO$_2$ catalysts.[59]

Preparation method	[Au] (wt.%)	Activation	d_{Au} (nm)	CO order	p_{CO}:p_{O2} (kPa)	O$_2$ order	p_{CO}:p_{O2} (kPa)	T(K)	E_a (kJ mol^{-1})	References
DP	3.3	C	3	0.05	0.5–20; 20	0.24	1; 0.7–20	273	34	39
IMP	2.3	HTR/C/LTR	30	0.24	2–25; 5	0.4	5; 2–19	313–333	9.6	19
IMP	1	HTR/C/LTR	30	0.4	n.r.; 5		5; n.r.	273–313	29	52
IMP	2	HTR/C	30	0.45	0.5–2; 4	0.19	0.3–2, 4	320	32	60
IMP (phosphine)	1	C	3	0.25	0.25–3.5	0.41	0.25–3.5	263	16.3	51
MP (phosphine)	1	HTR/C	5	0.2	3.67–16.5; 3.67	0.46	3.67–17.0, 3.67	293	16.3	43
DP	2.4	LTR/C	2.7	0.34	0.02–2; 0.1	0.32	1; 0.2–3	353	—	59
DP	3	C	3.6	0.85	n.r.	0.07	>10% O$_2$ <10% O$_2$	n.r.	25	57

DP, deposition–precipitation; IMP, impregnation with chloroauric acid, unless specified; LTR, low-temperature reduction under H$_2$ at 473 K; HTR, high-temperature reduction at 773 K; C, air treatment at 673 K.

strongly adsorbed as to inhibit reaction. Of course measurements need to be made well within the kinetic regime (Section 1.4), and it is possible for the reaction to pass into the area of mass-transport limitation as the pressure of one reactant is changed.

The only penetrating study of the kinetics of this reaction has been conducted by Al Vannice and his associates[19,52] (Table 6.3). The kinetics obtained using Au/TiO_2 prepared by impregnation, and thus having somewhat large gold particles, were tested against six possible mechanisms by comparing observed rates with those calculated using entropies and enthalpies of chemisorption.[19] Mechanisms of the Eley–Rideal type (Section 1.4) and those involving a rate-limiting adsorption of one of the reactants were at once eliminated, as both would require a first-order dependence for one of them. Langmuir–Hinshelwood mechanisms (Section 1.4) involving competitive (one site) or noncompetitive (two sites) adsorption of carbon monoxide and oxygen atoms were rejected because calculated entropy losses for dissociation of oxygen were larger than the entropy of the gaseous molecule. Entropy and enthalpy values were however consistent with either competitive or noncompetitive mechanisms in which the reaction between carbon monoxide and oxygen molecules appeared in the slow step. The noncompetitive route was later confirmed[52] by the DRIFTS observation that oxygen had no effect on coverage by carbon monoxide; it presumably therefore adsorbed on the support close to the interface with the metal, and this kinetic study therefore resolved the main features of the reaction for this catalyst and this temperature range (273–333 K). Other kinetic results[39,43,49,51] (see Table 6.3) can be interpreted in the same way, but quite different orders of reaction have also been found.[57]

Activation energies measured in the kinetic regime above about 273 K are frequently in the region of $30 \, kJ \, mol^{-1}$, although lower values are known (Table 6.3). Measurements over a wide range of temperature (90–400 K) have revealed three regions[3,5] (Figure 6.2): (I) above 300 K, where the reaction is clearly mass-transport limited, because the activation energy is very low ($\sim 2 \, kJ \, mol^{-1}$) (although strictly the Arrhenius equation (Equation 1.6) does not apply here); (II) 200–300 K, where its value is normal ($\sim 30 \, kJ \, mol^{-1}$); and (III) below 200 K, where it is again very low ($\sim 0 \, kJ \, mol^{-1}$). It was suggested that below 200 K, the titania surface was inactivated by unreactive carbonate ions,[50,58] causing reaction to proceed only on the metal, but the very efficient reaction might also be due to a particularly reactive form of the oxygen molecule, stable only at low temperature.

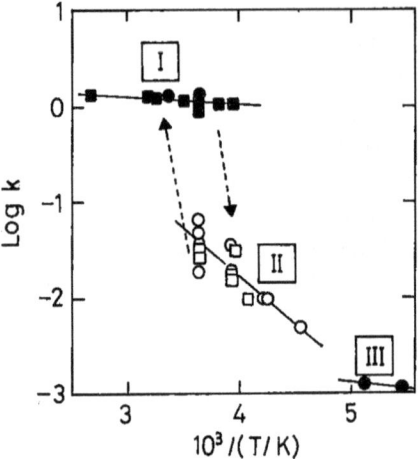

Figure 6.2: Arrhenius plot for the reaction over Au/TiO_2 catalysts.[3]

6.2.6. Deactivation and regeneration

Discordant results are also reported on rates of deactivation and on means of reactivation; again, in view of the variety of catalysts and conditions used, this should not be a surprise. However, it is necessary to explore the origins of these differing results, and of the explanations applied to them.

As to possible reasons for deactivation, growth of metal particles by sintering occurred with model catalysts ($Au/TiO_2(110)$,[61] $Au/SiO_2/Si(100)$,[62]) but not with 'real' catalysts (Au/TiO_2,[32] Au/Al_2O_3[30]). Water may deactivate interfacial titania sites[52] (but see Section 6.2.3); so may 'carbon species'.[63] More usually, activity loss is attributed to unreactive carbonate species blocking active sites,[57,64,65] and perhaps also to depletion of surface hydroxyl groups.[33] Activity loss by Au/Al_2O_3 activated in a dry atmosphere was restored by treatment with either hydrogen or water,[30,35] probably associated with the gold. Some gold catalysts seem however not to deactivate: $Au/Fe(OH)_3$ made from impregnation with $[Au(PPh_3)]NO_3$ and calcined at $673\,K$ was stable for $8\,h$,[20] as was Au/TiO_2 made from $[Au_6(PPh_3)_6](BF_4)_2$ followed by treatment under hydrogen at $773\,K$ then under air at $673\,K$;[43] the latter lost activity later, and this suggests that stability tests need to be conducted for much longer periods. About $10\,wt.\%$ Au/MnO_x actually gained activity during use and maintained its activity for at least $4\,days$.[66]

6.3. Oxidation of Carbon Monoxide on Various Supported Gold Catalysts

6.3.1. Introduction

The activity of a supported gold catalyst for the oxidation of carbon monoxide is dependent principally on three characteristics: (i) particle size of the gold, (ii) chemical nature of the support and (iii) its physical structure. In making this generalisation, we assume that other factors such as the presence of chloride ion or other toxins, and of water vapour, are of lesser importance although they may in fact influence the observed rate. The particle size, and in some cases the physical and chemical nature of the support, are determined by the chosen method of preparation, including whatever pretreatment is applied before the rate is measured. For example, what starts out as hydroxide support may end up as an oxide, and hydrothermal sintering of a microporous support will diminish its surface area. It is therefore not easy to assign the cause of activity to a specific feature of the catalyst, as the attempt to alter the level of one variable, may accidentally affect the level of another. In the following, we first consider the choice of the support as it enters the preparation (Section 6.3.2); then the influence of the preparative method on each of a number supports (Section 6.3.3); and finally, we review the oxidation state of the gold in the active condition (Section 6.4), the effect of particle size, and the metal–support interaction (Section 6.5).

6.3.2. Influence of the nature of the support

From the very first observations on catalysts made by coprecipitation and calcined at 673 K, it became clear that the chemical nature of the support plays a dominant role in determining activity[39,67] (Table 6.4). Best are the oxides of the base metals of Groups 8–10 and the hydroxides of alkaline earth metals; then semiconducting oxides of *sp*-metals (not however d^0), followed by oxides of other transition metals; least effective are silica and alumina. Gold particle sizes present in most catalysts evaluated were small $(\bar{d}_{Au} < 5\,\text{nm})$ except for the least active oxides (CdO, Cr_2O_3, SiO_2) whose $\bar{d}_{Au}$ was larger than 20 nm. A similar picture emerged[20,68] with catalysts prepared by impregnating freshly precipitated hydroxides with a solution of [Au(PPh$_3$)]NO$_3$ (Table 6.5), but the hydroxides of lanthanum, vanadium and chromium joined the least active group. It is very likely that

Table 6.4: Activity of gold catalysts prepared by coprecipitation and calcined at 673 K.[39,67]

Type of supports	Oxides or hydroxides	$T_{50\%}$ (K)
Oxides of Group VIII 3d transition metals	Fe_2O_3, NiO, Co_3O_4	<266
Hydroxides of alkaline earth metal	$Be(OH)_2$, $Mg(OH)_2$	<266
Semi-conductive oxides	CuO, ZnO, In_2O_3, SnO_2	266–298
	CdO	479
Oxides of other transition metals	Sc_2O_3, La_2O_3, TiO_2, ZrO_2, Cr_2O_3	333–433
Insulating oxides	Al_2O_3	357
	SiO_2	477

$T_{50\%}$: temperature for 50% conversion, 5 wt.% Au except with NiO (10 wt.%). Reactions conditions: 200 mg of catalyst, CO = 1% in air, 66 ml s^{-1}, SV = 20 000 ml h^{-1} g$_{cat}^{-1}$.

Table 6.5: Activity of gold catalysts prepared by impregnation of hydroxide supports with $Au(PPh_3)NO_3$, and calcined at 673 K.[20,68]

Support precursor	T_s(K)	$T_{50\%}$(K)	$T_{100\%}$(K)
$Mn(OH)_2$	<203	<203	273
$Co(OH)_2$	<203	<203	273
$Ni(OH)_2$	<203	230	273
$Fe(OH)_3$	<203	206	273
$Zn(OH)_2$	<203	248	273
$Ce(OH)_4$	243	263	283
$Mg(OH)_2$	203	250	>373
$Ti(OH)_4$	253	304	433
$Cu(OH)_2$	223	334	443
$La(OH)_3$	283	335	503
$Al(OH)_3$	373	606	>633
$V(OH)_3$	~383	649	>773
$Cr(OH)_3$	~473	735	>773

T_s, $T_{50\%}$ and $T_{100\%}$: temperatures for reaction start, 50 and 100% conversion, 3 wt.% Au.
Reactions conditions: 200 mg of catalyst, CO = 1% in air, 66 ml s^{-1}, SV = 20 000 ml h^{-1} g$_{cat}^{-1}$.

hydroxide supports used above about 400 K became partially or completely dehydrated to the oxide.

The use of various methods (coprecipitation, impregnation and deposition–precipitation) confirmed the superiority of the transition metal oxides as supports,[69] these being more easily reducible than the ceramic oxides that gave low activities (Table 6.6). With mixed Fe_2O_3–MgO supports activity increased with iron content, not withstanding a growth in gold particles size.[69]

With these chemical methods of preparation, the oxidation state of the gold in the finished catalysts is somewhat uncertain of preparation, so it is of interest to see whether physical methods of preparation, which give gold particles that are at least initially metallic, arrive at the same conclusions. Preparation of catalysts by laser vaporisation of gold clusters (Section 4.4.5) gave particles of about the same size on several supports (Table 6.7), of which titania gave highest activity.[70] On the other hand, the use of chemical vapour deposition (CVD) with Me_2Au^{III}(acac) followed by treatment in air at 673 K[41,42] gave Au/TiO_2 that was less active than the catalyst made by DP (Table 6.9), while with alumina as support the CVD method gave an active product (Table 6.8); in these cases the activation energy was almost unchanged, but in the case of silica, where CVD gave a catalyst of comparable activity to those on the other supports, the activation energy

Table 6.6: Rates and TOFs for gold catalysts calcined at 673 K (573 K for $Au/Mg(OH)_2$).[69]

Catalysts	Preparation method	$\bar{d}_{Au}$(nm)	Rate $(mol_{CO}\, g_{Au}^{-1}\, s^{-1})$	TOF at 353 K$(s^{-1})^a$
Au/Fe_2O_3	DP	2.3–7	39	1.3–3.0
	COPPT	5.5–7	43	3.2–3.4
Au/NiO	COPPT	3.2	20	1.3
Au/Co_3O_4	IMP	3.4	22	1.8
Au/TiO_2	IMP	2.7	33	1.6
$Au/Mg(OH)_2$	COPPT	<4	13	0.5–0.9
Au/MgO	COPPT	6.0	3.8	0.3
Au/Al_2O_3	IMP	4.4	6.0	0.35

DP, deposition–precipitation; COPPT, coprecipitation; IMP, impregnation at pH $\sim$ IEP of the supports.

Reaction conditions: 100 mg diluted in Al_2O_3 (conversion <20%) 1 kPa CO, 1 kPa O_2 in N_2 (30 ml min^{-1}).

^{a}TOF normalised to the number of surface atoms (hemi-spherical particle shape).

Table 6.7: Activities of gold catalysts prepared by laser vaporisation.[70]

Support	$\bar{d}_{Au}$(nm)	[Au](wt.%)	$T_{50\%}$(K)	TOF at 393 K(s^{-1})
Al$_2$O$_3$	2.6	0.08	603	2–7 × 10^{-2}
ZrO$_2$	2.9	0.05	463	2.81 × 10^{-1}
TiO$_2$	2.9	0.02	413	1.1

Reaction conditions: 800–1200 mg of catalyst, 2% CO, 2% O$_2$ in He, 50 ml min^{-1}. TOF normalised to the number of surface atoms (spherical particle shape).

Table 6.8: TOFs and activation energies for gold catalysts prepared by CVD of Me$_2$Au(acac).[41,42]

Support	$\bar{d}_{Au}$(nm)	TOF at 273 K(s^{-1})	E_a(kJ mol^{-1})
Al$_2$O$_3$	3.5	1 × 10^{-2}	36
SiO$_2$	6.6	2 × 10^{-2}	17
TiO$_2$	3.8	2 × 10^{-2}	41

Activation conditions: 473–773 K in air.
Reactions conditions: 100 mg of catalyst, CO = 1% in air, 66 ml s^{-1}, SV = 20 000 ml h^{-1} g$_{cat}^{-1}$.
TOF normalised to the number of surface atoms.

Table 6.9: Activity of Au/TiO$_2$ prepared by different methods.

Preparation Methods	[Au] (wt.%)	$\bar{d}_{Au}$ (nm)	$T_{reaction}$ (K)	Rate (mol · s^{-1} · g$_{cat}^{-1}$)	TOF (s^{-1})	E_a (kJ · mol^{-1})	References
CVD	4.7	3.8	273	1.2 × 10^{-6}	2 × 10^{-2}	41	42
DP	—	1.7	273	3.3 × 10^{-6}	6 × 10^{-2}	37	42
DP	0.5	3.5	300	3.8 × 10^{-7}	3.7 × 10^{-2}	27	40
DP	0.8	3.1	300	6.9 × 10^{-7}	3.4 × 10^{-2}	19	40
DP	1.8	2.7	300	5.5 × 10^{-6}	1.2 × 10^{-1}	18	40
DP	2.3	2.5	300	4.5 × 10^{-6}	6.8 × 10^{-2}	20	40
DP	3.1	2.9	300	2.0 × 10^{-5}	2.6 × 10^{-1}	27	40
PD	1.0	4.6	300	1.5 × 10^{-10}	9.6 × 10^{-6}	56	40
PD	3.6	6.0	300	3.6 × 10^{-10}	8.3 × 10^{-6}	57	40
IMP	1.0	n.r.	300	1.7 × 10^{-10}	n.r.	58	40

CVD: chemical vapour deposition; DP: deposition–precipitation; PD: photodeposition; IMP: impregnation with HAuCl$_4$.
Activation conditions: 473 K in air.
Reaction conditions: 1% vol. CO in air (p_{CO} = 3.67 kPa, p_{O_2} = 3.67 kPa), SV = 20 000 ml h^{-1} g$_{cat}^{-1}$.
TOF calculated from the number of surface atoms.

was lower (see Table 6.8). The reason for the better activity given by the CVD method with the ceramic supports is by no means clear.

From this, we may conclude that while reducible supports perform best, gold on irreducible ceramic oxides still shows a modicum of activity providing that the particles are small enough; the reaction then proceeds solely on the gold without any assistance from the support.

We must now consider how important is the physical structure of the support, starting with the porosity. Au/TiO_2 catalysts have been made with both microporous and mesoporous supports,[71] the latter being the freshly precipitated hydroxide; the catalysts were prepared by impregnation using $[Au(PPh_3)]NO_3$. With the microporous support, reaction occurred only near the external surface, and TAP measurements showed weak reversible adsorption of both reactants and product. With the mesoporous support, however, molecular oxygen was more strongly adsorbed. The particle size of the support is also important. With Au/ZrO_2 prepared by DP and calcined at 673 K, the rate increased sixfold as the zirconia's particle size was decreased from 40–200 to 5–10 nm.[72] Even larger effects ($\times 10^2$) have been observed with ceria[73] (see Section 6.3.3.6) and with yttria,[74] when the particles reach a few nm in size.

The crystalline form of the support also has an effect. Titania exists in three closely-related modifications: anatase, rutile and brookite; the commonly used Degussa P-25 is a mixture of about $\sim 70\%$ anatase and $\sim 30\%$ rutile. Comparison of Au/TiO_2 catalysts made by DP with each of these forms as supports showed that gold on brookite suffered least sintering on reduction in hydrogen at 573 K, and therefore provided the most active catalyst;[25,75] after reduction at 423 K, however, the differences were small.[26] At 473 K gold particles were fully reduced, and in the form of thin rafts, the activity of which was independent of the underlying support structure, but oxidation at 573 K caused them to change into three-dimensional particles, which remained smaller in brookite, and led to a more active catalyst than with the other supports.[25]

6.3.3. Influence of the method of preparation

The method of preparation plays a dominant role in determining the structure and composition of the finished catalyst, and in the COPPT method the support is formed during the preparation. We now examine how the method adopted controls the structure of the catalyst, and hence its activity for carbon monoxide oxidation, and since the effects vary from one support to another, we look at each of the most popular supports in turn.

6.3.3.1. *Iron oxides*

The iron oxides are among the most effective supports for gold (see Tables 6.5 and 6.6), but much depends on the method of preparation;[76] catalysts prepared by impregnation of ferric oxide with $HAuCl_4$ are less active ($\times 10^{-2}$) than those prepared by COPPT, and preparations employing impregnation of freshly precipitated ferric hydroxide ($Fe(OH)_3$) with a solution of $[Au(PPh_3)]NO_3$ are even better ($\times 3$).[77] The system is complicated by the fact that the presence of gold can affect the nature of the iron oxide and temperature of its phase transitions: with Au/Fe_2O_3 prepared by COPPT, the proportion of ferrihydrite ($Fe_5(OH)_8 \cdot 4H_2O$) increases with gold content, probably because dissolution of Au^{3+} ions into the iron oxide inhibits the growth of haematite (α-Fe_2O_3) or goethite during ageing.[76]

Au/Fe_2O_3 prepared by COPPT,[39,67,78] by DP[69] and by impregnation with $[Au(PPh_3)]NO_3$[20,56] was active after heating in air at 673 K. Gold particles are small and fully reduced after heating in air at 673 K ($\bar{d}_{Au} = 3$–5 nm). Only in the case of impregnation with gold phosphine, has the catalyst also been tested after drying, and found to be inactive. However, in other research, Au/Fe_2O_3 prepared by COPPT or DP was found to be active after drying at $\sim$360 K, high temperature calcination (673 K) significantly decreasing the activity.[33,34,36,47,76,79,80] Mössbauer spectroscopy of dried samples has revealed the presence of either Au^{III} as the oxyhydroxide ($AuO(OH \cdot xH_2O)$)[81] or both Au^0 and Au^{III},[33,76,80] both states being also seen by XPS.[34] The Au^{III} state has been suggested as an active species for oxidation of carbon monoxide, especially when present as the oxyhydroxide,[34,36,80,81] although the mechanism by which it might work has not been studied. The ferrihydrite phase may also be responsible for activating the oxygen molecule,[76] and with Au/Fe_2O_3 prepared by COPPT, the combination of Au^0, $AuO(OH)$ and ferrihydrite phases may be highly active because of the high concentration of hydroxyl groups.[80] It is noteworthy that calcined Au/Fe_2O_3 was less active than an 'as prepared' sample, but it did not deactivate to the same extent.[34]

6.3.3.2. *Titania*

Many preparations of Au/TiO_2 have been made by the DP method (Section 4.2.3); they are more active than those made by COPPT[39] or by CVD using $Me_2Au(acac)$, because the particle size is smaller[42] (Table 6.9). Photodeposition and IMP are not suitable methods, although the former produces quite small particles (Table 6.9); this may be because they are the wrong shape, i.e. spherical rather than hemispherical, as needed to maximise the metal–support interface. Au/TiO_2 catalysts made by DP with either NaOH or urea, having different gold contents but similar particle sizes exhibit the same specific rates after drying or calcination.[32]

Au/TiO_2 made from freshly precipitated $Ti(OH)_4$ and $[Au(PPh_3)]NO_3$ was more active than when preformed titania was used:[68] the T_{50} being 280 K instead of 500 K after treatment in air at 673 K, but activity depended on the temperature of treatment. With $[Au_6(PPh_3)_6](BF_4)_2$, the particle size and the activity changed greatly with the thermal treatment used[43] (Table 6.10), the best being comparable to the best made by DP and better than that made with the monomeric gold phosphine complex.

The necessity for using thermal activation is controversial. Catalysts prepared in the dark and therefore initially lacking any Au^0 need to be heated in air to activate them; use of the reactant mixture at low temperature does not reduce the Au^{III},[32] but preparations made in the light may contain some Au^0, which is enough to trigger the reaction, and to produce sufficient heat to complete the reduction under reaction conditions and make the catalyst active without any other activation treatment.[17,31] Treatment in air at a temperature of 473 K[12,32] or 573 K[13,50,82] is often recommended, and in such catalysts neither XAFS,[25,32,82] nor XPS,[68,83] nor FT-IR[84] have revealed any cationic gold. Lower temperatures have failed to achieve complete reduction.

6.3.3.3. *Alumina*

Although the activities reported for Au/Al_2O_3 are very variable, these catalysts are usually stated to be less active than the corresponding Au/TiO_2

Table 6.10: Influence of the conditions of activation of Au/TiO_2 prepared by impregnation with $[Au_6(PPh_3)_6](BF_4)_2$ on catalytic activity.[43]

Activation conditions	$\bar{d}_{Au}(nm)$	$T_{50\%}(K)$
LTC: air/673 K/1 h	—	443
HTC: air/773 K/1 h	7.9	408
HTR/LTC: H_2/773 K/30 min then air/673 K/30 min	4.7	313
HTC/LTR: air/773 K/30 min then H_2/673 K/30 min	8.5	343

Reaction conditions: CO (3.67 kPa), O_2 (3.67 kPa) in He, $SV = 20\,000\,ml\,h^{-1}\,g_{cat}^{-1}$.

catalysts made by the same method and having similar particle sizes, e.g.
DP,[12,47,85] or laser vaporisation[70] (Table 6.7). In the case of CVD the differences are smaller[42] (Table 6.8). 'As prepared' Au/Al_2O_3 showed no activity
at 298 K,[86,87] but on exposure to the reactants at 373 K, it proved more
active than after calcination at 623 K; thus unlike the case of ferric oxide,
Au^{III} was inactive and not reduced under the conditions of reaction at 298 K
(see also Section 6.2.2).

After heating to 623 K in helium, which also effected reduction according
to XAFS, treatment with sodium cyanide solution removed the Au^0, and
left $\sim$10% of gold as cationic Au^{III}.[54] The specific rate of carbon monoxide
oxidation was constant, irrespective of the treatment with sodium cyanide,
as were activation energy and orders of reaction. It appeared that the Au^{III}
was reduced under reaction conditions, and was not an active species.

When considering results for Au/Al_2O_3 catalysts, it has to be remembered that the oxides, oxyhydroxides and hydroxides of aluminium can exist
in many crystalline forms and can have a wide range of surface areas, these
can both change in response to pretreatments, structure, surface area and
changes during preparation are not always reported. There has been no
systematic study of the importance of these variables on catalytic activity.

6.3.3.4. *Microporous and mesoporous silica*

Silica is probably the least used support for gold, because small gold particles are difficult to obtain (Chapter 4), and they sinter easily because
the metal–support interaction is weak; reported activities are low, and
much lower than for corresponding Au/TiO_2 catalysts.[56] Small particles
can however be formed in the channels of mesoporous silica (Section 4.5.2),
although even these showed lower activities than gold in mesoporous titania,
but they were remarkably stable on heating to 773 K.[88] Using CVD with
$Me_2Au(acac)$ resulted in similar activities in both Au/SiO_2 and Au/TiO_2
(Table 6.8); applied to mesoporous MCM-41, however, not all the gold man-
aged to enter the pores, and the catalyst was not found to be more active
than on regular silica[41] (Table 6.11). By including colloidal gold during the
synthesis of MCM-41, gold contents of up to 36 wt.% could be obtained
(Section 4.5.2), nanosized support particles giving much higher activities
than micron-sized particles; even so, activities were low (reaction temperature 353 K).[89]

Au/SBA-15 prepared by anion adsorption onto SBA-15 functionalised
by cationic groups also had large gold particles ($\sim$8 nm) and low activity,

Table 6.11: Activity of gold catalysts prepared by CVD using $Me_2Au(acac)$.[41]

Support	SA $(m^2 g^{-1})^a$	[Au] (wt.%)	$\bar{d}_{Au}$ (nm)	$T_{50\%}$ (K)	TOF at $273 K (s^{-1})^c$
SiO_2	180	6.6	6.6 ± 3.8	227	2×10^{-2}
MCM-41 (2.2 nm)^b	1010	4.2	4.2 ± 1.4	263	2×10^{-2}
MCM-41 (2.7 nm)	1036	2.9	4.0 ± 1.5	259	5×10^{-2}
MCM-41 (3.1 nm)	979	3.5	4.9 ± 1.9	264	4×10^{-2}
SiO_2–Al_2O_3 (14% Al_2O_3)	200	3.0	6.9 ± 1.7	543	—
SiO_2–Al_2O_3 (27% Al_2O_3)	240	3.0	5.8 ± 1.6	618	—
Activated carbon	2000	5.2	5.0 ± 1.6	<573	—

^{a}SA: surface area. ^{b}Mean pore diameter. ^{c}At 253 K.
Activation conditions: 673 K in air.
Reaction conditions: 100 mg catalyst, 1% vol. CO in air (p_{CO} = 3.67 kPa, p_{O_2} = 3.67 kPa), 33 ml min^{-1}, SV = 20 000 ml h^{-1} g$_{cat}^{-1}$.

but modification of its surface by titania enabled small gold particles (~1 nm) to be formed by DP, and the catalyst had comparable activity with Au/TiO_2 made by DP.[90]

Moisture has a drastic effect in increasing the activity of Au/SiO_2 catalysts[45] (Section 6.2.3 and Figures 6.1c and 6.3); water is a prerequisite for low-temperature activity, but since silica is not expected to contribute to the reaction it was suspected that water was adsorbed on the gold and helped the adsorption of oxygen.

6.3.3.5. *Zirconia*

Au/ZrO_2 catalysts are less active than Au/TiO_2 catalysts, whatever method of preparation is used: deposition of colloidal gold,[83,91] DP[12] or laser vaporisation.[70] Activity depends on the method used (Table 6.12), and appears to be due only to the presence of Au^0. The reason for the difference between zirconia and titania is not understood; Zr^{4+} is more difficult to reduce than Ti^{4+}, so anion defects may be harder to form. The lattice structures also differ: in monoclinic zirconia (baddleyite) the Zr^{4+} ion is unusually seven coordinate, and phase transitions into tetragonal and cubic structures occur at >1370 and >2570 K, respectively. However, the

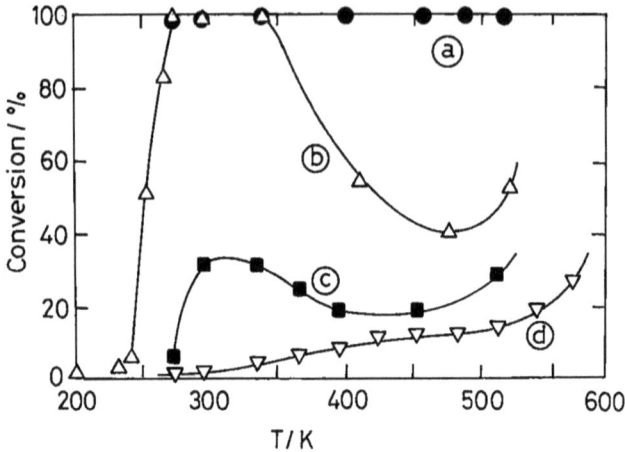

Figure 6.3: Conversion of carbon monoxide over Au/SiO_2 as a function of temperature for various moisture concentrations (a) ~6000 ppm, (b) ~150 ppm, (c) ~1 ppm and (d) ~0.3 ppm.[45]

Table 6.12: Activity of Au/ZrO_2 prepared by different methods.

Preparation	[Au] (wt.%)	$\bar{d}_{Au}$ (nm)	p_{O_2} (mbar)	p_{CO} (mbar)	Rate (mmol$_{CO}$ (g$_{Au}$ s)$^{-1}$)	References
Laser vaporisation	0.05	2.9	20	20	7.8×10^{-2}	70
Colloid deposition	1.7	2	2.5	2.5	0.8×10^{-2}	83
Co-precipitation	1.9	4	2.5	2.5	2.7×10^{-2}	92
Oxidation of ZrAu alloy	61.5	>7	20	20	2.5×10^{-2}	93
Deposition–precipitation	0.77	5.4	200	10	9.4×10^{-2} (nano ZrO_2)	72
	0.74	4.0	200	10	1.5×10^{-2}	72

Reaction temperature: 343–350 K.

effects of phase structure and of surface area have not yet been systematically evaluated, as with titania.

6.3.3.6. *Ceria*

There are certain similarities between ceria and ferric oxide: the cations Ce^{4+} and Fe^{3+} are both quite easily reduced, and in the case of ceria, the surface lattice oxide ions are readily mobilised, so that cation vacancies are common. Oxide ions inside the lattice are also removable, and a whole range of nonstoichiometric oxides between CeO_2 and Ce_2O_3 are known; they have been much studied in the context of vehicle exhaust treatment,[94] for which ceria is an important oxygen storage component (see Section 11.3).

Reported activities for Au/CeO_2 catalysts for oxidation of carbon monoxide are variable, sometimes low[95] and sometimes as high as a similar Au/Fe_2O_3.[96] With Au/CeO_2–Al_2O_3 the activity is associated with the ceria.[97]

Au/CeO_2 (the support being a commercial product) prepared by DP and dried at 393 K, was as active as a DP Au/TiO_2 calcined at 673 K (Table 6.13), but use of the SMAD technique (Section 4.4.7) that leads to Au^0 ($\bar{d}_{Au} = 5\,nm$), gave a less active catalyst.[98] XPS showed that the most active materials had Au^I (80%) and Au^{III} (20%) held in a fluorite-type solid solution $Au_xCe_{1-x}O_{2-\delta}$, in which the Ce-O bond is weaker than in ceria itself, resulting in greater oxygen mobility. However, the way in which carbon monoxide reacts was not explained, but a Mars–van Krevelen mechanism is possible (Section 1.4). There is other evidence for the modification of the ceria lattice by gold; for example, the temperature of reduction of ceria is considerably lower in the presence of gold. The formation of such a $Au_xCe_{1-x}O_{2-\delta}$, solid solution has also been suggested in the case of Au/CeO_2 prepared by combusting of a mixture of cerium ammonium

Table 6.13: Activity of Au/CeO_2 catalysts at 263 K.

Preparation	[Au] (wt.%)	$\bar{d}_{Au}$ (nm)	Au species[a]	Rate ($\mathrm{mmol_{CO}}$ $(\mathrm{g_{cat} \cdot s})^{-1}$)	References
DP at pH 10 on CeO_2 (3.3 nm), drying at 373[b]	2.8	4	Au^0, Au^+, Au^{3+}	2.2×10^{-3}	73, 99
DP on TiO_2, calcination at 673 K[c]	3.3	3.6	Au^0	7.4×10^{-3}	101
DP on CeO_2 (23 nm), drying at 393 K[d]	3.0	—	Au^+ (80%) and Au^{3+} (20%)	6.0×10^{-3}	98

[a]Characterisation before reaction.
[b]Reaction conditions: CO/air/He 0.2/19.8/80, $W/F = 94\,\mathrm{g_{cat}\,h\,mol_{CO}^{-1}}$.
[c]Reaction conditions: CO/air 1/99, VS $= 20\,000\,\mathrm{ml\,h^{-1}\,g_{cat}^{-1}}$.
[d]Reaction conditions: $CO/O_2/He$ 1/1/98, 50 mg, 50 ml min^{-1}, VS $= 60\,00\,\mathrm{ml\,h^{-1}\,g_{cat}^{-1}}$.

nitrate, $HAuCl_4$ and oxalyl-dihydrazide (ignition $T \sim 1273\,K$ for 5 min);[95] after heating to $1073\,K$ for 100 h, the amount of Au^{III} was increased (30–43%) and the gold particles were smaller (5–10 nm to undetectable by XRD), as was the activity.

Gold on 'nanosized' ceria prepared by DP and dried at $373\,K$, and containing metal gold particles ($\bar{d}_{Au} = 3.3\,nm$), is $\sim 10^2$ times more active than Au/CeO_2 prepared by COPPT followed by thermal treatment in air;[73,99,100] this compares favourably with the activity of Au/TiO_2 (Table 6.13). Raman spectroscopy showed the presence of superoxide (O_2^-) and peroxide (O_2^{2-}) species only on the nanosized ceria support, held at one-electron defect sites,[99] and they are thought to be the species that react with carbon monoxide at the gold–support interface.[73] Activity was also associated with Au^{III} as a result of attributing an FT-IR band for carbon monoxide at $2148\,cm^{-1}$ to Au^{III}–CO,[99] but the assignment is doubtful (see Section 5.3.4).

The involvement of the ceria and in particular of its oxide ions in the oxidation of carbon monoxide could be clarified by the use of $^{18}O_2$. The extent and manner of this involvement may well depend on the genesis of the ceria. Measurement of the oxidation state of the gold during reaction would also be of interest.

6.3.3.7. *Magnesia*

Moderately active gold catalysts can be made by DP on magnesium hydroxide if the gold particles are small (<1 nm), but they are less active than Au/TiO_2[102,103] (Table 6.6). After calcination at $623\,K$, the most active catalysts were shown by EXAFS to be fully reduced,[104] although ^{197}Au Mössbauer spectroscopy showed two gold species Au^0 and Au^I, the latter presumably located at the metal–support interface.[105] However, the interpretation of the ^{197}Au Mössbauer spectra is not straightforward, since a similar spectrum obtained on Au/TiO_2 ($\bar{d}_{Au} = 5\,nm$) has been interpreted[106] in terms of two types of metallic gold atoms, one in the bulk and one on the surface. Small gold particles ($\bar{d}_{Au} = 3\,nm$) can also be formed on magnesia from CVD of $Me_2Au(acac)$, followed by reduction at $573\,K$.[107] XANES and TPO/TPR showed that they can be re-oxidised to the Au^{III} state by oxygen at $\sim 485\,K$, and be reduced again by hydrogen;[108] this is the first observation of the full oxidation of Au^0 particles. XANES showed that during re-oxidation of carbon monoxide both Au^0 and Au^I were present, the latter being thought to be at the Au^0–support interface.[107] The

Au^0/Au^I ratio decreased as the carbon monoxide pressure increased,[103] the rate being maximum when this ratio was about unity. In line with this, modification of Au/MgO by ascorbic acid raised the activity, possibly because it favourably alters the Au^{x+}/Au^0 ratio.[109] Iron and manganese also act as promoters, perhaps by forming $M^{n+}-Au_n^0$ (M = Fe, Mn) ensembles on magnesia.[110]

The Au/MgO system is very sensitive to the water content of the feed gas (Figure 6.4).[55] The occurrence of a negative activation energy at low temperature (<373 K) under very dry conditions (80 ppb of water) suggested a different mechanism from that operating at high temperature (>373 K). A possible explanation is that under very dry conditions a highly reactive but weakly adsorbed form of the oxygen molecule can exist at low temperature, but it gradually desorbs on heating, giving the apparently negative activation energy; the presence of traces of water inhibits the adsorption of this species, but allows an hydroxyl-mediated reaction to occur above room temperature.

Au/Be(OH)$_2$ catalysts are also active for carbon monoxide oxidation when the gold particles are 1 nm in size.[2]

6.3.3.8. *Acidic supports*

CVD of Me$_2$Au(acac) onto activated carbon produced small gold particles, but the activity of the resulting catalyst was very low ($T_{50} > 500$ K).[41]

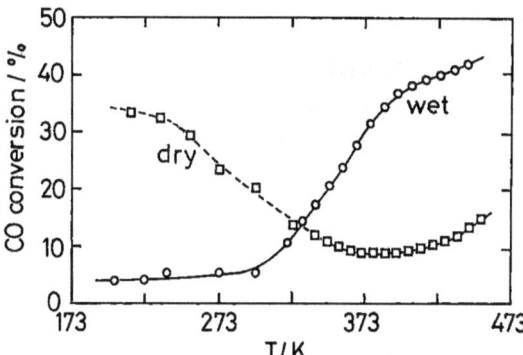

Figure 6.4: Conversion of carbon monoxide over Mg(OH)$_2$ prepared by deposition–precipitation as a function of temperature for various moisture concentrations, wet: 10 ppm; dry: 80 ppb.[55]

Modifying activated carbon fibres (ACF) by ferric oxide improved the gold dispersion and their activity at room temperature, but they deactivated quickly.[111]

While Au/Al_2O_3–SiO_2 catalysts have only low activity,[41] results of considerable interest have been obtained with zeolites containing gold. With gold deposited in HY and USY zeolites by DP, heating in nitrogen to 373 and 723 K, respectively, produced optimum activity (100% conversion at 273 K) associated with the presence of both Au^{III} and Au^0 clusters.[112] Pretreating Au/HY in hydrogen gave better activity. In contrast, Au/NaY was almost inactive because the gold particles aggregated. Adsorption of gold species from $HAuCl_4$ solution at pH 6 onto Y and FeY zeolites gave after washing and drying, small gold particles ($\sim 2\,nm$) comprising Au^{III} and Au^0 species.[113] While Au/FeY was immediately active at 273 K, but was unstable due to carbonate formation, Au/Y required a 4 h induction period, after which its activity was stable. Au/mordenite and Au/β-zeolite gave much lower conversions than Au/Y, and these rapidly decreased due to gold particle growth.[114] XAFS and FTIR characterisation of $Me_2Au(acac)$ adsorbed on calcined NaY zeolite showed that gold remained mononuclear during reaction, but that the Au^{III} initially present was reduced to Au^I and not to Au^0 during oxidation of carbon monoxide.[115,116] The Au^{III} species was said to be tenfold more active than Au^I, but these conclusions were based on assigning an IR vibration at $2169\,cm^{-1}$ to Au^{III}–CO, which is a doubtful assignment (see Section 5.3.4).

6.3.3.9. *Conclusions*

It is very difficult and indeed perhaps impossible to derive any general conclusions as to how the method of preparation affects the structure and activity of supported gold catalysts, as so much depends on the gold precursor and the support; each system needs to be considered on its own. Only a few strong correlations are apparent, and there is much speculation: it is easier to list the unresolved questions than the definitive conclusions.

Activity is strongly associated with methods that give small gold particles rather than large ones, and is more generally found with supports that are more or less easily reducible than with the ceramic oxides. Chloride ion is generally regarded as a poison, and water as a promoter or co-catalyst, especially when the support is a ceramic oxide (Table 6.2). Au^0 is often regarded as essential to activity, but Au^{III} is frequently encountered, especially with reducible oxide supports and is occasionally regarded as the only active species, acting either directly (by a mechanism not specified) or more probably indirectly by disturbing the support. These matters will be discussed in more detail in the following sections.

6.4. Oxidation State of Gold in Active Catalysts

Much of the effort deployed by scientists in their research on gold catalysts is directed to the attempted determination of the active oxidation state (Au^0, Au^I and Au^{III}), or to establishing whether some combination of them is needed. This turns out to be a very difficult problem because of the extreme sensitivity of supported gold catalysts to their surroundings, and a kind of catalytic uncertainty principle operates because the mere act of examining a material may change its composition. Examination of its state before (or after) use does not necessarily reveal its state during use, and there are few techniques (XAFS is one) that can give direct information on a working catalyst.

The positions adopted by investigators are often based on experience with only one system, although they may be claimed to have general relevance; it seems more realistic to seek an answer for each individual system. A further problem is that it is not sufficient to state that a given procedure of activation makes an active catalyst; it is necessary to know how active it is, and how its activity compares with that found after use of other procedures of activation. Without such information we cannot fully understand the factors that contribute to activity, or to find the conditions for obtaining the best catalytic properties.

These positions fall into one of four categories.

(1) 'As prepared' samples, simply dried and not further activated, are inactive so that Au^{III} is not of itself active and is not reduced by a reactant mixture; this has been observed with Au/Al_2O_3[11] and with Au/TiO_2,[16] prepared in the dark and vacuum-dried without heating, without detected trace of Au^0.

(2) 'As prepared' samples are more active than activated samples; their activity could be due to Au^{III}, but because of sensitivity to heat, light and adventitious reductants, it is difficult to ensure that only Au^{III} is present; sometimes, it is clearly reported that gold is partially reduced.[33,73,98] Reduction with the reactant mixture can even start at quite a low temperature (Section 6.2.2); perhaps at least a trace of Au^0 is needed to start things off. In one case with Au/La_2O_3 catalyst, it is clearly claimed that Au^{III} is the active site.[117] Moreover, comparison between 'as prepared' and activated samples is difficult since thermal activation may affect particle size and morphology, and metal–support interaction.

(3) Gold must be fully reduced to be active. This usually requires the catalyst to have been thermally activated, and there are cases where only Au^0 has been detected by XPS,[43,53,118] XAFS[27,32] or Mössbauer spectroscopy,[106] and the catalyst found to be active.

(4) Cationic gold species are inactive by themselves, but are necessary in combination with Au^0 to obtain activity. The search for traces of unreduced species after thermal activation is not an easy task; FT-IR of adsorbed carbon monoxide is probably the most sensitive technique, provided the bands are correctly assigned, since Au^{III} is reduced by carbon monoxide, and Au^I–CO may be confused with the molecule adsorbed on a cation of the support (Section 5.3.4). It is not always easy to distinguish between Au^I and Au^{III}, even by XANES, where interpretation is uncertain.[25–27,107,108,119] As for the location of the cationic gold species, they may be at the metal–support interface[107] or atomically dispersed on or in the support;[50] an XPS study[27] has indicated that Au^{III} and Au^0 form separate phases on titania.

From the suggestions found in the literature as summarised in Section 6.3.3, one might conclude that with ceramic oxides, including titania and zirconia, the dominant active species is Au^0, either alone or with some cationic species. With ferric oxide, it may be Au^0 associated with Au^{III}, with magnesia, it may be Au^0 associated with Au^I, and with ceria, it is not clear which combination of the three species is active.

6.5. Particle Size and Metal–Support Interaction

There is general agreement that small gold particles are more active than large ones for the oxidation of carbon monoxide,[15,32,40,61,91,120] and that the TOF[32,39,120] and the *specific* rate (the rate per unit amount of gold)[121] start to increase rapidly as the size falls below about 4 nm (Figure 6.5). From the consistency of the results gathered in Figure 6.5, one can conclude that particle size is more important than the kind of support in determining activity. This plot shows rates increasing without limits, but there is one clear exception to this: with 'model' Au/TiO_2 (110), the TOF shows a maximum at $\sim$3 nm (Figure 6.6).[61]

There have understandably been numerous attempts to account for this rise in activity at very small sizes in terms of the changes in physical properties described in Section 3.4. These efforts have not resulted in a unanimous view, and the reasons for this are not hard to find.

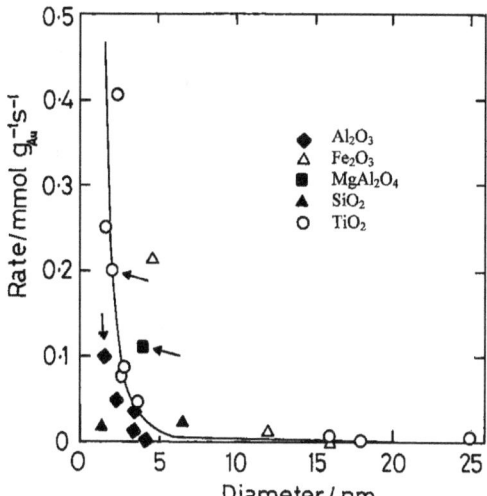

Figure 6.5: Rates at 273 K over different Au catalysts as a function of average gold particle diameter. The points are collected from Ref. 121 (indicated by arrows) and from other work.[18,19,42,56,69,78,122,123]

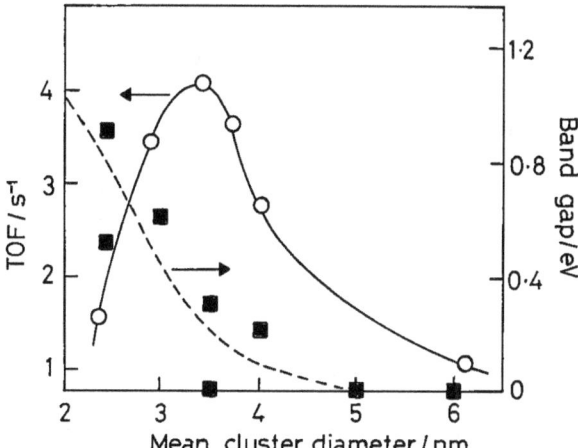

Figure 6.6: Turnover frequency (TOF) at 350 K as a function of the diameter of the gold particles on Au/TiO$_2$ (110) and the band gap of the bidimensional clusters.[61]

First, the various changes are all connected: the loss of metallic character (which in Figure 6.6 appears to cause a *loss* of activity), change in particle morphology and in the proportion of surface atoms of low coordination, shortening of Au–Au distance and larger interaction with the support are all linked (see Sections 3.4 and 3.5). It is likely to be the summation of these and many other factors that account for the rise in activity.

Second, the rate is a function of an apparent activation energy and a pre-exponential factor (see Chapter 1). These two terms depend on the extent to which the active surface is covered by reactants. The strengths of adsorption of the reactants determine not only the number of functioning centres but also the energetics of reaction at each centre. If the rate is measured as a function of temperature: the apparent activation energy indicates the intrinsic activity of each active centre and the pre-exponential factor suggests how many of these active centres there are. Information on the strength of adsorption of the reactants under reaction conditions is obtainable from the dependence of rate on their concentrations (partial orders of reaction), but as we have seen in Section 6.2.4, there has only been one penetrating analysis of reaction kinetics, and that with a catalyst having somewhat large gold particles.[52] There has been no systematic study of how kinetics depends on particle size. The many values of activation energy range widely (Tables 6.3 and 6.9), but many fall between 30 and $40\,\text{kJ}\,\text{mol}^{-1}$. A provisional assessment of the cause of variation of rate with particle size might therefore be that it is chiefly due to the number of active centres rather than the energetics of the transition state.

Although as shown in Figure 6.5, particle size is chiefly responsible for activity, there is nevertheless a major role for the support to play, especially when it is an oxide of a transition metal. The influence of the support on gold and the reciprocal influence of gold on the support has been surveyed in Section 3.5, but of the several physical effects identified, none has been definitely connected with catalytic activity. Perhaps the most important consequence of a metal–support interaction is the finding that with reducible supports (e.g. ceria and ferric oxide), gold cations can become associated with the support by 'dissolving' in it, and can be the major source of catalytic activity (e.g. in water gas shift, see Chapter 10). So it may be that the special quality of small gold particles is their readiness to relinquish some of their atoms to become ions; this would be a very indirect manifestation of the particle size effect.

6.6. Mechanisms of Carbon Monoxide Oxidation

6.6.1. Introduction

At the outset we must be ready to accept that there is not a single unique mechanism for this reaction; that in particular the mechanism may differ from one type of catalyst to another, that different mechanisms may operate at the same time on a given catalyst, and as a consequence that a particular mechanism may be predominant under a given set of reaction conditions (temperature (Section 6.2.5), moisture level (Section 6.3.3.7), and probably partial pressures). In some systems, e.g. Au/TiO_2, it is observed that below $\sim 220\,K$, the reaction has a very low, almost zero activation energy (see Figure 6.2, Region III); clearly as temperature is raised, this will be overtaken by another process having a much higher activation energy, and proceeding by a different mechanism (Section 6.2.5).

We may distinguish four main types of mechanism: (1) the reaction proceeds only on metallic gold particle, (2) it requires the simultaneous availability of metallic gold and cationic gold species, (3) it involves collaboration between metal and support, with the possible assistance of cationic species and (4) it proceeds solely *via* cationic species on the support. In the next section we examine mechanisms 1 and 2; then in Section 6.6.3, we look at mechanisms 3 and 4.

6.6.2. Mechanisms on gold particle only

The mechanisms of carbon monoxide oxidation that have been devised to involve only the gold component should be less complicated than those where the support is also implicated, and so indeed they are. The reaction on gaseous clusters has been mentioned in Section 5.3.1. Coadsorption of carbon monoxide and oxygen on size-selected anionic gold clusters (Au_2^- to Au_{20}^-) prepared by laser vaporisation showed that their reaction takes place on these clusters,[124,125] and proceeds through a co-adsorption that is cooperative and not competitive. A metastable intermediate product was isolated at $100\,K$ and assigned to the structure $Au_2CO_3^-$ (peroxyformate or carbonate),[125] indicating that this complex might represent a key step in the catalytic oxidation of carbon monoxide. The results of number of DFT calculations on small unsupported particles[126−129] are in agreement that co-adsorption of the reactants without dissociation of oxygen, followed by their reaction in a Langmuir–Hinshelwood mode, is a distinct possibility for the slow step; one paper[126] using a neutral Au_{10} rigid particle, suggests two possible routes, the one mentioned above and one involving oxygen dissociation. Neutral Au_{13} and Au_{55} particles of both icosahedral and cubo-octahedral forms have been studied by unrestricted hybrid DFT;[127] an active site is created by dynamic charge polarisation, the carbon monoxide molecule being adsorbed at a positively polarised gold atom, with an oxygen molecule at a neighbouring atom negatively polarised. This idea bears some resemblance to the concept of cooperative coadsorption mentioned in Section 5.3.1.

Little assistance is expected from the support in the case of the ceramic oxides (SiO_2, Al_2O_3 and MgO),[69] and the type of mechanism described above may apply. Another mechanism thought to occur only on a gold particle was proposed for Au/Al_2O_3 as is shown in Figure 6.7(A);[11,35,86] but it requires an Au^+ cation at the edge of the particle, carrying an OH^- group (Figure 6.7(B)). An oxygen molecule then adsorbs dissociatively on steps or defect sites of metallic gold atoms. A carbon monoxide molecule then arrives and reacts *via* an hydroxycarbonyl ion as shown, liberating carbon dioxide and restoring the initial centre. No kinetic evidence was advanced in support of this mechanism. The same intermediate has also been proposed in the reaction over Au/Fe_2O_3 containing both Au^0 and Au^{III} [33] (Section 6.3.3). The existence of the $Au^{I}OH$ entity at the gold–alumina interface was deduced from observations on the catalyst's deactivation, the positive influence of water in the feed,[35] the effect of chloride ion,[28] and a TOF-SIMS experiment that detected AuO^- and AuO_2^-;[85] these species might however have arisen during transfer of catalyst through the air to the instrument.

The principal difficulty with 'metal only' mechanisms is to find a plausible way of activating the oxygen molecule. The problem is discussed in Section 5.2.2; formation of the O_2^- ion, in which the O–O bond is considerably weakened (see Table 5.2), appears possible on small particles,[130,131] and the suggestion that the adsorption of oxygen and of carbon monoxide is mutually supportive opens a possible route for a 'metal-only' mechanism.

The Au/MgO and $Au/Mg(OH)_2$ systems have received considerable experimental attention, both with 'real' (Section 6.3.3.7) and 'model' catalysts,[132–135] and theoretical calculation[129,136–139] due no doubt to the structural simplicity of the support. DFT calculations on Au_{12} and Au_{34} particles and on stepped $Au(211)$ and (221) surfaces lead to the proposal of

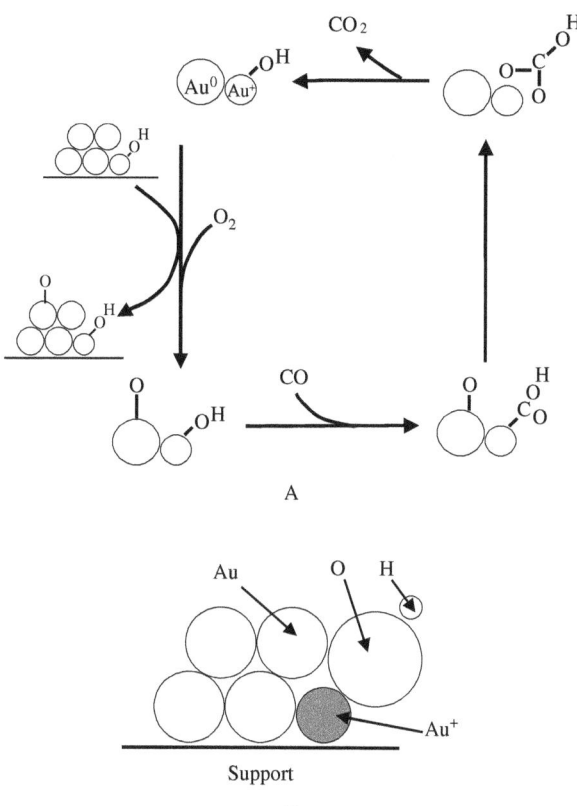

Figure 6.7: (A) Mechanism and (B) model ensemble of Au^0 and Au^+-OH sites proposed by Kung *et al.*[11] (Reprinted with permission from *Appl. Catal. A: Gen.* **343** (2003) 15.)

an Eley–Rideal type of mechanism,[129] in which gaseous oxygen reacts with carbon monoxide adsorbed on gold, through the formation of a metastable O–O–CO intermediate complex:

$$O_2 + CO^* \rightarrow CO_2 + O^* \text{ (slow reaction)}$$

the extra oxygen atom adsorbed on gold then reacting quickly with a second CO molecule:

$$O^* + CO \rightarrow CO_2 \text{ (fast reaction)}.$$

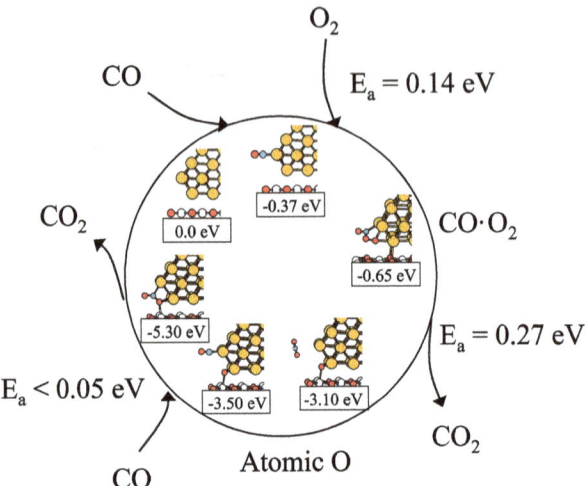

Figure 6.8: Catalytic cycle for oxidation of carbon monoxide at Type-II Au/MgO interface proposed by Molina and Hammer.[139] (Reprinted with permission from *Appl. Catal. A: Gen.* **291** (2005) 21.)

A mechanism for reaction at the metal–support interface in Au/MgO has also been proposed (Figure 6.8);[136,137] three gold atoms and a Mg^{2+} ion combine to form a metastable O–O–CO intermediate with the reactants. Other calculations[138] have however thrown doubt on the probability of formation of this intermediate.

Anionic gold sites formed by electron transfer from the support have also been invoked in the case of Au/MgO,[134,140] and a correlation between the concentration of F centres in magnesia, formed by heating an ultrathin film of MgO/Mo(100) to high temperature, with the activity of Au/MgO has also been advanced.[133] A 'metal-only' mechanism has also been suggested for 'model' Au/TiO$_2$(110) catalysts[61] (Figure 6.6). An activity maximum was found for a gold layer two atoms thick, and a role for anionic gold was again put forward.[141,142]

The reaction with gold catalysts supported ceramic oxides seems particularly sensitive to the presence of water in the feed (Figure 6.1, Figure 6.3 and Figure 6.4); its beneficial effect implies a role for hydroxyl groups either on the support, close to gold particles, or attached to Au^+ ions at their edge

(Figure 6.7). A role for water in aiding chemisorption of reactants on gold particles in Au/MgO(100) has been anticipated in DFT calculations.[143]

6.6.3. Mechanisms involving the support

We turn finally to consider the mechanisms that have been proposed for the reaction on catalysts where the support is assigned an active role. The presence of the support provides an escape route from the problem of oxygen chemisorption, and the mechanisms to be discussed have the common feature that oxygen is in some way 'activated' by the support, while carbon monoxide is chemisorbed on metallic gold particles. This concept focuses attention on the periphery of the particle, where the two adsorbed reactants can meet, i.e. on a reaction at the gold-support interface. Several groups consider that the gold particle is fully metallic[3,18,39,69,70,144,145] (Figure 6.9), while others propose that gold cations are also present[4] (Figure 6.10). All the mechanisms focus on the opening reaction, i.e. reaction between adsorbed carbon monoxide and an oxygen molecule, assumed to be rate-determining, and the supposition (if considered at all) that reaction with the residual oxygen atom will be fast.

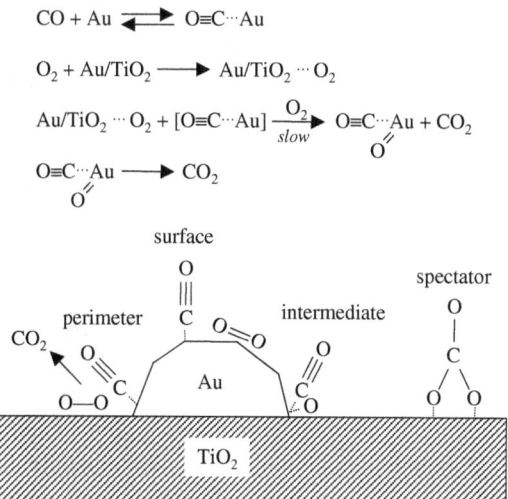

Figure 6.9: Mechanism proposed by Haruta *et al.*[8]

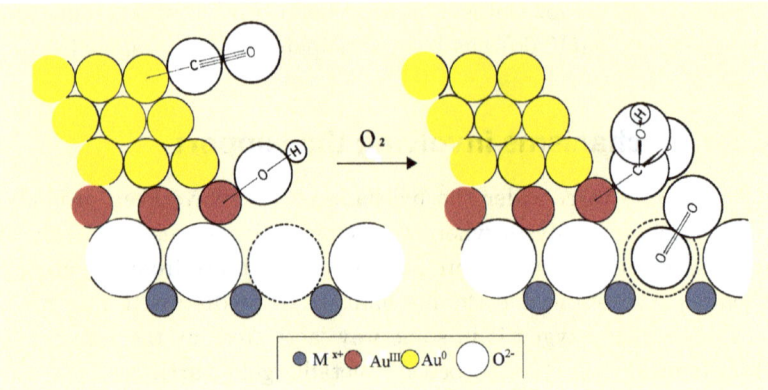

Figure 6.10: Mechanism proposed by Bond and Thompson.[4] (Reprinted from G.C. Bond and D.T. Thompson, *Gold Bull.* **33** (2000) 41, with permission from World Gold Council.)

The manner in which the support 'activates' the oxygen molecule is sometimes left unspecified. Very careful kinetic analysis of the reaction on Au/TiO_2 shows that only a Langmuir–Hinshelwood mechanism involving noncompetitive adsorption of the reactants fits the observed dependence of rate on their pressures;[19] thus two different but closely adjacent sites at the particle edge must be involved, and the importance of this edge region is highlighted by the superiority of supports and methods of preparation that give particles that maximise the length of the interfacial region, such as small particles of hemispherical shape.

More specific proposals for 'activating' the oxygen molecule have also been made. The supports in question are all to some extent reducible, and the oxide ion vacancies are presumed to exist on their surfaces, either as a result of thermal desorption or by reduction with carbon monoxide spilling over from the metal.[146] Such vacancies might be more common near gold particles than elsewhere (Section 3.5.3). Oxygen molecules may adsorb at these vacancies as O_2^- which has been identified by EPR[51,69,147] and then, if not already close to a gold site, might migrate to it;[148] the mean distance of travel would become less as the particle size decreases for a given gold loading, and hence the number of activated oxygen species close to the gold particles would increase. The O_2^- superoxide ion might dissociate at the particle edge before reacting.[69] There is no evidence from experiments with $^{18}O_2$ that lattice oxide ions can react with carbon monoxide on Au/TiO_2.[50,51,59,69,71,149]

The specific role played by surface anion vacancies is reinforced by the observation, frequently made, that a common cause of deactivation is unreactive 'spectator' carbonate ions[3,5,38,58,69,92] that block these sites, thus forbidding them to oxygen molecules. The carbonates are readily decomposed on heating in air, and activity partially if not wholly restored. Other by-products such as bicarbonate[33] and formate ions[59,92] have been also detected. The stability of the carbonate at low temperature, when the titania surface becomes saturated with them,[58,150] has led to the suggestion that reaction then occurs uniquely on the gold[3,5,151] (see also Section 6.2.5). The toxicity of chloride ion can also be understood by its occupation of anion vacancies.

A further mechanism[4] attributes importance to cationic gold at the interface between the metal and the support, and appearing also at the particle edge (Figure 6.10). It is worth explaining how this idea arose. At the time it was advanced, there was much confusion in the literature about the importance on even the necessity of calcination in preparing active catalysts, an uncertainty that still exists, and it was felt that this might be understood if the gold entities were responsive to the reaction environment. If they started as metal, they might become somewhat oxidised, or if they started as oxide, they might become partly reduced. The most likely location for any cations was thought to be at the interface, since with supported metals of Groups 8–10 a layer of cations forming a 'chemical glue' was well-established as being responsible for the stability of small particles. Such species, shown as Au^{3+} ions in that publication,[4] but they could just as well have been Au^+, would have to bear an anion, such as the hydroxyl ion (Figure 6.10), and the reaction might then proceed as shown, by oxidation of an hydroxycarbonyl ion by O_2^-. This mechanism therefore resembles that discussed in the last section, but no support involvement was invoked then (Figure 6.7).

Other supports are more easily reducible than titania, and with them a role for surface anion vacancies is even more plausible. Based on a TAP study, an elaborate mechanism has been proposed for the reaction on Au/Fe_2O_3[33] (Figure 6.11); this appears to employ lattice oxide ions adjacent to the gold particle, and O_2^- ions occupying vacancies created by their use, the product being formed by decomposition of -CO_3H. Extensive use was made of hydroxyl groups on both the gold and the support, as well as water molecules on the latter.

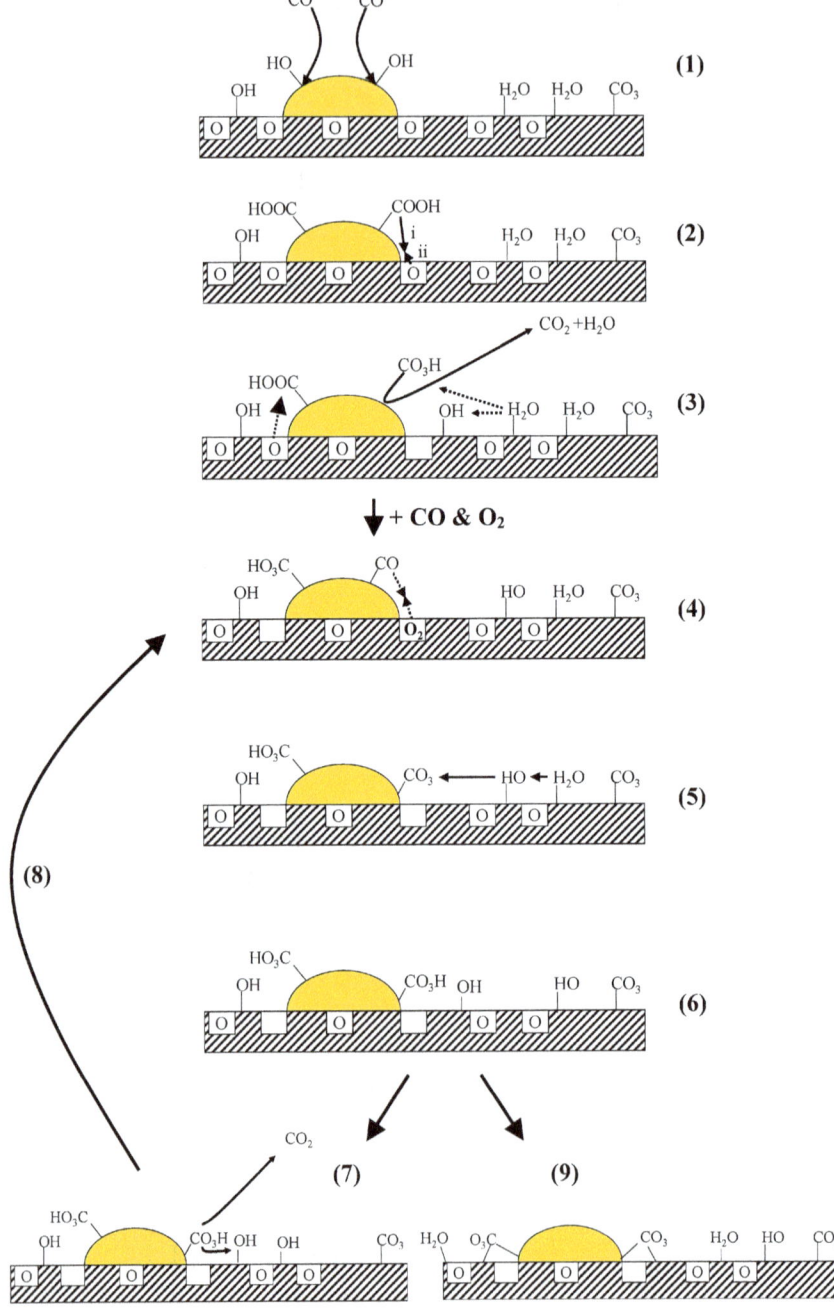

Figure 6.11: Schematic model of oxidation of carbon monoxide on as-prepared (dried) Au/Fe_2O_3 proposed by Makkee *et al.*[33] (Reprinted with permission from *J. Catal.* **230** (2005) 52.)

On ceria, unreduced gold seems to play a major role in the reaction (Section 6.3.3.6). Ceria may provide sites for the formation of superoxide and peroxide–type species,[100] or act as an oxygen supplier for a reaction of Mars and van Krevelen-type.[97] It is occasionally suggested that the reaction might go entirely on the support itself modified by gold ions, and forming a solid solution of type $Ce_{1-x}Au_xO_{2-\delta}$.[98]

There have been a number of DFT studies of the reaction on Au/TiO_2 supports.[129,138,139,152,153] Four publications in particular support the idea of a Langmuir–Hinshelwood mechanism;[129,138,139,152] oxygen molecules adsorb at the particles edge, the O–O bond being stretched by electron transfer from gold, initiated by the presence of Ti cations and oxygen vacancies at the interface. Other calculations[153,154] indicate lattice expansion in the gold particle due to mismatch with the lattice of the support, as was in fact observed with $Au/MgO(100)$[155] and with $Au/anatase-TiO_2(110)$;[156] this was thought to increase the effectiveness of gold in catalysing carbon monoxide oxidation. On the other hand, several EXAFS studies have shown that supported gold particles exhibit a contraction of the gold–gold lattice distance when the particle size becomes smaller than 3 nm,[25,27,29] i.e. in the range of size where they are most active.

6.7. Retrospect and Prospect

It is evident that a very great deal of work has been carried out on the oxidation of carbon monoxide over supported gold catalysts, and it is pertinent to enquire what else remains to be done. The reaction will doubtless continue to be used as a handy way of assessing the effectiveness of a preparation method or a new catalyst composition, but since catalysts are now available that complete the reaction at or below 273 K at high space velocity (at least with a high O_2/CO ratio), there would seem to be little need to search for yet more active catalysts. Two cords of caution are however necessary: (i) the usual reaction mode for measuring activity by a single increasing-temperature scan does not reveal anything about a catalysts stability. This as we have seen in Section 6.2.6, is a problem with almost all catalysts, and more work is needed to connect deactivation rate with operating conditions, and to find ways of maintaining activity for longer times. This is particularly important for developing environmental applications, such as in air quality improvement and respirators (see Section 14.2.1). (ii) High activity for carbon monoxide oxidation does not guarantee the same for any other reaction catalysed by gold; indeed there is good evidence now that there are other requirements such as an easily reducible support for the water-gas shift (Chapter 10) and larger gold particles for the synthesis of hydrogen peroxide (Chapter 8).

As to the mechanisms, it has to be stated that most of the proposals are unsupported by the rigorous kinetic analysis. While spectroscopic, TAP and other techniques can provide vital information on certain aspects of mechanism, reaction kinetics alone lead us to the composition of the transition state, and with the sole exception of the work of Al Vannice,[19] there has been no attempt at comprehensive mathematical modelling of the reaction.

References

1. M. Haruta, T. Kobayashi, H. Sano and N. Yamada, *Chem. Lett.* **2** (1987) 405.
2. M. Haruta, K. Saika, T. Kobayashi, S. Tsubota and Y. Nakahara, *Chem. Express* **3** (1988) 159.
3. M. Haruta and M. Daté, *Appl. Catal. A* **222** (2001) 427.
4. G.C. Bond and D.T. Thompson, *Gold Bull.* **33** (2000) 41.
5. M. Haruta, *Cattech* **6** (2002) 102.
6. M. Haruta, *Chem. Record* **3** (2003) 75.
7. R. Meyer, C. Lemire, S.K. Shaikhutdinov and H.J. Freund, *Gold. Bull.* **37** (2004) 72.
8. M. Haruta, *J. New Mater. Electrochem. Systems* **7** (2004) 163.
9. G.J. Hutchings, *Catal. Today* **100** (2005) 55.
10. M.S. Chen and D.W. Goodman, *Catal. Today* **111** (2006) 22.
11. C.K. Costello, J.H. Yang, H.Y. Law, Y. Wang, J.N. Lin, L.D. Marks, M.D. Kung and H.H. Kung, *Appl. Catal. A: Gen.* **243** (2003) 15.
12. A. Wolf and F. Schüth, *Appl. Catal. A: Gen.* **226** (2002) 1.
13. M. Daté, Y. Ichihashi, T. Yamashita, A. Chiorino, F. Boccuzzi and M. Haruta, *Catal. Today* **72** (2002) 89.
14. T. Akita, P. Lu, S. Ichikawa, K. Tanaka and M. Haruta, *Surf. Interf. Anal.* **31** (2001) 73.
15. B. Schumacher, V. Plzak, K. Kinne and R.J. Behm, *Catal. Lett.* **2003** (2003) 109.
16. R. Zanella and C. Louis, *Catal. Today* **107–108** (2005) 768.
17. W.-C. Li, M. Comotti and F. Schüth, *J. Catal.* **237** (2006) 190.
18. M. Haruta, *Catal. Today* **36** (1997) 153.
19. S.D. Lin, M. Bollinger and M.A. Vannice, *Catal. Lett.* **17** (1993) 245.
20. Y. Yuan, A.P. Kozlova, K. Asakura, H. Wan, K. Tsai and Y. Iwasawa, *J. Catal.* **170** (1997) 191.
21. J.M.C. Soares, P. Morrall, A. Crossley, P. Harris and M. Bowker, *J. Catal.* **216** (2003) 17.
22. C. Sze, E. Gulari and B.G. Demcyzyk, *Mater. Res. Soc. Proc.* **286** (1993) 143.
23. B. Solsona, M. Conte, Y. Cong, A. Carley and G. Hutchings, *Chem. Commun.* (2005) 2351.
24. D. Cunningham, S. Tsubota, N. Kamijo and M. Haruta, *Res. Chem. Interm.* **19** (1993) 1.

25. V. Schwartz, D.R. Mullins, W. Yan, B. Chen, S. Dai and S.H. Overbury, *J. Phys. Chem. B* **108** (2004) 15782.

26. W. Yan, B. Chen, S.M. Mahurin, V. Schwartz, D.R. Mullins, A.R. Lupini, S.J. Pennycook, S. Dai and S.H. Overbury, *J. Phys. Chem. B* **109** (2005) 10676.

27. J.H. Yang, J.D. Henao, M.C. Raphulu, Y. Wang, T. Caputo, A.J. Groszek, M.C. Kung, M. Scurrell, J.T. Miller and H.H. Kung, *J. Phys. Chem. B* **109** (2005) 10319.

28. H.S. Oh, J.H. Yang, C.K. Costello, Y.M. Wang, S.R. Bare, H.H. Kung and M.C. Kung, *J. Catal.* **210** (2002) 375.

29. J.T. Miller, A.J. Kropf, Y. Zha, J.R. Regalbuto, L. Delannoy, C. Louis, E. Bus and J.A. van Bokhoven, *J. Catal.* **240** (2006) 222.

30. M.C. Kung, C.K. Costello and H.H. Kung, in *Specialist Periodical Reprints: Catalysis*, J.J. Spirey and G.W. Roberts, (eds.), Roy. Soc. Chem., London, Vol. 17, p. 152, 2004.

31. F. Moreau, G.C. Bond and A.O. Taylor, *J. Catal.* **231** (2005) 105.

32. R. Zanella, S. Giorgio, C.-H. Shin, C.R. Henry and C. Louis, *J. Catal.* **222** (2004) 357.

33. S.T. Daniells, A.R. Overweg, M. Makkee and J.A. Moulijn, *J. Catal.* **230** (2005) 52.

34. M. Khoudiakov, M.-C. Gupta and S. Deevi, *Appl. Catal. A: Gen.* **291** (2005) 151.

35. C.K. Costello, M.C. Kung, H.-S. Oh and H.H. Kung, *Appl. Catal. A: Gen.* **232** (2002) 159.

36. A.M. Visco, A. Donato, C. Milone and S. Galvagno, *React. Kinet. Catal. Lett.* **61** (1997) 219.

37. Q. Xu, K.C.C. Kharas and A.K. Datye, *Catal. Lett.* **85** (2003) 229.

38. A.K. Tripathi, V.S. Kamble and N.M. Gupta, *J. Catal.* **187** (1999) 332.

39. M. Haruta, S. Tsubota, T. Kobayashi, H. Kageyama, M.J. Genet and B. Delmon, *J. Catal.* **144** (1993) 175.

40. G.R. Bamwenda, S. Tsubota, T. Nakamura and M. Haruta, *Catal. Lett.* **44** (1997) 83.

41. M. Okumura, S. Tsubota and M. Haruta, *J. Molec. Catal. A: Chem.* **199** (2003) 73.

42. M. Okumura, S. Nakamura, S. Tsubota, T. Nakamura, M. Azuma and M. Haruta, *Catal. Lett.* **51** (1998) 53.

43. T.V. Choudhary, C. Sivadinarayana, C.C. Chusuei, A.K. Datye, J.P. Fackler Jr. and D.W. Goodman, *J. Catal.* **207** (2002) 247.

44. M. Daté and M. Haruta, *J. Catal.* **201** (2001) 221.

45. M. Daté, M. Okumura, S. Tsubota and M. Haruta, *Angew. Chem. Int. Ed.* **43** (2004) 2129.

46. J.T. Calla and R.J. Davis, *J. Phys. Chem. B* **109** (2005) 2307.

47. E.D. Park and J.S. Lee, *J. Catal.* **186** (1999) 1.

48. A.P. Kozlova, A.I. Kozlov, S. Sugiyama, Y. Matsui, K. Asakura and Y. Iwasawa, *J. Catal.* **181** (1999) 37.

49. H. Liu, A.I. Kozlov, A.P. Kozlova, T. Shido and Y. Iwasawa, *Phys. Chem. Chem. Phys.* **1** (1999) 2851.

50. F. Boccuzzi, A. Chiorino, M. Manzoli, P. Lu, T. Akita, S. Ichikawa and M. Haruta, *J. Catal.* **202** (2001) 256.

51. H. Liu, A.I. Kozlov, A.P. Kozlova, T. Shido, K. Asakura and Y. Iwasawa, *J. Catal.* **185** (1999) 252.

52. M.A. Bollinger and M.A. Vannice, *Appl. Catal. B: Env.* **8** (1996) 417.

53. J.-D. Grunwaldt, C. Kierner, C. Wogerbauer and A. Baiker, *J. Catal.* **181** (1999) 223.

54. J.T. Calla and R.J. Davis, *Catal. Lett.* **99** (2005) 21.

55. D.A.H. Cunningham, W. Vogel and M. Haruta, *Catal. Lett.* **63** (1999) 43.

56. Y. Yuan, K. Asakura, H. Wan, K. Tsai and Y. Iwasawa, *Catal. Lett.* **42** (1996) 15.

57. P. Konova, A. Naydenov, C. Venkov, D. Mehandjiev, D. Andreeva and T. Tabakova, *J. Molec. Catal. A: Chem.* **213** (2004) 235.

58. F. Boccuzzi, A. Chiorino, S. Tsubota and M. Haruta, *J. Phys. Chem.* **100** (1996) 3625.

59. B. Schumacher, Y. Denkwitz, V. Plzak, M. Kinne and R.J. Behm, *J. Catal.* **224** (2004) 449.

60. N.W. Cant and N.J. Ossipoff, *Catal. Today* **36** (1997) 125.

61. M. Valden, X. Lai and D.W. Goodman, *Science* **281** (1998) 1647.

62. L. Guczi, G. Peto, A. Beck, K. Frey, O. Geszti, G. Molnar and C.S. Daroczi, *J. Am. Chem. Soc.* **125** (2003) 4332.

63. G. Srinivas, J. Wright, C.S. BAi and R. Cook, *Stud. Surf. Sci. Catal.* **101** (1996) 427.

64. M.M. Schubert, V. Plzak, J. Garche and R.J. Behm, *Catal. Today* **76** (2001) 143.

65. P. Konova, A. Naydenov, T. Tabakova and D. Mehandjiev, *Catal. Commun.* **5** (2004) 537.

66. S.D. Gardner, G.B. Hoflund, M.R. Davidson, H.A. Laitinen, D.R. Schryer and B.T. Upchurch, *Langmuir* **7** (1991) 2140.

67. M. Haruta, H. Kageyama, N. Kamijo, T. Kobayashi and F. Delannay, *Stud. Surf. Sci. Catal.* **44** (1988) 33.

68. Y. Yuan, K. Asakura, H. Wan, K. Tsai and Y. Iwasawa, *Catal. Today* **44** (1998) 333.

69. M.M. Schubert, S. Hackenberg, A.C.v. Veen, M. Muhler, V. Plzak and R.J. Behm, *J. Catal.* **197** (2001) 113.

70. S. Arii, F. Mortin, A.J. Renouprez and J.L. Rousset, *J. Am. Chem. Soc.* **126** (2004) 1199.

71. M. Olea and Y. Iwasawa, *Appl. Catal. A: Gen.* **275** (2004) 35.

72. X. Zhang, H. Wang and B.-Q. Xu, *J. Phys. Chem. B* **109** (2005) 9678.

73. S. Carrettin, P. Concepcion, A. Corma, J.M.L. Nieto and V.F. Puntes, *Angew. Chem.* **43** (2004) 2538.

74. J. Guzman and A. Corma, *Chem. Commun.* (2005) 743.

75. W. Yan, B. Chen, S.M. Mahurin, S. Dai and S.H. Overbury, *Chem. Commun.* (2004) 1918.

76. F.E. Wagner, S. Galvagno, C. Milone, A.M. Visco, L. Stievano and S. Calogero, *J. Chem. Soc. Faraday Trans. I* **93** (1997) 3403.

77. A.I. Kozlov, A.P. Kozlova, H. Liu and Y. Iwasawa, *Appl. Catal. A: Gen.* **182** (1999) 9.

78. M. Haruta, N. Yamada, T. Kobayashi and S. Iijima, *J. Catal.* **115** (1989) 301.

79. S.K. Tanielyan and R.L. Augustine, *Appl. Catal. A: Gen.* **85** (1992) 73.

80. N.A. Hodge, C.J. Kiely, R. Whyman, M.R.H. Siddiqui, G.J. Hutchings, Q.A. Pankhurst, F.E. Wagner, R.R. Rajaram and S.E. Golunski, *Catal. Today* **72** (2002) 133.

81. R.M. Finch, N.A. Hodge, G.J. Hutchings, A. Meagher, Q.A. Pankhurst, M.R.H. Siddiqui, F.E. Wagner and R. Whyman, *Phys. Chem. Chem. Phys.* **1** (1999) 485.

82. J. Guzman, J.C. Fierro-Gonzalez, S. Kuba and B.C. Gates, *Catal. Lett.* **95** (2004) 77.

83. J.D. Grunwaldt, M. Maciejewski, O.S. Becker, P. Fabrizioli and A. Baiker, *J. Catal.* **186** (1999) 458.
84. M.A.P. Dekkers, M.J. Lippits and B.E. Nieuwenhuys, *Catal. Lett.* **56** (1998) 195.
85. L. Fu, N.Q. Wu, J.H. Yang, F. Qu, D.L. Johnson, M.C. Kung, H.H. Kung and V.P. David, *J. Phys. Chem. B* **109** (2005) 3704.
86. H.H. Kung, M.C. Kung and C.K. Costello, *J. Catal.* **216** (2003) 425.
87. C.K. Costello, J. Guzman, J.H. Yang, Y.M. Wang, M.C. Kung, B.C. Gates and H.H. Kung, *J. Phys. Chem. B* **108** (2004) 12529.
88. S.H. Overbury, L. Ortiz-Soto, H. Zhu, B. Lee, M.D. Amiridis and S. Dai, *Catal. Lett.* **95** (2004) 99.
89. J.-H. Liu, Y.-S. Chi, H.-P. Lin, C.-Y. Mou and B.-Z. Wan, *Catal. Today* **93–95** (2004) 141.
90. W. Yan, V. Petkov, S.M. Mahurin, S.H. Overbury and S. Dai, *Catal. Commun.* **6** (2005) 404.
91. M. Maciejewski, P. Fabrizioli, J.D. Grunwaldt, O.S. Becker and A. Baiker, *Phys. Chem. Chem. Phys.* **3** (2001) 3846.
92. A. Knell, P. Barnickel, A. Bauker and A. Wokaun, *J. Catal.* **137** (1992) 306.
93. M. Lomello-Tafin, A.A. Chaou, F. Morfin, V. Caps and J.-L. Rousset, *Chem. Commun.* (2005) 388.
94. *Catalysis by Ceria and Related Materials*, A. Trovarelli, (ed.), Imperial College Press, London (2002).
95. P. Bera and M.S. Hegde, *Catal. Lett.* **79** (2002) 75.
96. W. Liu and M. Flytzani-Stephanopoulos, *J. Catal.* **153** (1995) 304.
97. A.C. Gluhoi, H.S. Vreeburg, J.W. Bakker and B.E. Nieuwenhuys, *Appl. Catal. A: Gen.* **291** (2005) 145.
98. A.M. Venezia, G. Pantaleo, A. Longo, G.D. Carlo, M.P. Casaletto, F.L. Liotta and G. Deganello, *J. Phys. Chem. B* **109** (2005) 2821.
99. J. Guzman, S. Carrettin and A. Corma, *J. Am. Chem. Soc.* **127** (2005) 3286.
100. J. Guzman, S. Carrettin, J.C. Fierro-Gonzalez, Y. Hao, B.C. Gates and A. Corma, *Angew. Chem. Int. Ed.* **44** (2005) 4778.
101. S. Tsubota, T. Nakamura, K. Tanaka and M. Haruta, *Catal. Lett.* **56** (1998) 131.
102. S. Tsubota, M. Haruta, T. Kobayashi, A. Ueda and Y. Nakahara, *Stud. Surf. Sci. Catal.* **72** (1991) 695.
103. J. Guzman and B.C. Gates, *J. Am. Chem. Soc.* **126** (2004) 2672.
104. D.A.H. Cunningham, W. Vogel, H. Kageyama, S. Tsubota and M. Haruta, *J. Catal.* **177** (1998) 1.
105. Y. Kobayashi, S. Nasu, S. Tsubota and M. Haruta, *Hyperfine Interactions* **126** (2000) 95.
106. A. Goossens, M.W.J. Crajé, A.M. van der Kraan, A. Zwijnenburg, M. Makkee, J.A. Moulijn and L.J. de Jongh, *Catal. Today* **72** (2002) 95.
107. J. Guzman and B.C. Gates, *J. Phys. Chem. B* **106** (2002) 7659.
108. J. Guzman and B.C. Gates, *J. Phys. Chem. B* **107** (2003) 2242.
109. J.L. Margitfalvi, A. Fasi, M. Hegedüs, F. Lonyi, S. Göbölös and N. Bogdanchikova, *Catal. Today* **72** (2002) 157.
110. J.L. Margitfalvi, M. Hegedus, A. Szegedi and I. Sajo, *Appl. Catal. A: Gen.* **272** (2004) 87.
111. D.A. Bulushev, L. Kiwi-Minsker, I. Yuranov, E.I. Suvorova, P.A. Buffat and A. Renken, *J. Catal.* **210** (2002) 149.
112. K. Okumura, K. Yoshino, K. Kato and M. Niwa, *J. Phys. Chem. B: Env.* **109** (2005) 12380.

113. J.-N. Lin, J.-H. Chen, C.-Y. Hsiao, Y.-M. Kang and B.-Z. Wan, *Appl. Catal. B: Env.* **36** (2002) 19.

114. J.-H. Chen, J.-N. Lin, Y.-M. Kang, W.-Y. Yu, C.-N. Kuo and B.-Z. Wan, *Appl. Catal. A: Gen.* **291** (2005) 162.

115. J.C. Fierro-Gonzalez and B.C. Gates, *J. Phys. Chem. B* **108** (2004) 16999.

116. J.C. Fierro-Gonzalez, B.G. Anderson, K. Ramesh, C.P. Vinod, J.W. Niemantsverdriet and B.C. Gates, *Catal. Lett.* **101** (2005) 265.

117. J.C. Fierro-Gonzalez, V.A. Bhirud and B.C. Gates, *Chem. Commun.* (2005) 5275.

118. R.J.H. Grisel, C.J. Weststrate, A. Goossens, M.W.J. Crajé, A.M. van der Kraan and B.E. Nieuwenhuys, *Catal. Today* **72** (2002) 123.

119. N. Weiher, E. Bus, L. Delannoy, C. Louis, D.E. Ramaker, J.T. Miller and J.A. van Bokhoven, *J. Catal.* **240** (2006) 100.

120. A.I. Kozlov, A.P. Kozlova, K. Asakura, Y. Matsui, T. Kogure, T. Shido and Y. Iwasawa, *J. Catal.* **196** (2000) 56.

121. N. Lopez, T.V.W. Janssens, B.S. Clausen, Y. Xu, M. Mavrikakis, T. Bligaard and J.K. Nørskov, *J. Catal.* **223** (2004) 232.

122. M. Haruta, *Stud. Surf. Sci. Catal.* **110** (1997) 123.

123. S.-J. Lee and A. Gavriilidis, *J. Catal.* **206** (2002) 305.

124. W.T. Wallace and R.L. Whetten, *J. Am. Chem. Soc.* **124** (2002) 7499.

125. L.D. Socaciu, J. Hagen, T.M. Brenhardt, L. Wöste, U. Heiz, H. Häkkinen and U. Landman, *J. Am. Chem. Soc.* **125** (2003) 10437.

126. N. Lopez and J.K. Norskov, *J. Am. Chem. Soc.* **124** (2002) 11262.

127. M. Okumura, Y. Kitagawa, M. Haruta and K. Yamaguchi, *Appl. Catal. A: Gen.* **291** (2005) 37.

128. A. Shiga and M. Haruta, *Appl. Catal. A: Gen.* **291** (2005) 6.

129. Z.-P. Liu, P. Hu and A. Alavi, *J. Am. Chem. Soc.* **124** (2002) 14770.

130. T. Fukushima, S. Galvagno and G. Parravano, *J. Catal.* **57** (1979) 177.

131. H. Berndt, I. Pitsch, S. Evert, K. Stuve, M.-M. Pohl, J. Radnik and A. Martin, *Appl. Catal. A: Gen.* **244** (2003) 169.

132. O. Meerson, G. Sitja and C.R. Henry, *Eur. Phys. J. D* **34** (2005) 119.

133. Z. Yan, S. Chinta, A.A. Mohamed, J.P. Fackler Jr. and D.W. Goodman, *J. Am. Chem. Soc.* **127** (2005) 1604.

134. A. Sanchez, S. Abbet, U. Heiz, W.D. Schneider, H. Häkkinen, R.N. Barnett and U. Landman, *J. Phys. Chem. A* **103** (1999) 9573.

135. B. Yoon, H. Häkkinen, U. Landman, A.S. Wörz, J.-M. Antonietti, S. Abbet, K. Judai and U. Heiz, *Science* **307** (2005) 40.

136. L.M. Molina and B. Hammer, *Phys. Rev. Lett.* **90** (2003) 206102.

137. L.M. Molina and B. Hammer, *Phys. Rev. B* **69** (2004) 155424.

138. Z.-P. Liu, X.-Q. Gong, J. Kohanoff, C. Sanchez and P. Hu, *Phys. Rev. Lett.* **91** (2003) 266102.

139. L.M. Molina and B. Hammer, *Appl. Catal. A: Gen.* **291** (2005) 21.

140. H. Häkkinen, S. Abbet, A. Sanchez, U. Heiz and U. Landman, *Angew. Chem. Int. Ed.* **42** (2003) 1297.

141. M.S. Chen and D.W. Goodman, *Science* **306** (2004) 252.

142. D.C. Meier and D.W. Goodman, *J. Am. Chem. Soc.* **126** (2004) 1892.

143. A. Bongiorno and U. Landman, *Phys. Rev. Lett.* **95** (2005) 106102.

144. L. Guczi, D. Horvath, Z. Paszti, L. Toth, Z.E. Horvath, A. Karacs and G. Peto, *J. Phys. Chem. B* **104** (2000) 3183.

145. R.J.H. Grisel and B.E. Nieuwenhuys, *Catal. Today* **64** (2001) 69.

146. G.C. Bond, M.J. Fuller and L.R. Molloy, *Proc. 6th Intern. Congr. Catal.*, G.C. Bond, P.B. Wells and F.C. Tompkins, (eds.) The Chemical Society, London (1976), p. 356.

147. M. Okumura, J.M. Coronado, J. Soria, M. Haruta and J.C. Conesa, *J. Catal.* **203** (2001) 168.

148. J.-D. Grunwaldt and A. Baiker, *J. Phys. Chem. B* **103** (1999) 1002.

149. M. Olea, M. Kunitake, T. Shido and Y. Iwasawa, *Phys. Chem. Chem. Phys.* **3** (2001) 627.

150. F. Boccuzzi, A. Chiorino, M. Manzoli, D. Andreeva and T. Tabakova, *J. Catal.* **188** (1999) 176.

151. M. Haruta, M. Daté, Y. Iisuka and F. Boccuzzi, *Shokubai (Catalysts and Catalysis)* **43** (2001) 125.

152. L.M. Molina, M.D. Rasmussen and B. Hammer, *J. Chem. Phys.* **120** (2004) 7673.

153. M. Mavrikakis, P. Stoltze and J.K. Norskov, *Catal. Lett.* **64** (2000) 101.

154. Y. Xu and M. Mavrikakis, *J. Phys. Chem. B* **107** (2003) 9298.

155. S. Giorgio, C. Chapon, C.R. Henry, G. Nihoul and J.M. Penisson, *Phil. Mag. A* **64** (1991) 87.

156. S. Giorgio, C.R. Henry, B. Pauwels and G.P. Tenderloo, *Mater. Sci. Eng. A* **297** (2001) 197.

CHAPTER 7

The Selective Oxidation of Carbon Monoxide in Hydrogen

7.1. Introduction: The Nature of the Problem

7.1.1. The chemical industry's need for hydrogen

The chemical industry has need of enormous quantities of hydrogen for a multiplicity of processes, most notably for ammonia synthesis, petroleum reforming and a number of hydrogenations, including fat-hardening.[1,2] For many years now the principal source of hydrogen has been the steam-reforming of methane and higher alkanes, which produces a product containing carbon oxides as well as hydrogen and unreacted compounds. Now with some adjustments this product is used as a *synthesis gas* for the synthesis of methanol and for making other alcohols and hydrocarbons by the Fischer–Tropsch synthesis. However, for those processes that employ metal catalysts for hydrogenations the presence of carbon monoxide is immensely harmful, so the initial product of steam-reforming is first subjected to the water-gas shift

$$CO + H_2O \rightarrow CO_2 + H_2 \qquad (7.1)$$

(see Chapter 10), initially by a high-temperature operation using a catalyst containing the oxides of iron and chromium, and then by a low-temperature operation that employs a copper–zinc oxide-alumina catalyst.[1] It is this latter stage where gold catalysts have some prospect of being used (see Chapter 10). However, at the moment the best quality of hydrogen obtainable after water-gas shift treatment still contains of the order of 1% carbon monoxide,[3] which is too much for many applications, so that a yet further treatment is needed. This can be methanation:

$$CO + 3H_2 \rightarrow CH_4 + H_2O, \qquad (7.2)$$

but this consumes too much hydrogen, and the best conceivable option would be to oxidise the carbon monoxide to the dioxide, if a catalyst could be found to perform this selectively, i.e. without simultaneously oxidising any hydrogen.

7.1.2. Pure hydrogen for use in fuel cells

There is a very large potential for further use of pure hydrogen in the operation of low-temperature *fuel cells*[3-5] (Chapter 14), where power is generated from the electrochemical oxidation of hydrogen at a platinum/carbon electrode on a polymer base. This polymer electrolyte fuel cell (PEFC; also termed the proton exchange membrane (PEM) fuel cell) represents a highly efficient means of directly converting chemical energy into electrical energy, which can then be used as the motive power for vehicles and for powering portable computers and other devices. The greatest hurdle to be overcome is finding a suitable source of hydrogen of sufficient purity. The expectation is that it will be obtained from steam-reforming methanol or ethanol:

$$CH_3OH + H_2O \rightarrow CO + 3H_2, \tag{7.3}$$

$$CH_3OH + 2H_2O \rightarrow CO_2 + 4H_2, \tag{7.4}$$

$$C_2H_5OH + 3H_2O \rightarrow 2CO_2 + 6H_2, \tag{7.5}$$

$$C_2H_5OH + H_2O \rightarrow 2CO + 4H_2. \tag{7.6}$$

However, the removal of carbon monoxide by water-gas shift to a low level still demands its selective oxidation to the minimum concentration possible. Much research and development has been conducted during the past decades to find a gold catalyst that can do this; the target is usually described by the acronym PROX (preferential oxidation), but sometimes as SCO (selective catalytic oxidation). The task is somewhat simplified by the constraints that are externally imposed: the preferred feed gas, often termed '*idealised*' *reformate*, has the composition 1.0% CO, 1.0% O_2, 75.0% H_2, balance nitrogen or other inert gas, and while of course variations to this composition can be made to explore the kinetics and mechanism, and the effects of the products water and carbon dioxide can be added to observe their effects, the successful catalyst must remove almost all the carbon monoxide (to <10 ppm) and less than 0.5% hydrogen. This requirement is expressed as a *selectivity* based on the percentage of the oxygen consumed that is taken by the carbon monoxide; this should exceed 50%, under conditions where the conversion of carbon monoxide is above 99.5%.[5]

There is a further constraint in that the working of the fuel cell means that the temperature of the oxidation should be close to 353 K, and this is normally taken as the standard value,[6] although establishing the temperature dependence of the selectivity in this region is clearly of interest.

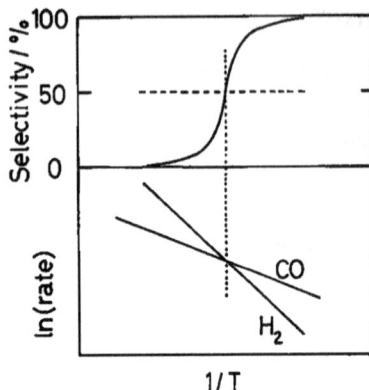

Figure 7.1: Schematic diagram showing how dependence of selectivity on reciprocal temperature follows from the difference in activation energy of the component reactions.

This limitation is not a trivial one, because all research shows that the oxidation of hydrogen has the higher activation energy,[7] so that, while very good selectivity is obtainable at temperatures in the region of ambient, it invariably falls as temperature rises, and is often unacceptably low above 373 K (Figure 7.1). Since there is almost no leeway in the choice of operating conditions, the whole emphasis of research is laid on the catalyst, and ways of lowering the rate of hydrogen oxidation relative to that of carbon monoxide are being eagerly sought.

The kinetics and mechanism of hydrogen oxidation to water by gold has been little examined, except in the context of PROX, but the following section focuses on what has been done. Its *selective* oxidation to hydrogen peroxide is treated in Chapter 8. The simultaneous oxidations are discussed in Section 7.3, but in view of the large and continually growing literature it is impossible to give a comprehensive coverage, so that emphasis is placed on the principles so far revealed, with illustrative examples, drawn especially from the more promising systems.

7.2. The Oxidation of Hydrogen to Water

The gold-catalysed oxidation of hydrogen was first studied by Bone and Wheeler as long ago as 1906;[8] using gold gauze in a recirculatory system at 523 K, they found the reaction to be first order in hydrogen and zero order in oxygen. This is the earliest publication concerning catalysis by gold. Some years later, Benton and Elgin[9] followed the reaction on solid gold in a dynamic system at 403 K, and found orders of respectively unity and 0.5. Both these studies therefore suggested that oxygen was more strongly chemisorbed than hydrogen, which was perhaps not adsorbed at all. More recently it was noted[10] with Au/SiO_2 that large particles obtained with loadings above 0.1% were active for water formation, while small ones at loadings below 0.05% were chiefly active for hydrogen–deuterium equilibration (Section 9.2). Very recently a lengthy and detailed paper has described the reaction on the same type of catalyst, as well as on gold supported on solids of the silicalite type.[11]

This paper,[11] which is a model of its kind, reported a study of the reaction on two Au/SiO_2 catalysts, having respectively 0.15 and 5% gold; unfortunately both had somewhat broad particle size distributions, namely 3–9 nm (0.15% Au) or 3–7 nm (5% Au), with a significant number of very large (>10 nm) particles. This complicated the interpretation of the results, as no clear particle size effect could be seen. However, silicalite-1 (Si-MFI) and TS-1 (a titanium-containing silicalite, Ti-MFI) were also used, and the size of the channels constrained the particle size to be less than 3 nm in both cases. These size differences accounted for the marked variations in activity observed at 433 K:

$$0.15\% \text{ Au/MFI supports} \gg 0.15\% \text{ Au/SiO}_2 > 5\% \text{ Au/SiO}_2.$$

Reaction kinetics were similar for all catalysts, orders in hydrogen being 0.69–0.82 and in oxygen 0.08–0.19; activation energies were remarkably consistent at $39 \pm 2 \text{ kJ mol}^{-1}$, showing that the cause of the activity differences lay entirely in the number of active sites. Added water did not affect the rate. Low coverages of gold by molecular oxygen were detected calorimetrically, but the heat released by addition of hydrogen after evacuation was due only to reaction with residual oxygen, giving hydroxy and hydroperoxy groups (see Chapter 5). A notable feature of this work is the extensive use that was made of DFT calculations to model the process and to assign energies to component steps, and hence to identify the rate-limiting step. So intimately are observations and calculations combined that distinguishing the two is not always easy.

Several mechanisms were considered and rejected as being inconsistent with the observed kinetics,[11] but finally a two-site mechanism was found acceptable, on one of which both reactants adsorb competitively and on another of which only hydrogen is dissociatively adsorbed. Reaction led to a hydroperoxy radical on the first type of site, and this was converted by hydrogen peroxide by molecular hydrogen. Its decomposition to water was considered to occur quickly and therefore not to be kinetically significant. An optimum particle size for the reaction was to be expected because large ones (Au_{55} or greater) would not chemisorb oxygen (Section 5.2), while for very small ones (Au or smaller) the formation of the peroxy intermediate was predicted by DFT calculation to be energetically unfavourable. Once again, as with so many DFT calculations,[12] the exact mode of bonding of adsorbed intermediates is irretrievably buried in the procedure, so that for example the manner of bonding the hydrogen peroxide molecule is not made clear. These results have also to be reconciled with the knowledge that similar catalysts are able to synthesise hydrogen peroxide in three-phase systems, or at least are unable to decompose it completely (Section 8.5).

Hydrogen oxidation in the absence of carbon monoxide is occasionally studied as part of a PROX programme. Au/Al_2O_3, Au/ZrO_2 and Au/TiO_2 were prepared by low-energy cluster beam deposition into powder supports[13]; gold loadings were low (0.02–0.08%) and particles were small (~ 3 nm). Temperatures for 50% conversion were respectively 394, 423 and 450, but the specific rates at 353 K were all very similar, and activation energies were about $30 \, \text{kJ} \, \text{mol}^{-1}$. The World Gold Council reference catalyst Au/TiO_2 (1.47% Au) showed a *specific* rate 10 times greater, perhaps because it had a more intimate contact between metal and support.

A multi-component catalyst prepared by coprecipitation of Fe_2O_3 and SnO_2, followed by deposition of $Pd(OH)_2$ onto the tin component, calcination and deposition of gold onto the iron component by appropriate selection of pH values, afforded a catalyst gave that 100% oxidation of hydrogen at 300 K.[14]

7.3. Selective Oxidation of Carbon Monoxide in Hydrogen

7.3.1. Tour d'horizon

Although the target conditions for this process are closely delimited, the system is in fact quite complex because of the greater variety of reactants and possible products. It may be helpful if before examining the literature we attempt to define those qualities that a successful catalyst ought to have.

1. There should not be independent adsorption sites for each reactant; they should adsorb competitively, carbon monoxide much more strongly than hydrogen.

2. The hydrogen and water should not inhibit the reaction of carbon monoxide; it would be beneficial if they promoted it, for example by limiting formation of carbonate ion.

3. The oxidation of hydrogen should not be assisted by either of the carbon oxides or other intermediate products (CO_3^{2-}, $HCOO^-$, etc.).

4. Carbon monoxide should not be re-formed by the water-gas shift. If all these conditions were met, high selectivity should be attained.

Let us first try to narrow down the types of support that have been found beneficial from the large number tried. Table 7.1 summarises the results obtained with a variety of supports at 353 K;[6,15,16] taking the target selectivity to be greater than 50% and the rate of carbon monoxide oxidation to exceed $30 \times 10^{-4}\,\mathrm{mol\,g_{Au}^{-1}\,s^{-1}}$, the compounds of interest are the oxides of iron, ceria and titania, and indeed we find much of the published work has been with them. Many of the listed oxides when used as supports for gold show acceptably high selectivities, but rates that are too low, so that the required high conversion of carbon monoxide would not be obtained (see Section 7.3.3). Silica does not seem to be in the running, and carbon has only been used under very limited conditions.[17] Despite the low activity reported for $Au/\gamma-Al_2O_3$ reported in Table 7.1, it has in fact been quite widely studied.[6,13,18–20] On the whole, however, perhaps not surprisingly, the supports that feature most often are those that perform well in the oxidation of carbon monoxide alone.

As noted above, the reaction mixture most often used contains only the amount of oxygen required to oxidise *either* the carbon monoxide *or* some of the hydrogen, or a modest excess. This, together with knowledge of the relative activation energies (Figure 7.1) helps to explain the temperature profile of conversions and selectivity, as shown in a typical case in Figure 7.2. The maximum in the carbon monoxide conversion arises because

Table 7.1: Supported gold catalysts for PROX: rate of CO oxidation and selectivity (S) to CO_2 at 353 K.[18]

Support	Method	Au (%)	d_{Au} (nm)	r^a	S (%)
Fe_2O_3	DP	0.94	2.3	61	64
MgO	COPPT	0.96	5.8	3.8	64
CeO_2	DP	2.2	2.2	45	58
TiO_2	IMP	1.8	2.4	33	48
Al_2O_3	IMP	1.7	4.4	6	59

[a]Rate in $\mathrm{mol_{CO}\,g_{Au}^{-1}\,s^{-1}} \times 10^4$. *Note*: All catalysts pretreated in air at 673 K.

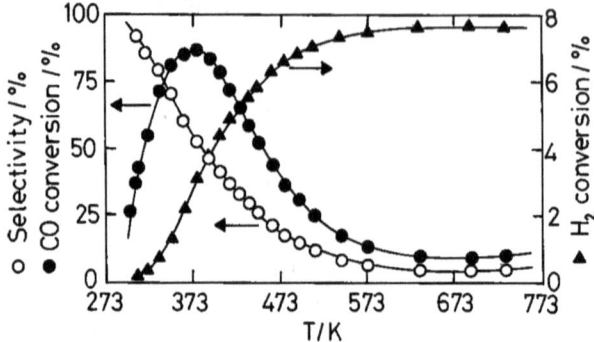

Figure 7.2: Temperature dependence of conversion of carbon monoxide and hydrogen, and selectivity, over the World Gold Council 1.4% Au/TiO$_2$ ($d_{Au} = 3.7 \pm 1.5$ nm); reaction mixture had [CO] = [O$_2$] = 2%; [H$_2$] = 4.8%; [He] = 48%.[13]

the hydrogen by reacting faster above the maximum consumes most of the limited amount of oxygen. With rising temperature the relative strengths of adsorption may also move in favour of hydrogen, and this may be why it reacts faster at higher temperatures; above about 473 K the reverse water-gas shift may also contribute to the increase in the carbon monoxide level:

$$CO_2 + H_2 \rightarrow CO + H_2O. \tag{7.7}$$

Exploration of the effects of adding the reaction products and determination of the kinetics also help in appreciating the factors favouring high selectivities, as well as hinting at the reaction mechanism. Information of this type is most comprehensively available for the Au/α–Fe$_2$O$_3$ catalyst, which we will consider first. Features of interest shown by other catalysts will be presented afterwards.

7.3.2. Catalysis by gold on ferric oxide

The selective oxidation of carbon monoxide in hydrogen on Au/α–Fe$_2$O$_3$ (haematite) catalysts shows characteristics that differ from those of the oxidation of carbon monoxide by itself (Chapter 6). The presence of hydrogen, or perhaps more importantly, of the water formed from it, decreases the rate of activity loss by inhibiting formation of carbonate ion, and causes the rate of carbon monoxide oxidation to increase a little.[7,18,21,22]

When carbon dioxide was added to the feed gas, the IR band for carbon monoxide moved to higher wave numbers (2110.5–$2114\,\text{cm}^{-1}$), and its intensity decreased; this implied decrease in the C–O stretching frequency may have been due to adjacently adsorbed molecules of the carbon dioxide acting as electron acceptors and hence lowering the electron density beneath the molecules of the monoxide. This effect was counteracted by adding water to the feed, which also increased selectivity since it tended to suppress hydrogen oxidation, especially at the lower end of the temperature range. Adding water could also reactivate catalysts poisoned by carbonate ion. Increasing the pressure of carbon dioxide lowered the selectivity, because it interfered with the oxidation of the monoxide more than that of hydrogen. Catalysts experienced slight deactivation of both reactions, the selectivity remaining constant at ∼65%. Orders of reaction (Table 7.2) indicated that the gold surface was far from fully covered by carbon monoxide at $353\,\text{K}$, the same conclusion having been reached in other work,[23] but it was higher at $313\,\text{K}$; increasing its pressure from 0.07 to $1\,\text{kPa}$ raised the selectivity from 38 to 60%. The activation energy for carbon monoxide oxidation in idealised reformate (Table 7.2) was hardly altered by adding carbon dioxide, although naturally the rate was lowered, but including water as well decreased it to $20\,\text{kJ}\,\text{mol}^{-1}$ since its beneficial effect lessened as temperature went up. Other points to note from these extensive and detailed studies of PROX[7,18,21,22] include the partial reduction of the support to haematite (Fe_3O_4) and formation of some carbonate ($FeCO_3$), and the strongly inhibiting effect of sodium on carbon monoxide oxidation at concentrations above 1%.[6] Bands in the 1200–$1800\,\text{cm}^{-1}$ region of the IR spectrum (see Section 5.3.4) increased in intensity with pressure of carbon dioxide, but the species responsible for them (e.g. bicarbonate) were not in the mainstream of the reactions.

Catalysts used in this work[7,18,21,22] were prepared either by deposition–precipitation or coprecipitation, and were pretreated in air at $673\,\text{K}$. Depending on the way in which the α-Fe_2O_3 support was made, the precursor before calcination may have been either ferrihydrite ('2-line' or '6-line'[6]) or γ-Fe_2O_3 (magehmite), the former giving the better results.[6]

Oxidations of the two molecules appear to proceed almost but perhaps not quite wholly independently of each other; the dissociative chemisorption of hydrogen is not much affected by the carbon monoxide, which, in view of what was said in Chapter 5, is somewhat surprising, as there it seemed that both interacted with lowly coordinated gold atoms. The dependence of kinetic parameters on the hydrogen concentration has not been

Table 7.2: Au/Fe_2O_3 and Au/TiO_2 catalysts for PROX: CO conversion C (%) and selectivity to CO_2 (S) $r \propto P_{CO}^x P_O^y$.

Support	Method	Au (%)	d_{Au} (nm)	BET area ($m^2 g^{-1}$)	x	y	E (kJ mol^{-1})	C (%)	S (%)	Reference
Fe_2O_3	DP	2.3–3.1	2.2–2.7	47–63	—	—	29 (CO)	—	~40	7
Fe_2O_3	DP	~1	2.3	63	—	—	—	—	64	18
Fe_2O_3	COPPT	3.1	6.5	54	0.55[a]	0.23	31[b]	—	—	21,22
Fe_2O_3	COPPT	5	6.7	26	—	—	—	99.8	51	5,26
TiO_2	DP	2.4	~3	—	0.82	0.36	~25	—	48	18,27
TiO_2[c]	DP	1.5	3.7	—	—	—	—	80	70	13

[a]0.33 at 313 K. [b]Decreases as P_{CO} decreases. [c]World Gold Council catalyst. *Note:* All catalysts pretreated in air at 673 K.

examined, because its concentration needs to be high to fulfil the PROX requirements. Unlike the mechanism of the oxidation of carbon monoxide by itself, which has been frequently discussed (Section 6.6), that of the twin reaction has not yet been formulated in a comprehensive way. By a stroke of good luck, the presence of the hydrogen actually helps, perhaps because its oxidation involves hydroxyl ions that occupy anionic sites in the surface of the support close to gold particles, which otherwise would be blocked by carbonate ions, and which have been proposed as the location for the adsorption of oxygen.[24] There has been no suggestion that the kind of mechanism proposed in Section 7.2 for the oxidation of hydrogen, proceeding *via* hydrogen peroxide, has any role to play in the PROX system.

Throughout all this work[6,7,18,21,22] on $Au/\alpha-Fe_2O_3$ catalysts, the removal of carbon monoxide is given as a *rate* and not as a fractional *conversion*. It is therefore not easy to decide how close this type of catalyst, calcined at 673 K, approaches to the target performance (Section 7.1), but it appears[5] that it is not wholly satisfactory. Other work with this catalyst drew near but did not reach the target[25] because an excess of oxygen over the stoichiometric ratio was needed to obtain optimum performance. It was therefore somewhat of a relief to find that an Au/Fe_2O_3 catalyst made by coprecipitation and calcined first at 673 K and then at 823 K did in fact meet the target,[5,26] but with little to spare (CO conversion 99.8%; selectivity 51%); the particle size was then 6.7 nm. The additional calcination was thought to eliminate gold cations that otherwise would have catalysed the water-gas shift; this implies that oxidation of carbon monoxide can conveniently proceed at about the desired temperature of 353 K without the assistance of cationic species.

Au/Fe_2O_3 has been formed within the pores of alumina pellets to facilitate their use in a prototype reactor.[6]

7.3.3. Gold on other supports

Au/TiO_2 catalysts made by deposition–precipitation have been examined for the PROX reaction;[27] they show some differences from Au/Fe_2O_3 catalysts. Hydrogen now interferes with the oxidation of carbon monoxide, perhaps by competing for the same adsorption sites; this was shown by a marked increase in the order of reaction when hydrogen was present, but not in the oxygen order (see Table 7.2). However, there is no reason to expect that gold particles on these two supports differ in any fundamental way, although subtle metal-support interactions may be present (see Section 3.5). Activation energies were similar for both reactions in the absence of the other reactant, but with the mixed reactants it decreased to about zero for carbon monoxide pressures below 0.1 kPa, the oxygen pressure remaining constant. Selectivity was therefore a steep function of carbon monoxide concentration, making it difficult to obtain high values at high conversion. Hydrogen caused changes to the IR spectrum of carbon monoxide, and after some time a new asymmetric peak having maximum intensity at 2032 cm^{-1} was formed; it may have been caused by an H–Au–CO complex, but it was not present in a *reacting* hydrogen + carbon monoxide atmosphere. As with Au/Fe_2O_3, hydrogen also suppressed the formation of inhibiting but unreactive species such as carbonate and formate.[27]

Au/TiO_2 formed by laser vaporisation of gold onto anatase suffered from low activity because the gold content was only 0.02%.[13,28] The World Gold Council Au/TiO_2 catalyst performed better, showing 70% selectivity at 80% conversion (353 K).

Au/CeO_2 catalysts have not so far performed well. Despite some early promise,[18] in later work with catalysts prepared by coprecipitation, including some containing lanthanum or gadolinium in the support, high conversions were only achieved at 385 K, although selectivities of $\sim$65% were then obtained.[29,30]

Au/Al_2O_3 catalysts tend to show low activity, and sometimes deactivate quickly[19,20] (but not always[18]); lost activity could, however, be recovered by treating the catalyst either with hydrogen or water.[4,20] This underlined the importance of hydroxyl groups in the oxidation of carbon monoxide, these perhaps being attached to gold cations at the edges of particles, as had been previously suggested[24] (see Section 6.6.2). More success has been obtained with Au/Al_2O_3 promoted by magnesia and manganese oxide; selectivities above 90% were recorded below 373 K, but high conversion of carbon monoxide (99%) was only found when oxygen was present in fourfold excess over the stoichiometric ratio.[31,32]

A number of other oxides (including zirconia[7,28]) have been used as supports for gold in the PROX reaction,[18] but none have shown any particular promise. Au/C and Au/TiO_2 (World Gold Council) dispersed in an aqueous solution of a polyoxometalate compound such as $H_3PMo_{12}O_{40}$ efficiently oxidised carbon monoxide admixed with hydrogen, showing a reasonable selectivity ($\sim$90%) at room temperature.[17]

7.3.4. Conclusion

Comparison of supported gold catalysts for the selective oxidation of carbon monoxide in hydrogen have a distinct advantage over platinum catalysts,[21] and this is mainly due to their much greater activity for the target reaction. This in turn is a consequence of the weaker chemisorption of the reactant, which ensures that it is more easily disengaged from its site, a conclusion that has been supported by DFT calculations.[15] The best gold catalyst (Au/α–Fe_2O_3 suitably prepared) meets the practical requirements for use in association with hydrogen-based fuel cells.[4] Application of gold catalysts to the larger-scale purification of hydrogen for the chemical industry will require much further development.

References

1. G.C. Bond, *Heterogeneous Catalysis: Principles and Applications*, 2nd edition Clarendon Press, Oxford, 1987.
2. A.J. Farrauto and C.H. Bartholomew, *Fundamentals of Industrial Catalytic Processes*, Chapman and Hall, London, 1997.
3. D.L. Trimm, *Appl. Catal. A: Gen.* **296** (2005) 1.
4. D. Cameron, R. Holliday and D. Thompson, *J. Power Sources* **118** (2003) 298.
5. P. Landon, J. Ferguson, B.E. Solsona, S. Garcia, S. Al-Sayari, A.F. Carley, A.A. Herzing, C.J. Kiely, M. Makkee, J.A. Moulijn, A. Overweg, S.E. Golunski and G.J. Hutchings, *J. Mater. Chem.* **15** (2005) 1.
6. V. Plzak, J. Garche and R.J. Behm, *Eur. Fuel Cell News* **10**(2) (2003) 8.
7. M.M. Schubert, A. Venugopal, M.J. Kahlich, V. Plzak and R.J. Behm, *J. Catal.* **222** (2004) 32.
8. A.W. Bone and R.V. Wheeler, *Phil. Trans.* **206A** (1906) 1.
9. A.F. Benton and J.C. Elgin, *J. Am. Chem. Soc.* **49** (1927) 2426.
10. S. Naito and M. Tanimoto, *J. Chem. Soc. Chem. Commun.* (1988) 832.
11. D.G. Barton and S.G. Podkolzin, *J. Phys. Chem. B* **109** (2005) 2262.
12. D.H. Wells Jr., W.N. Delgass and K.T. Thompson, *J. Catal.* **225** (2004) 69.
13. C. Rossignol, S. Arrii, F. Morfin, L. Piccolo, V. Caps and J.-L. Rousset, *J. Catal.* **230** (2005) 476.
14. M. Okumura, T. Akita, M. Haruta, X. Wang, O. Kajikawa and O. Okada, *Appl. Catal. B: Env.* **41** (2003) 43.
15. S. Kandoi, A.A. Gokhale, L.C. Grabow, J.A. Dumesic and M. Mavrikakis, *Catal. Lett.* **93** (2004) 93.
16. M.M. Schubert, S. Hackenberg, A.C. van Veen, M. Mihler, V. Plzak and R.J. Behm, *J. Catal.* **197** (2001) 113.
17. Won Bae Kim, T. Voiti, G.J. Rodriguez-Rivera and J.A. Dumesic, *Science* **305** (2004) 1280.
18. M.M. Schubert, V. Plzak, J. Garche and R.J. Behm, *Catal. Lett.* **76** (2001) 143.
19. S.S. Pansare, A. Sirijaruphan and J.G. Goodwin Jr., *J. Catal.* **234** (2005) 151.
20. C.K. Costello, J.H. Yang, H.Y. Law, J.-N. Lin, L.D. Marks, M.C. Kung and H.H. Kung, *Appl. Catal. A: Gen.* **243** (2003) 15.

21. M.M. Schubert, M.J. Kahlich, H.A. Gasteiger and R.J. Behm, *J. Power Sources* **84** (1999) 175.

22. M.J. Kahlich, H.A. Gasteiger and R.J. Behm, *J. Catal.* **182** (1999) 430.

23. O. Meerson, G. Sitja and C.R. Henry, *Eur. J. Phys. D* **34** (2005) 119.

24. G.C. Bond and D.T. Thompson, *Gold Bull.* **32** (2000) 41.

25. G. Avgouropoulos, T. Ioannides, Ch. Papadopoulou, J. Batiste, S. Hocevar and H.K. Matralis, *Catal. Today* **75** (2002) 157.

26. P. Landon, J. Ferguson, B.E. Solsona, T. Garcia, A.F. Carley, A.A. Herzing, C.J. Kiely, S.E. Golunski and G.J. Hutchings, *Chem. Commun.* (2005) 3385.

27. B. Schumacher, Y. Denkwitz, V. Plzak, M. Kinne and R.J. Behm, *J. Catal.* **224** (2004) 449.

28. M. Lomello-Tafin, A.A. Chaou, F. Morfin, V. Caps and L. Rousset, *Chem. Commun.* (2005) 388.

29. A. Lucngnarucmitchai, S. Osuwan and E. Gulari, *Internat. J. Hydrogen Energy* **29** (2004) 429.

30. W.-L. Deng, J. de Jesus, H. Saltsburg and M. Flytzani-Stephanopoulos, *Appl. Catal. A: Gen.* **291** (2005) 126.

31. R. Grisel, K.-J. Weststrate, A. Gluhoi and B.E. Nieuwenhuys, *Gold Bull.* **35** (2002) 39.

32. R.J.H. Grisel, C.J. Weststrate, A. Goossens, M.W.J. Coajé, A.M. van der Kraan and B.E. Nieuwenhuys, *Catal. Today* **72** (2002) 123.

CHAPTER 8

Selective Oxidation

8.1. Introduction

8.1.1. A survey of selective oxidation

The selective oxidation of organic molecules is one of the most important processes used in the chemical industry. Its reactions fall into two broad categories: (i) gas-phase oxidation of hydrocarbons (alkanes, alkenes) to oxygenated products and (ii) further oxidation of molecules containing one or more oxygen atoms (mainly in the liquid phase).

The first set of reactions is the mainstay of the petrochemical industry;[1] outstanding examples are the oxidation of propene to propenal (acrolein) catalysed by bismuth molybdate, and of ethene to oxirane (ethylene oxide) catalysed by silver. In general these processes work at high but not perfect selectivity, the catalysts having been fine-tuned by inclusion of promoters to secure optimum performance. An especially important reaction is the oxidation of ethene in the presence of acetic (ethanoic) acid to form vinyl acetate (ethenyl ethanoate) catalysed by supported palladium–gold catalysts; this is treated in Section 8.4. Oxidation reactions are very exothermic, and special precautions have to be taken to avoid the catalyst over-heating.

The second set of reactions is more related to the fine chemicals and pharmaceutical industries, although some of them are carried out industrially on a very significant scale. Temperature-control in three-phase systems is easier, and is rarely a problem, but adequate mixing of the phases is essential to avoid mass-transport limitation. Selectivity here is more directed towards securing the desired product, which may be one of several closely related ones.

Turning to current work employing gold catalysts, the reaction that has attracted most attention is the oxidation of propene to methyloxirane (propene oxide or 1,2-epoxypropane).[2] While silver catalysts work very well for epoxidising ethene, they fail utterly with propene, because the selective intermediate is more reactive than the reactant, and carbon dioxide alone

217

is formed. The situation with gold catalysts is somewhat more promising (Section 8.2.1).

The most dramatic results obtained so far with gold catalysts have been with the liquid phase processes. They are conducted with oxygen or air, often using water as solvent, and are therefore felt to be environmentally benign. Particular success has been obtained with reducing sugars (Section 8.3.2) and other aldehydes (Section 8.3.3), and with alcohols and other hydroxy-compounds (Sections 8.3.4–8.3.7). Reactions that use soluble gold complexes to catalyse selective oxidation are reported in Chapter 12.

Selective oxidation in inorganic chemistry is unusual, but it seems appropriate to consider the synthesis of hydrogen peroxide as the selective oxidation of hydrogen (Section 8.5). The more usual (non-selective) process yielding water was considered in Chapter 7.

8.1.2. Reaction mechanisms and the use of bimetallic catalysts

While the mechanisms of certain gas-phase oxidations are understood, sometimes quite completely, those of reactions proceeding in the liquid phase have been less thoroughly investigated, perhaps because they are used by organic chemists more interested in what is made than how it is made. Proper kinetic measurements are also harder to obtain. A particular feature of liquid-phase oxidations is that they often seem to go better with a bimetallic catalyst,[2] the combination of gold with palladium being especially effective, as with the synthesis of vinyl acetate and of hydrogen peroxide, and the oxidation of aldehydes and diols. There appears to be little understanding of why bimetallic catalysts are preferable and even the physical structure of the catalyst is often unknown. The following notes may serve as a background to the presentation in the following sections of work that does not focus on reaction mechanism.

Reactions of organic molecules sometimes follow a complex path through various intermediates. Where a reaction can be broken down into two or more steps, each may prefer to proceed on a different type of atom or on an active centre of a size and composition that uniquely fits it. Migration from one site or atom to another could proceed by surface migration or might require desorption and movement through the fluid phase (Scheme 8.1A). If this can happen, there is no necessity for both elements to occupy the same particle. Where carbon monoxide is a product of the reaction, as with

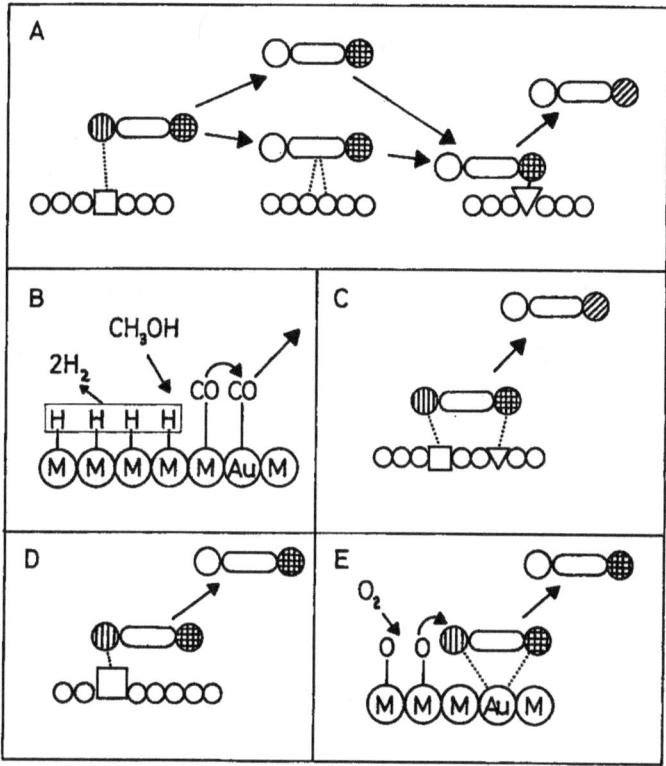

Scheme 8.1: Modes of action of bimetallic catalysts in selective oxidations: in **A**, **C** and **D** the squares and triangles represent atoms of either gold or the Group 10 metal involved in the reaction in a matrix of non-participating atoms (small circles); see text for explanation.

methanol decomposition (Section 9.6.2), the presence of gold provides an escape route for molecules formed on a more active component, thereby liberating it from their otherwise toxic effect (Scheme 8.1B). If an organic molecule possesses two or more functional groups that may react catalytically at the same time, accelerated rate may result from each function interacting with a different beneficial site (Scheme 8.1C); this requires both metals to be in the same particle. Equally, dilution of the active Group 10 metal in the surface by an inactive Group 11 metal, resulting in the formation of small ensembles or even single atoms of the active metal, may permit selective operation on one of the functions while the other remains unaffected (Scheme 8.1D). When oxidation of an organic molecule proceeds faster on a bimetallic catalyst, this may be because the oxygen molecule is dissociated in one component, while the organic molecule is held at the other component (Scheme 8.1E).

It is generally thought that the oxygen molecule has to dissociate into atoms before reacting, although this is probably not the case with hydrogen peroxide synthesis. In view of the known reluctance of gold to manage this, its efficacy in selective oxidations has come as somewhat of a surprise. We have reviewed the evidence concerning the interaction of oxygen with gold surfaces in Section 5.2; it appeared that dissociative chemisorption was indeed possible on very small particles, and this may be how selective oxidations proceed, but there is evidence in some cases that larger particles work better, in which case some other mode of activation of the molecule must apply. It may be possible for dissociation to happen by the molecule colliding with the adsorbed state of the other reactant in a kind of Rideal-Eley mechanism (Section 1.4); in this case the rate would be directly proportional to the oxygen pressure.

8.2. Selective Oxidation of Hydrocarbons

8.2.1. Oxidation of propene

Early work on attempts to achieve selective oxidation of propene (and ethene) on gold has been reviewed.[3] Gold film proved to be totally non-selective,[4] but Au/SiO_2 gave mainly acetone (3%) and propenal (50%);[5] reaction kinetics were determined. Lower selectivities were found with Au/Al_2O_3 and gold sponge. This mode of reaction, in which the methyl group is the locus of reactivity, therefore resembles that shown by binary oxide combinations (Bi–Mo, Sb–Sn, etc.); loss of a hydrogen atom from the methyl group gives a symmetrical π-allyl radical, as shown by isotopic tracer studies. Acquisition of an oxygen atom then leads to the unsaturated aldehyde.[5] Use of gold catalysts for this type of reaction has only recently been followed up,[6] and attention has turned instead to the epoxidation of propene to give methyloxirane (propene oxide).

Methyloxirane is an important building block for the manufacture of polyurethane, of various organic intermediates and solvents[7–13] (Section 14.3.6). It is currently made in one of two ways: (i) the chlorohydrin process and (ii) the hydroperoxide process.[7–10,12] The former is not environmentally friendly, due to the formation of calcium chloride and hazardous chlorinated organic compounds as by-products. The latter route has a demand–supply gap for its co-products. There is therefore growing interest in the development of new routes for the direct gas-phase synthesis of methyloxirane, analogous to that for the silver-catalysed epoxidation of ethene.[7,8,10,14] Attempts to use silver catalysts promoted by organic or inorganic chlorides or potassium, or by palladium or palladium + platinum supported by a titanosilicate such as TS-1, have met with only limited success. Yields have been low, with numerous by-products, and the complex structure of the catalysts has prevented their use. An industrially acceptable catalyst and process is therefore still awaited.

The first steps in this direction were taken by the observation[7,15] that, when the epoxidation of propene was performed over Au/TiO_2 in the presence of hydrogen, the selectivity to methyloxirane was very high (~99%), but only at very low conversion (~1%). Typical reaction conditions were[16] 323 K with a feed of $H_2:O_2:C_3H_6:Ar$ = 1:1:1:7 and a flow-rate of $2000\,cm^3\,min^{-1}$. A number of papers have appeared subsequently in which Au/TiO_2 has been used,[7,17–20] and high selectivities to methyloxirane have been routinely found, but the support adsorbs and further oxidises the desired product, so that yields were invariably low. Subsequent work has attempted to overcome these limitations in a variety of ways, especially by modification of the support.

Industrial use of gold catalysts could become possible if high selectivity could be maintained to a conversion of 10%, but a number of problems would still remain. Selectivity decreases at higher temperatures and catalysts deactivate with time-on-stream;[1] the hydrogen efficiency is also low (<30%). Conversion of propene has been raised to 5–10% by using higher temperatures, without too much loss of selectivity:[8,14] with Au/TS-1 (a titanium-substituted silicalite) a selectivity of 80% has been obtained at 473 K,[21] although in other work better results were obtained at 413 K.[22] Only small improvements in deactivation rates have so far been reported; Au/TiO_2–SiO_2 catalysts lost activity less quickly than Au/TiO_2,[2] perhaps because the product adheres less strongly to the support.

As with so many other gold-catalysed reactions, the particle size is important; it has been claimed[9,20] that only hemispherical particles between 2 and 10 nm in size are suitable and early work recommended 2.3 nm particles as the best.[16] The deposition–precipitation (DP) method is favoured because of the intimate contact between metal and support that it gives. Particles smaller than 2 nm appear to shift the reaction mode towards formation of propane,[7,9,10,15] which is clearly a waste of both propene and hydrogen. This may be a result of the particles having lost their metallic character,[10] or of easier chemisorption of hydrogen (see Chapters 3 and 5). On the other hand, specific activity for methyloxirane formation (i.e. the rate per unit weight of gold) has been shown to increase as gold loading was lowered, catalysts having only 0.05 or even 0.01% gold giving high productivities at 473 K ($350\,g\,h^{-1}\,g_{Au}^{-1}$).[22] This was obtained with a sample of TS-1 having a very high Si/Ti ratio (500), and it was noted that using the DP method the gold loading obtained correlated with the titanium content, suggesting that gold is preferentially attached to a titanium site.

Hydrogen can of course also be lost if it is oxidised to water instead of the hydrogen peroxide that is needed for epoxidation. Its consumption can be lowered by 90% and selectivity raised to 97% at 1.7% conversion by admixing Au/Ti-MCM-41 with caesium chloride,[23] but it is unclear how it acts. Water has however been found to decrease the rate of deactivation of Au/TiO$_2$.[24,25]

Proper choice of support is critical. In a wide-ranging study, it was shown that using oxides of main-group elements (Mg, Si, Sn) only aldehydes were formed, while oxides of Transition Metals (Fe, Ce) gave only carbon dioxide; titania was the only support to give epoxidation,[26] and it has been present as a component in all other work. The crystalline and pore structure of titania and of TiO$_2$–SiO$_2$,[7,10,12,18,19,27] where the titanium is dispersed as isolated species in a silica matrix, greatly affect selectivity to methyloxirane. Only the anatase form of titania gives high selectivity; rutile and the amorphous form do not.[12,28] The activity of Au/Ti-MCM-48 is higher than that of Au/Ti-MCM-41, indicating that three-dimensional branched pore structure is preferable.[8] The wide pores of supports of this type (TS-1, TS-2, Ti-β, Ti-MCM-41 and -48)[1,2,8,9,12,14,23,28–32] encourage the egress of the product from the catalyst, thus minimising its further reaction. The accumulation of product on the catalyst surface is thought to be the main factor in limiting selectivity and in causing deactivation.[8,18,19,33] FTIR spectroscopy has shown[18] that methyloxirane adsorbs irreversibly as propoxy groups on acidic Ti–OH groups on titania, and on both Ti–OH and Si–OH groups on TiO$_2$–SiO$_2$, especially when water is present. Modification of the hydrophilic character of support surfaces may therefore adjust the adsorption and desorption of reactants and products.[34] Thus silylation of Au/Ti-MCM-41 improved the deactivation rate and increased selectivity and hydrogen utilisation.

While these considerations are of great importance, it is first necessary to form the methyloxirane. Investigation of a number of titanium-containing

Table 8.1: Selected results for the selective oxidation of propene on titania- and silica-titania catalysts.

Support	Au (%)	SV[a]	T (K)	Conv. (%)	S^b	Reference
TiO$_2$	1.0	~4000	323	<1	99	16
TiO$_2$/SiO$_2$ (46)[c]	0.3	4000	423	5	91	36
TS-1 (48)[c]	0.52	7000	443	6.2	71	31
TS-1 (36)[c]	0.05	7000	473	9	81	22

[a]Space velocity in cm^3 h^{-1} g$_{cat}^{-1}$. [b]Selectivity with respect to propene. [c]Si/Ti ratio.

supports has led to a correlation between distance between titanium species and catalytic performance. The best was Au/TS-1, which gave ~95% selectivity at 473 K, although at a conversion of only 1.4%. Too high a concentration of titanium in the support led to unwanted side-reactions and further oxidation. Table 8.1 gives a small selection of the results achieved by using gold supported on various titania-containing supports. The industrial significance of this reaction is evident from the high level of patent activity shown by major companies such as Dow, which claim the use of Au/TiO$_2$ and gold on various TiO$_2$–SiO$_2$ supports, although high selectivities were only reported at conversions below 0.5%.[35] Silver and other metals have also been named as promoters.

There has as yet been no clear statement of reaction mechanism or any determination of kinetics, but certain broad features are clearly apparent.[36] Well-dispersed (but not atomically dispersed) titanium cations in silica are very effective for the epoxidation of propene by hydrogen peroxide.[28] Since this type of material cannot synthesise hydrogen peroxide, the principal role of the gold must be to make it, and of course this is well known to occur (Section 8.5). A possible mechanism therefore entails the migration of hydrogen peroxide or a peroxy radical from the gold to a suitable titanium site[37] where it reacts with a propene molecule that may never have been chemisorbed by the gold. In support of this view, inelastic neutron scattering has been used to study hydrogen peroxide formation from hydrogen and oxygen under conditions relevant to propene epoxidation, and has found evidence for hydroperoxy species at 523 K.[38] DFT calculations have also been made in this reaction with a neutral Au$_3$ cluster;[39] the results suggested that particles smaller than 1 nm could be important in epoxidation, although this is unlikely in view of their tendency to form propane.

Calculations have also been made on the binding of propene to small gold particles (Section 5.5.2); using neutral and charged particles it was inferred that chemisorption would be stronger if the support took electron density *from* the particle, becoming somewhat positive; it could then accept charge from the propene in order to bind it.[40] It was also noted that adsorption of oxygen involves donation of charge *from* the particle *to* the molecule (Section 5.2.2), so that the reaction might constitute another example of collaborative chemisorption. Such arguments do not however provide a role for the support in binding hydroperoxy species, as suggested above. The mechanism of propene epoxidation will no doubt receive further attention. In view of the success of the gold + palladium combination in synthesising hydrogen peroxide (Section 8.5), it is strange that it has not yet been tried for the selective oxidation of propene.

Some years ago a novel means for activating the surface of gold was developed in the Battelle Geneva Research Centre.[41] It involved producing an electric current carrying electrons with energies above the Fermi level ('hot' electrons) by means other than temperature, namely, by a thin-layer device well known in solid state electronics. A sandwich was formed comprising a supporting base metal covered by 5–150 nm of alumina, which was in turn coated by a 10 nm layer of gold. When a small voltage is applied across the device, the Fermi level E_F of the positively biased gold is lowered, and electrons flow through the insulator by direct tunnelling; they lose energy in doing so, but if the gold layer is not too thick some will reach the gold surface with energies greater than E_F. These 'hot' electrons lose energy as they fall back to E_F and the excess energy can be used to activate adsorbed species. At a current density greater than $1\,A\,cm^2$, the rate of propene oxidation increased markedly, giving methanoic acid, methyloxirane and carbon dioxide. No further interest has been shown in this interesting way of stimulating a catalytic reaction, but it is important because the unsupported metal gave some of the desired product by direct oxidation (i.e. without hydrogen being present).

8.2.2. Oxidation of other alkenes

It has recently been found[42] that Au/C and Au/graphite catalysts are effective for the aerobic oxidation of other alkenes at 353 K; small amounts of either hydrogen peroxide or *tert*-butylhydroperoxide (^tBuOOH) are needed as oxygen chain initiators. The products of the reactions of cyclohexene,

Z-cyclo-octene, *Z*-stilbene (*Z*-diphenylethene) and styrene (phenylethene) were very dependent on the type of solvent used; thus with toluene as solvent, cyclohexene gave no epoxide, but only cyclohexen-3-one and cyclohexen-3-ol, while with 1,2,3,5-tetramethylbenzene an Au/C catalyst gave 50% of the epoxide and 26% of the ketone at 30% conversion. Modification of the catalyst with bismuth, using the nitrate, gave higher total selectivity (98%) due to an increase in the amount of the ketone formed. *Z*-Cyclo-octene without a solvent gave 81% of the epoxide at 8% conversion, while *Z*-stilbene gave mainly or only the *E*-epoxide. With styrene, toluene was a satisfactory solvent, but the main product was benzaldehyde; this plus the epoxide (23%) and phenylmethylketone (12%) accounted for 7% of the styrene converted. Clearly much thought is needed to understand the basis of the solvent effect.

Some interest has been shown in the epoxidation of styrene using ^{t}BuOOH as oxidant. Using gold on several types of mesoporous alumina, selectivities of ~70% were obtained at 355 K and moderate to high conversions; benzaldehyde was the other major product.[43] Best results were obtained at low gold loading (1%), but with Au/TiO$_2$ made by DP selectivities of ~50% were found at the same temperature, independent of gold content.[44] Other supports have been explored, including the oxides of gallium, indium and thalium, and selectivities of ~60% reported for Au/Tl$_2$O$_3$;[45] other oxide supports including cupric oxide have also been examined.[44,46]

Epoxidation of styrene with ^{t}BuOOH catalysed by Au/SiO$_2$ in the presence of zinc and tetrabutylammonium bromides and carbon dioxide led directly to styrene carbonate; at 1 MPa pressure and 353 K, conversion was ~90% and selectivity 35%.[47] Gold on functionalised polymer also catalyses the reaction of carbon dioxide with various epoxides to give lactones, and with amines to give carbonates.[48]

1-Butene has given some 1,3-butadiene as well as 2-butene isomers by oxidative dehydrogenation on several supported gold catalysts.[49] Gold supported on V$_2$O$_5$–SiO$_2$ and on MoO$_3$–SiO$_2$ catalysed the oxidation of propene to propenal and acetone above 473 K.[6]

8.2.3. Oxidation of cyclohexane

Cyclohexanone and cyclohexanol formed by aerobic oxidation of cyclohexane are essential intermediates needed to make *eta*-caprolactam and adipic acid for the manufacture respectively of Nylon-6 and Nylon-6,6, and the global production exceeds 10^6 tonnes per year (Section 14.3.3). The process operates at $\sim$430 K with a cobalt naphthenate catalyst, giving selectivities of 70–85% but only at 4% conversion. Attempts to find better ways of making it have been briefly reviewed,[50] but a satisfactory direct oxidation of cyclohexane has yet to be found. Several studies have been reported on the use of gold catalysts.[42,50–53] With gold on graphite, reaction without solvent but with a halogenated benzene additive gave quite good selectivities to cyclohexanol + cyclohexanone (maximum 92%), the amounts of the two products being comparable;[50] conversions were however limited to about 1%. Similar results were found with graphite-supported platinum and palladium. Very good total selectivities ($\sim$95%) were obtained at 423 K with gold on mesoporous silica catalysts, conversions being 20–30%.[52] Using Au/ZSM-5 catalysts at 423 K, good total selectivities were found at conversions of $\sim$10–15%.[51,53] Assuming that cyclohexanol can be dehydrogenated to the ketone, the use of mesoporous supports to allow easy diffusion of the products away from the active centres would seem to offer promise for large-scale development.

8.2.4. Oxidation of alkanes

This will be a very short section, because attempts to obtain selectively oxidised products from alkanes have, with two exceptions, been unsuccessful. On Au/La$_2$O$_3$–CaO above 970 K methane gave significant amounts of C$_2$ hydrocarbons, and with hydrogen peroxide present some benzene was formed.[54] Gold actually suppressed the activity of magnesia for this reaction.[55] In what must be the earliest application of Au/TiO$_2$ as a catalyst, an American patent[56] claimed the oxidative dehydrogenation of ethylbenzene at 973 K to give 94% selectivity to styrene at 53% conversion. At this temperature the gold must have been heavily sintered if in the metallic state.

8.3. Selective Oxidation of Oxygen-Containing Molecules

8.3.1. Introduction

The selective oxidation of multifunctional organic molecules to give products of greater value is an important target in the fine chemicals industry.

As an example we may cite the oxidation of the reducing sugar D-glucose, obtained by hydrolysis of sucrose, starch and cellulose, to D-gluconic acid or its salts, which find use as water-soluble cleansing agents and as additives to food and beverages. They are manufactured on a scale of some 6×10^4 tonnes per year by fermentation, despite problems with separation of the ferment, with control of by-products and disposal of waste water.[57-59] The development of an aerobic inorganic catalyst for this and related processes would therefore be of great interest.

One of the most surprising and potentially important aspects of the recent advances in catalysis by gold has been the discovery that many reactions of this type can be successfully accomplished under mild conditions.[60-62] It is clearly a requirement that only one of the functions of the molecule should be attacked; in the case of the reducing sugars and related alcohols (e.g. D-sorbitol), the aldehyde group or one of the terminal hydroxyl groups can be selectively oxidised by gold catalysts:

$$R-CH_2OH \rightarrow R-CHO \rightarrow R-COOH. \qquad (8.1)$$

Similar success has been obtained with smaller diols (e.g. 1,2-dihydroxyethane) and triols such as glycerol (1,2,3-trihydroxypropane) (Sections 8.3.5 and 8.3.6). However, the form of catalyst giving best results is unlike those so successful for oxidation of carbon monoxide and hydrogen; gold supported on activated carbon works well,[57-59,63-65] and even unsupported gold in colloidal form is effective.[59,66] Thus many of the questions posed by the gaseous oxidation of simple molecules are absent when we consider these other reactions; new questions do however arise. A further new feature is the finding that bimetallic catalysts (e.g. Au–Pt, Au–Pd) are even more effective than pure gold, although colloids of pure noble metals other than gold are not effective.[59,66]

8.3.2. Oxidation of reducing sugars

The reducing sugars D-glucose, D-lactose and D-maltose have been oxidised by molecular oxygen to the corresponding monocarboxylic acids under mild conditions (303–363 K) at slightly alkaline pH with near 100% selectivity; these acids find application in the detergent, food and pharmaceutical industries (Section 14.3.5). Although Au/TiO$_2$ has been used,[67] the preferred support is activated carbon,[68] of which a number of different types have been tried.[69-71] Gold is frequently introduced as the colloid:[72] in a typical procedure,[61,69] a dilute HAuCl$_4$ solution was reduced by NaBH$_4$ in the presence of PVA stabiliser, and after formation of the colloid the carbon was added. The mean particle size was typically ~12 nm.[71] Gold uptake was complete in 2 h[61] or 1–3 days.[69] The preparation of Au/C catalyst has been carried out on the 500 g scale, with results that were quite comparable to those obtained on a smaller scale;[70] particle sizes were 3–4 nm. Similar methods have been used to make bimetallic catalysts by the colloidal route.[61] Although platinum catalysts have been widely investigated, gold catalysts are much superior, particularly with respect to resistance to poisoning,[73] either by reaction products or other adventitious species. They are also preferable to homogeneous catalysts based on salts of copper, iron, cobalt or manganese, for which their separation from the product causes difficulty. The use of water as solvent has obvious environmental advantages; reactions proceed in it by an ionic pathway, whereas in organic solvents a radical route is favoured. In the oxidation of glucose, products other than gluconic acid are observed, particularly at temperatures above 323 K. These include fructose, formed by base-catalysed epimerisation of glucose, and mannose formed by further epimerisation.[69]

Two studies of the kinetics of glucose oxidation have been reported,[69,74] leading to different conclusions regarding the mechanism. One study[69] utilised catalysts made by the colloidal method, the gold particles (mean sizes 3.3–6.1 nm) being deposited onto two carbons differing in surface area. There was a maximum rate at 323 K (pH 9.5), and the rate increased logarithmically with gold surface area, the selectivity being unaffected. The rate also increased with catalyst weight to a power between zero and unity, and it and selectivity were both roughly independent of glucose concentration. These results were fitted to a Langmuir–Hinshelwood model, but the rate dependence on oxygen pressure was not determined. The other study[74] was performed with colloidal gold of 3.5 nm in size. The rate was proportional to the oxygen pressure, and increased with glucose concentration but showing no maximum; the activation energy was 47 kJ mol^{-1}. This kinetic form was compared with that shown by enzymatic reaction, and it was concluded that the gold catalyst was less active by a factor of 55, based on the mass of gold used; however, based on the probable number of active centres that can function simultaneously on a gold particle, the difference becomes much smaller, and almost zero within experimental error. This second study[74] favoured a 'Rideal-Eley' mechanism (Section 1.4) in which an oxygen molecule reacts directly with chemisorbed glucose: this mechanism allows the formation of hydrogen peroxide in parallel with gluconic acid, and this has been observed. The critical parameter that distinguishes the two mechanisms is the dependence of rate on glucose concentration. It has to be said that the amount of relevant experimental evidence is hardly sufficient to support the differentiation.

8.3.3. Oxidation of other aldehydes

The oxidation of glyoxal (ethane-di-al, OHC–CHO) provides a striking example of the selectivity that can be achieved with gold-containing catalysts, and the benefits that arise from using bimetallic palladium–gold catalysts. High selectivity to glyoxalic acid (OHC–COOH) by oxidation of one of the aldehyde groups was obtained in aqueous medium at 311 K,[75,76] but reasonable conversions were only found when using Au–Pd/C catalysts (Au:Pd = 1:1). Selectivity fell with increasing conversion as sequential oxidation to oxalic acid (HOOC–COOH) took place, but was still ∼60% at 37% conversion. High conversions of n-butanal to n-butanoic acid have been noted when gold in gel-type functional resins were used at 343 K.[77]

The importance of the support in the gold-catalysed liquid-phase oxidation of n-heptanal at 323 K has been emphasised by use of variously prepared ceria as supports.[78] Very small ceria particles, in samples described as either nano- or meso-structured,[79] make much better supports in terms both of activity and selectivity, than one prepared conventionally. The most active Au/CeO$_2$ gave 100% conversion in about 3 h, and 95% selectivity; aromatic aldehydes were also oxidised successfully.

8.3.4. Oxidation of a C$_6$ polyhydric alcohol: sorbitol

The oxidation of one of the two terminal –CH$_2$OH groups of D-*sorbitol* is an alternative route to D-gluconic acid (Scheme 8.2), but the reaction is complicated by the base-catalysed inversion of configuration about the two middle carbon atoms, giving D-gulonic acid; oxidation of both –CH$_2$OH groups affords D-glucaric acid (Scheme 8.2). Au/C, Au–Pd/C (Au:Pd = 1:8 to 8:1) and Au–Pt/C (Au:Pt = 1:1) catalysts have been investigated for this reaction,[61,80] using oxygen either at atmospheric pressure (pH 11) or at 3 atm pressure and a sorbitol/NaOH ratio of unity; the temperature was 323 K in both cases. The bimetallic catalysts performed best under both sets of conditions, giving selectivities to the monocarboxylic acids of ∼60–70%, which were almost independent of the gold: palladium ratio in the range investigated. These catalysts did not suffer from the deactivation that the monometallic catalysts experienced. It is

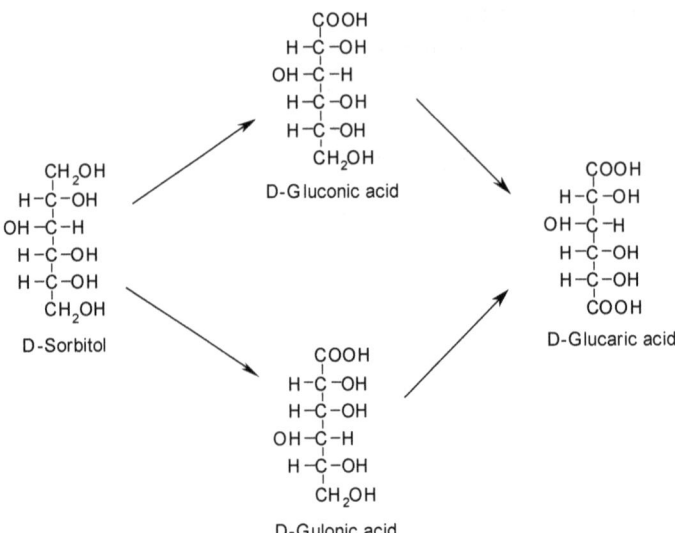

Scheme 8.2: The oxidation of sorbitol to gluconic, gulonic and glucaric acids.

unclear to a mere physical chemist why the secondary hydroxyl groups do not react.

8.3.5. Oxidation of diols

In line with the success obtained in oxidising selectively one of the terminal hydroxy-groups of sorbitol, it has been found that vicinal diols such as ethane-1,2-diol (ethylene glycol) and propane-1,2-diol can be oxidised with high selectivity by gold catalysts to the corresponding monocarboxylic acids under similar conditions (e.g. 343 K, high pH); the products are respectively glycolic acid ($HOCH_2$–$COOH$) and lactic acid (CH_3–$CH(OH)$–$COOH$).[81–84] Currently employed methods entail the use of toxic or corrosive reagents, as well as the use of high pressure; the alternative fermentation process for making lactic acid is subject to low productivity and severe problems of purification. In a number of publications,[58,60,70,84–90] Michele Rossi and his colleagues have shown that small gold particles are effective for these oxidations, being more active, selective and poison-resistant than other noble metal catalysts. Au/C catalysts can be recycled and used many times with only minimal loss of activity or of gold by leaching. With particle sizes between 7 and 12 nm,[87] selectivities to glycolate and lactate have exceeded 98%.[85-87]

Catalysts for these oxidations have been prepared by immobilising colloidal gold, suitably stabilised,[84,91] not only onto activated carbon,[86,87,90] but also onto alumina,[86] silica[86] and titania.[89] With Au/Al_2O_3, smaller particles were more active than larger ones,[92] but with Au/C the opposite trend, leading to a maximum rate of 7.5 nm,[86,88] was explained by supposing smaller particles became less available to reactants because they found their way into internal narrow micropores, whereas larger particles, being constrained to stay near the external surface, were more accessible. However, 3 nm gold particles could be fixed onto the outside of the carbon particles by using a highly polar stabiliser, giving much higher turnover frequencies.[84] Catalysts made this way were more active than those made by conventional methods;[91] they have been successfully prepared on a quite large scale (500 g).[70] With Au/Al_2O_3, selectivity for ethane-1,2-diol to glycolate (>98%) was independent of particle size.

The Nippon Shokubai Company has announced[93] the construction of a 50 tonnes per year plant to oxidise ethane-1,2-diol in the presence of methanol at 50 bar pressure and 373–473 K to make methyl glycolate; this finds use in, for example, semiconductor processing and metal cleaning (Section 14.3.4).

Other 1,2-diols have also been examined. Phenylethane-1,2-diol has been oxidised using a colloid-derived Au/C catalyst ($d_{Au} = 7$ nm) to mandelic acid ($C_6H_5CH(OH)CO_2H$), an important pharmaceutical intermediate.[84,85]

8.3.6. Oxidation of a triol: glycerol

Glycerol (propan-1,2,3-triol) is a readily available raw material from biosustainable sources such as rape-seed and sunflower; the many products that can be formed from it by oxidation find economic use as intermediates in the fine chemicals industry. However, its oxidation constitutes a complicated scenario by reason of the parallel and sequential reaction paths that can be followed (Scheme 8.3): obtaining a desired product therefore constitutes a considerable challenge. The mono-aldehyde readily isomerises under basic conditions to dihydroxyacetone, but fortunately it is less easily oxidised, so that glyceric acid ($HOCH_2–CH(OH)–CO_2H$) is frequently a major product. Gold catalysts, as in other reactions, do not suffer from deactivation, nor do

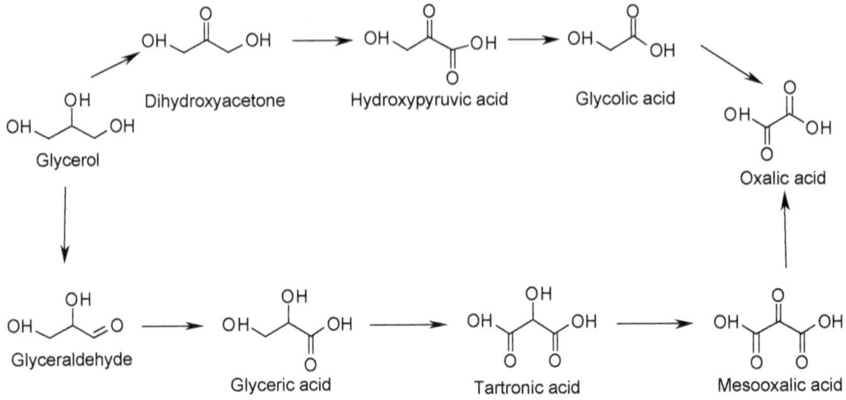

Scheme 8.3: General reaction pathways for the oxidation of glycerol.

they in general form products having only one or two carbon atoms;[94] glycolic acid ($HOCH_2$–CO_2H) is, however, sometimes a significant product,[95] and sequential oxidation of the second primary hydroxyl group can lead to the conversion of glyceric acid to tartronic acid (HOOC–CH(OH)–COOH). Much attention has therefore been paid to the effects of catalyst composition and operating conditions on product selectivities. These are closely connected in the sense that optimum conditions depend somewhat on the type of catalyst being used.

Glycerol has been oxidised to glyceraldehyde with 100% selectivity over 1% Au/C and 1% Au/graphite catalysts at 333 K using 3 bar oxygen pressure and high base/glycerol ratio; lower gold loadings, lower catalyst weights and higher oxygen pressures gave poorer results.[94,96–98] Catalysts were characterised by cyclic voltammetry in an attempt to ascertain what structural features gave best performance. Under similar conditions, use of various preparation methods and different types of carbon and graphite supports gave glycerate selectivities up to 90% at 90% conversion,[99] but it was possible to distinguish catalysts whose selectivities decreased with conversion ($d_{Au} = 6$ nm) from those whose selectivities remained constant ($d_{Au} > 20$ nm). Operating conditions giving maximum selectivity were identified; the presence of base (NaOH) was essential, and the rate increase on raising the oxygen pressure to 3 bar.[100] The lack of size effect on selectivity has been confirmed by the observation that catalysts having 5–50 nm mean particle sizes were comparably effective.[94,97] At both 313 and 333 K, palladium–gold and platinum–gold catalysts (Pd, Pt: Au = 1:1) supported on graphite were more active than the single metals; palladium–gold colloids have also been used to make supported bimetallic catalysts.[101,102]

The importance of choosing the right reaction conditions has been stressed, and the outstanding performance of carbon as a support (activated on 'black' or graphite) has been confirmed.[95] Large particles (23 and 42 nm) were however less effective than small ones (3.7 and 2.7 nm),[95] in disagreement with what has been said above, but this may be due to the use of different operating conditions. Rates were initially independent of catalyst loading, due perhaps to the formation of a foam that entrained some of the catalyst. They were independent of oxygen pressure (unlike the case of D-glucose, mentioned in Section 8.3.2), and increased with temperature according to an activation energy of about $50 \, kJ \, mol^{-1}$. Rates also increased with the NaOH/glycerol ratio, but at a molar ratio of 4 further oxidation to tartronic acid took place with progress of time. The dependence of rate and selectivity on gold particle size clearly deserves further study.

The poor performance of oxide-supported gold catalysts and the effectiveness of large gold particles clearly distinguishes these selective oxidations from that of carbon monoxide, and suggests very strongly that the metal is itself responsible; there has been no suggestion of a role for cationic species. The mode of activation of the oxygen molecule, however, remains obscure. The hydroxyl ion in solution may remove a proton from one of the terminal hydroxyl groups of the glycerol,[100] the glyceryl ion then adsorbing on the support by electron transfer; this might explain why quite large gold particles are satisfactory catalysts. The oxygen molecule, perhaps protonated as HO_2^+, may then react with this species, as was proposed for glucose (Section 8.3.2), forming hydrogen peroxide, which then decomposes.

8.3.7. Oxidation of monofunctional alcohols

It now remains for us to consider the oxidation of monofunctional alcohols and molecules containing the –OH group remote from other functions. The conversion of methanol to formaldehyde (methanal) can be performed either by dehydrogenation (difficult, see Chapter 9) or by oxidative dehydrogenation according to the equation:

$$CH_3OH + 0.5O_2 \rightarrow HCHO + H_2O. \qquad (8.2)$$

This process is performed very successfully on an industrial scale using either macrocrystalline silver or an iron molybdate type of catalyst. The hazard of course is over-oxidation, and some interest has been shown in what gold catalysts can do. The first study of the gold-catalysed reaction was reported[103] as long ago as 1920; reaction took place on a gold gauze between 790 and 1170 K, but it was concluded that silver was the better catalyst, and this may explain the lack of further attention to gold. A further study[104] in 1975 with the same type of catalyst showed that above 600 K methanal was formed, together with a small amount of hydrogen from decomposition of methanol. Platinum was more active ($T > 400$ K), but carbon dioxide formation occurred above 500 K.

This same reaction has recently been studied[105] with a different objective, namely, oxidation of methanol to produce hydrogen according to the equation

$$CH_3OH + 0.5O_2 \rightarrow 2H_2 + CO_2 \qquad (8.3)$$

although other reactions including oxidative dehydrogenation and steam-reforming

$$CH_3OH + H_2O \rightarrow 3H_2 + CO_2 \qquad (8.4)$$

were thought to participate. With Au/TiO_2 between 480 and 580 K, selectivity to hydrogen rose to 35%, although very high methanol conversion was taking place. Oxygen atoms on Au(111) also oxidise methanol.[106,107]

Alcohols such as propanol and *n*-butanol have been oxidised to the aldehydes by resin-supported gold in the liquid phase,[84] and they and other primary and secondary alcohols have been oxidised with very high selectivities (100% in most cases) in the vapour phase over 1% Au/SiO_2 between 373 and 573 K:[108] even prop-2-en-1-ol ($CH_2 = CH–CH_2OH$) gave 97% selectivity to propenoic acid at 523 K. *Iso*propanol has been oxidised to acetone over Au/Al_2O_3 and (better) $Au/CeO_2–Al_2O_3$ with high selectivity between 373 and 473 K,[109] and *o*-hydroxybenzyl alcohol (salicylic alcohol) to the corresponding acid with 90% selectivity at 90% conversion using gold supported on various oxides.[110,111] Amino-alcohols have also been converted to amino-acids;[59] for this reaction gold is preferable to palladium or platinum because the amino-group does not bind strongly to it.

The oxidation of benzyl alcohol to benzaldehyde has been detected[98] in solvent-free conditions at 373 K and 2 bar oxygen pressure using Au/SiO_2 prepared in several ways, Au/CeO_2 and Au/Fe_2O_3 (COPPT) and Au/TiO_2 and Au/C (IMP); high selectivities were noted, but conversions were low (0.6–7%), and with the more acidic supports (e.g. Fe_2O_3) further oxidation to the acid led to formation of the ester (benzyl benzoate). 1-Octanol was

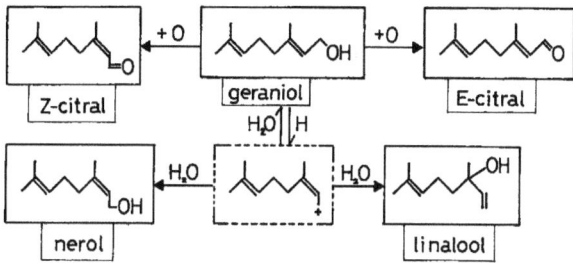

Scheme 8.4: Principal products in the oxidation of geraniol.

less reactive, but conversions were higher with geraniol (especially when using $Au/CeO2$); the main products were *E*- and *Z*-citral, their proportions varying greatly from one catalyst to another, but nerol and linalool were also formed by acid-catalysed isomerisation when the support had acidic character (see Scheme 8.4). Activities were stated to be comparable to those obtained with Group 10 metal catalysts. The same types of catalyst have also proved effective for the oxidation of 3-octanol,[112] but by far the highest activity was obtained when using gold supported on 'nano-crystalline' ceria; in many cases almost full conversions and high selectivities were found at 353 K and 1 atm oxygen pressure. A range of other primary and secondary alcohols were converted with similar success to the corresponding aldehyde or ketone or in some cases the acid or ester in less than 3 h with this catalyst.

8.4. Synthesis of Vinyl Acetate (Ethenyl Ethanoate)

It is convenient to consider the synthesis of vinyl acetate at this point, because it involves the aerobic oxidation of ethene to the unstable intermediate ethenol ($H_2C=CH(OH)$), followed by its immediate esterification with acetic (ethanoic) acid,

$$CH_2=CH_2 + CH_3COOH + 0.5O_2 \rightarrow CH_2=CH-CO-OCH_3 + H_2O \quad (8.5)$$

and although palladium by itself is capable of effecting the reaction, it goes much better when gold is also present (Table 8.2) and so it joins the several other cases treated above where the palladium–gold combination has distinct advantages over either metal singly (Section 14.3.1).

Table 8.2: Synthesis of vinyl acetate: comparison of palladium and palladium–gold systems.[112,114]

Catalyst	Space–time yield $(g\,l^{-1}\,h^{-1})$	Selectivity (%)
Pd	124	94.7
Pd/KOAc	100	95.4
Au/Pd	594	91.6
Au/Pd/KOAc	764	93.6

Notes: fixed-bed performance after 40 h on stream: test conditions, 438 K, 115 psig with feed of ethene, acetic acid, oxygen and nitrogen.

Vinyl acetate is an important intermediate for the production of paints, of adhesives for use in wallpaper paste and wood glue, and of surface coatings such as the protective laminate films in automotive safety glass.[113,114] The established fixed-bed processes may be assisted by many promoters, of which gold and potassium acetate are the most notable.[114,115] Typical results illustrating their promotional effects are shown in Table 8.2; selectivities are 90–95%. Some 80% of today's plants are more than 20 years old, and use fixed-bed processes, but in 2002 BP Chemicals commissioned a new plant in Hull (UK) that used a fluidised-bed process, for which a microspheroidal silica-supported palladium–gold catalyst was developed in collaboration with Johnson Matthey plc. This cost-saving route allowed the process to be simplified and intensified, and it required only a single reactor instead of the two needed for a fixed-bed process.

The mechanism of the reaction has not been entirely clarified and indeed it is not quite certain where it takes place. *"(It) is viewed to occur either on small clusters of palladium acetate dissolved in the 'supported liquid phase' or on palladium acetate dense surfaces"*.[112] The role of the gold is not unequivocally known, either. Its promoting influence has been assigned[112] to an electronic interaction, modifying the adsorption strength of various species on the palladium complexes, increasing the rate of vinyl acetate desorption. Other work[113,116] was based on the assumption that palladium metal was the active species: single-crystal studies showed that added gold promoted the formation of monodentate acetate, and favoured desorption of vinyl acetate over its decomposition.

The commercial fixed-bed catalyst loses activity with use, limiting its useful life to a few years, but the selectivity (~94%) remains high. The metal particle size was shown in one study[113] to have increased from

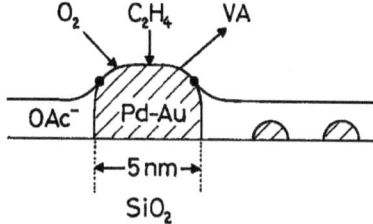

Scheme 8.5: Locus of reaction in the synthesis of vinyl acetate.

5 to 12 nm with use, but much of it was still highly dispersed. If this were caused by Ostwald ripening involving migration of palladium acetate species,[117] bimetallic particles should become richer in palladium, but this was not observed. Growth must therefore involve the migration of whole particles,[113] which would help to explain why selectivity remains constant while the rate declines. It was long ago established that very small palladium particles have low activity,[118] perhaps because they are totally covered by an aqueous acetic acid/acetate liquid film, in which ethene has small solubility. According to one scenario, reaction occurs at the interface of the three phases, on particles large enough to poke through the liquid film (Scheme 8.5).

8.5. Synthesis of Hydrogen Peroxide

We may close this chapter with a short account of recent work on the selective oxidation of hydrogen to hydrogen peroxide, for which the market is very large ($\sim 1.9 \times 10^6$ tonnes per year);[2] there is a great incentive to find means of making it where it is to be used, so as to avoid the heavy cost of transporting a hazardous chemical. Its production by the anthraquinone route is only economic on a large scale, but it is often required on a much smaller scale (Section 14.3.7). Theoretical calculations[119] and experiments[120,121] have both shown that its formation is favoured over gold surfaces. Au/SiO$_2$ prepared by impregnation and dispersed in water was effective at 283 K,[121] and Au/Fe$_2$O$_3$,[122] Au/Al$_2$O$_3$[120,123] and Au/TiO$_2$[124] gave high peroxide productivities when dispersed in methanol or methanol-water at 275 K. Higher temperature caused the hydrogen peroxide to decompose. Operating conditions were 3.7 Mbar with 1:2 hydrogen: oxygen ratio. In the case of Au/TiO$_2$ prepared either by DP or impregnation, calcination at 673 K was necessary for stability; gold in the uncalcined

Table 8.3: Synthesis of hydrogen peroxide: comparison of palladium, gold and bimetallic catalyst.[123]

Catalyst	Rate[a] of H_2O_2 formation (mmol $g_{cat}^{-1} h^{-1}$)
Au/Al_2O_3	1530
Pd/Al_2O_3	370
$Au\text{-}Pd(1:1)Al_2O_3$	4460

[a]At 275 K with mol ratio $O_2:H_2 = 1:2$, averaged over 30 min.

material readily dissolved, with disastrous consequences, so although activity after calcination was less than the *initial* activity before calcination the long-term performance was better.[124] While with all three supports (Fe_2O_3, Al_2O_3 and TiO_2), palladium gave useful productivities of the peroxide, in each case the combination of palladium + gold worked better; a 1:1 ratio was usually used (Table 8.3). Extensive characterisation involving STEM/EDX has been carried out on all three catalyst systems, and it has been concluded that in the working catalyst there was a partial or complete palladium shell overlying a gold core. In the case of Au–Pd/Al_2O_3[123] this structure appeared to be generated during the reaction; this may be the result of extractive chemisorption, because both reactants will chemisorb more strongly on the palladium, and hence draw it to the surface. It was noted that mean particle sizes for the pure gold catalysts were quite large (Au/Fe_2O_3, 48 nm; Au/Al_2O_3, ~10 nm), while for Au–Pd/TiO_2 the larger particles were bimetallic and the small ones gold. A quite precise inverse correlation between rates of hydrogen peroxide synthesis and carbon monoxide oxidation was established, showing that the mechanisms are entirely different. Speculation concerning the former has not yet been attempted.

8.6. Conclusion

The quite remarkable effectiveness of gold as a catalyst for the selective oxidation of organic molecules containing oxygenated groups has been a surprise comparable with that attending the discovery of its ability to oxidise carbon monoxide at low temperatures, but from what has been said above it should be clear that the requirements of catalysts for these two types of process are different. Good results have been obtained for selective oxidations irrespective of particle size, and certainly the necessity of having very small particles is absent for these reactions. A further clear point of difference is that a 'reactive' support is not needed in these cases; although ceria sometimes performs marvellously well as a support, in other cases the ceramic oxides work very satisfactorily. Understanding of reaction mechanisms has not advanced nearly as far as with carbon monoxide oxidation, but no doubt the continuing high level of interest being shown in gold as a catalyst foe selective oxidation, and therefore as a contributor to the field of 'green chemistry', will lead before long to the unravelling of many things that now remain a mystery.

References

1. Y.A. Kalvachev, T. Hayashi, S. Tsubota and M. Haruta, *J. Catal.* **186** (1999) 228.
2. M. Haruta, *Proc. Gold 2003*, Vancouver, Canada (2003).
3. G.C. Bond and D.T. Thompson, *Cat. Rev.-Sci. Eng.* **41** (1999) 319.
4. W.R. Patterson and C. Kemball, *J. Catal.* **2** (1963) 465.
5. N.W. Cant and W.K. Hall, *J. Phys. Chem.* **75** (1971) 2914.
6. M. Ruszel, B. Grzybowska, M. Gąsior, K. Samson, I. Gressel and J. Stoch, *Catal. Today* **99** (2005) 151.
7. T. Hayashi, K. Tanaka and M. Haruta, *J. Catal.* **178** (1998) 566.
8. B.S. Uphade, T. Akita, T. Nakamura and M. Haruta, *J. Catal.* **209** (2002) 331.
9. C. Qi, T. Akita, M. Okumura, K. Kuraoka and M. Haruta, *Appl. Catal. A: Gen.* **253** (2003) 75.
10. C. Qi, T. Akita, M. Okumura and M. Haruta, *Appl. Catal. A: Gen.* **263** (2004) 19.
11. G.U. Kulkarni, C.P. Vinod and C.N.R. Rao, in *Surface Chemistry and Catalysis*, A.F. Carley, P.R. Davis, G.J. Hutchings and M.S. Spencer, (eds.), Kluwer/Plenum, New York, 2002, p. 191.
12. B.S. Uphade, Y. Yamada, T. Akita, T. Nakamura and M. Haruta, *Appl. Catal. A: Gen.* **215** (2001) 137.
13. B.S. Uphade, Y. Tamada, T. Akita, T. Nakamura and M. Haruta, *Appl. Catal. A: Gen.* **215** (2001) 295.
14. A.K. Sinha, S. Seelan, T. Akita, S. Tsubota and M. Haruta, *Appl. Catal. A: Gen.* **240** (2003) 243.
15. M. Haruta, *Chem. Record* **3** (2003) 75.
16. M. Haruta, *Catal. Today* **36** (1997) 153.
17. T. Hayashi, L.B. Han, S. Tsubota and M. Haruta, *Ind. Eng. Chem. Res.* **34** (1995) 2298.
18. G. Mul, A. Zwijnenburg, B. van der Linden, M. Makkee and J.A. Moulijn, *J. Catal.* **201** (2001) 128.
19. A. Zwijnenburg, M. Saleh, M. Makkee and J.A. Moulijn, *Catal. Today* **72** (2002) 59.
20. E.E. Stangland, K.B. Stavens, R.P. Andres and W.N. Delgass, *J. Catal.* **191** (2000) 332.
21. L. Cumaranatunge and W.N. Delgass, *J. Catal.* **232** (2005) 38.
22. B. Taylor, J. Lauterbach and W.N. Delgass, *Appl. Catal. A: Gen.* **291** (2005) 188.
23. B.S. Uphade, M. Okumura, S. Tsubota and M. Haruta, *Appl. Catal. A: Gen.* **190** (2000) 43.
24. T.A. Nijhuis and B.T. Weckhuysden, *Chem. Commun.* (2005) 6002.
25. A. Nijhuis, T.Q. Gardner and B.M. Weckhuysen, *J. Catal.* **236** (2005) 153.

26. M. Gąsior, B. Grzybowska, K. Samson, M. Ruszel and J. Haber, *Catal. Today* **91–92** (2004) 131.

27. C. Qi, T. Akita, M. Okumura and M. Haruta, *Appl. Catal. A: Gen.* **218** (2001) 81.

28. E.E. Stangland, B. Taylor, R.P. Andres and W.N. Delgass, *J. Phys. Chem. B* **109** (2005) 2321.

29. Y.A. Kalvachev, T. Hayashi, S. Tsubota and M. Haruta, *Stud. Surf. Sci. Catal.* **110** (1997) 965.

30. B.S. Uphade, S. Tsubota, T. Hayashi and M. Haruta, *Chem. Lett.* (1998) 1273.

31. N. Yap, R.P. Andres and W.N. Delgass, *J. Catal.* **226** (2004) 156.

32. A.K. Sinha, S. Seelan, S. Tsubota and M. Haruta, *Angew. Chem. Int. Ed.* **43** (2004) 1546.

33. T.A. Nijhuis, B.J. Huizinga, M. Makkee and J.A. Moulijn, *Ind. Eng. Chem. Res.* **38** (1999) 884.

34. A. Zwijnenburg, M.M. Makkee and J.A. Moulijn, *Appl. Catal. A: Gen.* **270** (2004) 49.

35. A. Kuperman, R.G. Bowman, H.W. Clark, G.E. Hartwell and G.R. Meima, US Patent 2004/0176620, 9 September 2004 to Dow Chemical Co.

36. A.K. Sinha, S. Seelan, S. Tsubota and M. Haruta, *Topics Catal.* **29** (2004) 95.

37. D.H. Wells, H. David, W.N. Delgass and K.T. Thomson, *J. Am. Chem. Soc.* **126** (2004) 2956.

38. C. Sivadinarayana, T.V. Choudhary, L.L. Daemen, J. Eckert and D.W. Goodman, *J. Am. Chem. Soc.* **126** (2004) 38.

39. D.H. Wells, W.N. Delgass and K.T. Thomson, *J. Catal.* **225** (2004) 69.

40. S. Chrétien, M.S. Gordon and H. Metiu, *J. Chem. Phys.* **121** (2004) 3756.

41. J. Figar and W. Haidinger, *Chem. Phys. Lett.* **19** (1973) 564; *Gold Bull.* **7** (1974) 100.

42. M.D. Hughes, Y.-J. Xu, P. Jenkins, P. McMorn, P. Landon, D.I. Enache, A.F. Carley, G.A. Attard, G.J. Hutchings, F. King, E.H. Stitt, P. Johnston, K. Griffin and C.J. Kiely, *Nature* **437** (2005) 1132.

43. D.-H. Ying, L.-S. Qin, J.-F. Liu, C.-Y. Li and Y. Lin, *J. Molec. Catal. A: Chem.* **240** (2005) 40.

44. N.S. Patil, B.S. Uphade, P. Jana, R.S. Sonawane, S.K. Bhargava and V.R. Choudhary, *Catal. Lett.* **94** (2004) 89.

45. N.S. Patil, R. Jha, B.S. Uphade, S.K. Bhargava and V.R. Choudhary, *Appl. Catal. A: Gen.* **275** (2004) 87.

46. N.S. Patil, B.S. Uphade, D.G. McCulloh, S.K. Bhargava and V.R. Choudhary, *Catal. Commun.* **5** (2004) 681.

47. J.-M. Sun, S.-I. Fujita, F.-Y. Zhao, M. Hasegawa and M. Arai, *J. Catal.* **230** (2005) 398.

48. F. Shi, Q.-H. Zhang, Y. Ma, Y. He and Q.-Y. Deng, *J. Am. Chem. Soc.* **127** (2005) 4182.

49. F. Moreau, unpublished work.

50. Y.-X. Xu, P. Landon, D. Enache, A.F. Carley, M.W. Roberts and G.J. Hutchings, *Catal. Lett.* **101** (2005) 175.

51. R. Zhao, D. Ji, G. Lv, G. Qian, L. Yan, X. Wang and J. Suo, *Chem. Commun.* (2004) 904.

52. K.K. Zhu, J.C. Hu and R. Richards, *Catal. Lett.* **100** (2005) 195.

53. R. Zhao, D. Ji, G.-M. Lv, G. Qian, L. Yan, X.-L. Wang and J. Suo, *Catal. Lett.* **97** (2004) 115.

54. I. Eskandirov, N.J. Coville and V.D. Sokolovskii, *Catal. Lett.* **35** (1995) 33.
55. K. Blick, T.D. Mitrelias, J.S.J. Hargreaves, G.J. Hutchings, R.W. Joyner, C.J. Kiely and F.E. Wagner, *Catal. Lett.* **50** (1998) 211.
56. US Patent 374 2079 to Mobil Oil Corp., (1973); *Chem. Abs.* **79** (1973) 92807.
57. S. Schimpf, B. Kusserow, Y. Önal and P. Claus, *Proc. 13th Internat. Congr. Catal.*, Paris, July 2004.
58. S. Biella, L. Prati and M. Rossi, *J. Catal.* **206** (2002) 242.
59. S. Biella, G.L. Castiglioni, C. Fumagalli, L. Prati and M. Rossi, *Catal. Today* **72** (2002) 43.
60. L. Prati and G. Martra, *Gold Bull.* **32** (1999) 96.
61. N. Dimitratos and L. Prati, *Gold Bull.* **38** (2005) 73.
62. C. Basheer, S. Swaminathan, H. Lee and S. Valiyaveettil, *Chem. Commun.* (2005) 409.
63. S. Biella and M. Rossi, *Proc. Gold 2003*, Vancouver, Canada.
64. L. Prati, M. Rossi, C. Fumagalli, G. Castiglioni and C. Pirola, Italian Patent 99A 002611, Lonza S.p.A. (1999).
65. P. Claus, S. Schimpf and Y. Önal, *Proc. 18th Meeting North Am. Catal. Soc.*, Cancun, Mexico (2003), p. 365.
66. M. Comotti, C. Della Pina, M. Matarrese and M. Rossi, *Angew. Chem. Int. Ed.* **43** (2004) 5812.
67. A. Mirescu, U. Preusse and K.-D. Vorlop, *Proc. 13th Internat. Congr. Catal.*, Paris, July 2004, P5-059.
68. S. Biella, L. Prati and M. Rossi, *J. Molec. Catal. A: Chem.* **197** (2003) 207.
69. Y. Önal, S. Schimpf and P. Claus, *J. Catal.* **223** (2004) 122.
70. M. Comotti, C. Della Pina, R. Matarrese, M. Rossi and A. Siana, *Appl. Catal. A: Gen.* **291** (2005) 204.
71. C.L. Bianchi, S. Biella, A. Gervasini, L. Prati and M. Rossi, *Catal. Lett.* **85** (2003) 91.
72. A. Mirescu and U. Pruesse, *Catal. Commun.* **7** (2006) 11.
73. T. Mallat and A. Baiker, *Catal. Today* **19** (1994) 247.
74. P. Beltrame, M. Comotti, C. Della Pina and M. Rossi, *Appl. Catal. A: Gen.* **297** (2006) 1.
75. S. Hermans and M. Devillers, *Catal. Lett.* **99** (2005) 55.
76. S. Hermans, S. Vanderheyden and M. Devillers, *Proc. 13th Internat. Congr. Catal.*, Paris 1–30, 2004.
77. C. Burato, P. Centomo, G. Pace, M. Favaro, L. Prati and B. Corain, *J. Molec. Catal. A: Chem.* **238** (2005) 26.
78. A. Corma and M.E. Domine, *Chem. Commun.* (2005) 4042.
79. S. Carrettin, J. Guzman, A. Corma and H. Garcia, *Angew. Chem. Int. Ed.* **44** (2005) 2242.
80. N. Dimitratos, F. Porta, L. Prati and A. Villa, *Catal. Lett.* **99** (2005) 181.
81. S. Coluccia, G. Martra, F. Porta, L. Prati and M. Rossi, *Catal. Today* **61** (2000) 165.
82. S. Biella, L. Prati and M. Rossi, *Inorg. Chim. Acta* **349** (2003) 253.
83. L. Prati and M. Rossi, *J. Catal.* **176** (1998) 552.
84. L. Prati and F. Porta, *Appl. Catal. A: Gen.* **291** (2005) 199.
85. S. Biella, L. Prati and M. Rossi, *Inorg. Chim. Acta* **349** (2003) 253.
86. S. Coluccia, G. Martra, F. Porta, L. Prati and M. Rossi, *Catal. Today* **61** (2000) 165.

87. L. Prati and M. Rossi, *J. Catal.* **176** (1998) 552.
88. C. Bianchi, F. Porta, L. Prati and M. Rossi, *Topics Catal.* **13** (2000) 231.
89. F. Porta, L. Prati, M. Rossi and G. Scari, *J. Catal.* **211** (2002) 464.
90. F. Porta and M. Rossi, *J. Molec. Catal. A: Chem.* **204–205** (2003) 553.
91. S. Biella, F. Porta, L. Prati and M. Rossi, *Catal. Lett.* **90** (2003) 23.
92. H. Berndt, I. Pitsch, S. Evert, K. Struve, M.-M. Pohl, J. Radnik and A. Martin, *Appl. Catal. A: Gen.* **244** (2003) 169.
93. *Chem. Eng. NY* **111** (2004) 20.
94. G.J. Hutchings, *Catal. Today* **100** (2005) 55.
95. S. Demirel-Gülen, M. Lucas and P. Claus, *Catal. Today* **102–103** (2005) 166.
96. S. Carrettin, P. McMorn, P. Johnston, K. Griffin and G.J. Hutchings, *Chem. Commun.* (2002) 696.
97. S. Carrettin, P. McMorn, P. Johnston, K. Griffin, C.J. Kiely, G.A. Attard and G.J. Hutchings, *Topics Catal.* **27** (2004) 131.
98. D.I. Enache, D.W. Wright and G.J. Hutchings, *Catal. Lett.* **103** (2005) 41.
99. F. Porta and L. Prati, *J. Catal.* **224** (2004) 397.
100. S. Carrettin, P. McMorn, P. Johnston, K. Griffin, C.J. Kiely and G.J. Hutchings, *Phys. Chem. Chem. Phys.* **5** (2003) 1329.
101. C.L. Bianchi, P. Canton, N. Dimitratos, F. Porta and L. Prati, *Catal. Today* **102–103** (2005) 203.
102. N. Dimitratos, F. Porta and L. Prati, *Appl. Chem. A: Gen.* **291** (2005) 210.
103. M.D. Thomas, *J. Am. Chem. Soc.* **42** (1920) 609.
104. C.N. Hodges and L.C. Roselaar, *J. Appl. Chem. Biotechnol.* **25** (1975) 609.
105. F.-W. Chang, H.-Y. Yu, L.S. Roselin and H.-C. Yang, *Appl. Catal. A: Gen.* **290** (2005) 138.
106. D.A. Outka and R.J. Madix, *J. Am. Chem. Soc.* **109** (1987) 1709.
107. N. Salida, D.H. Parker and B.E. Koel, *Surf. Sci.* **410** (1998) 270.
108. S. Biella and M. Rossi, *Chem. Commun.* (2003) 378.
109. M.A. Centeno, M. Paulis, M. Montes and J.A. Odriozola, *Appl. Catal. A: Gen.* **234** (2002) 65.
110. C. Milone, R. Ingoglia, G. Neri, A. Pistone and S. Galvagno, *Appl. Catal. A: Gen.* **211** (2001) 251.
111. C. Milone, R. Ingoglia, A. Pistone, G. Neri and S. Galvagno, *Catal. Lett.* **87** (2003) 2.
112. W.D. Provine, P.L. Mills and J.J. Lerou, *Stud. Surf. Sci. Catal.* **101** (1996) 191.
113. A. Abad, P. Concepcion, A. Corma and H. Garda, *Angew. Chem. Int. Ed.* **44** (2005) 4066.
114. N. Macleod, J.M. Keel and R.M. Lambert, *Appl. Catal. A: Gen.* **261** (2004) 37.
115. W.J. Barteley, S. Jobson, G.G. Harkreader, M. Kitson and M. Lemanski, U.S. Patent 5274181 (1993).
116. R.D. Haley, M.S. Tikhov and R.M. Lambert, *Catal. Lett.* **76** (2001) 125.
117. R. Abel, G. Prauser and H. Tiltscher, *Chem. Eng. Technol.* **17** (1994) 112.
118. B. Samanos, P. Boutry and R. Montarnal, *J. Catal.* **23** (1971) 19.
119. P. Parades Olivera, E.M. Patrito and H. Sellers, *Surf. Sci.* **313** (1994) 25.
120. P. Landon, P.J. Collier, A.J. Papworth, C.J. Kiely and G.J. Hutchings, *Chem. Commun.* (2002) 2058; P. Landon, P.J. Collier, A.F. Carley, D. Chadwick, A.J. Papworth, A. Burrows, C.J. Kiely and G.J. Hutchings, *Phys. Chem. Chem. Phys.* **5** (2003) 917.
121. T. Ishihara, Y. Ohura, S. Yoshida, Y. Hata, H. Nishiguchi and Y. Takita, *Appl. Catal. A: Gen.* **291** (2005) 215.

122. J.K. Edwards, B. Solsona, P. Landon, A.F. Carley, A. Herzing, M. Watanabe, C.J. Kiely and G.J. Hutchings, *J. Mater. Chem.* **15** (2005) 4595.

123. P. Landon, P.J. Collier, A.F. Carley, D. Chadwick, A.J. Papworth, A. Burrows, C.J. Kiely and G.J. Hutchings, *Phys. Chem. Chem. Phys.* **5** (2003) 1917.

124. J.K. Edwards, B.E. Solsona, P. Landon, A.F. Carley, A. Herzing, C.J. Kiely and G.J. Hutchings, *J. Catal.* **236** (2005) 69.

CHAPTER 9

Reactions Involving Hydrogen

9.1. Introduction: The Interaction of Hydrogen with Gold

This chapter is concerned with reactions in which the hydrogen molecule is either a reactant (Sections 9.2–9.5) or a product (Section 9.6) and includes reactions in which it reacts incestuously with its close relatives and in which it is dissociated or its atoms recombine (Section 9.2). It is generally admitted that gold catalysts are markedly inferior to those containing the metals of Groups 8–10 for reactions requiring the dissociation of hydrogen, and this is undoubtedly a consequence of their much more limited ability to chemisorb it (Section 5.5). As Dr. Samuel Johnson remarked of a dog walking on its hind legs, it is not surprising that it is done badly — it is surprising it is done at all. Nevertheless, as we shall see, there are circumstances in which gold is able to show greater selectivity or specificity than other metals, and the conditions under which reactions do proceed throw valuable light on the ability of gold to act catalytically.

The ways in which hydrogen, either as a molecule or as an atom, can interact with gold are as follows: it may be either (i) chemisorbed without dissociation in a weak molecular state that is not stable above room temperature or (ii) chemisorbed dissociatively as atoms or (iii) dissolved within the lattice, either in sites that are immediately sub-surface or that lie deeper[1] or (iv) chemisorbed as atoms that have been provided to the surface as such. At high temperatures the molecule will dissociate and hydrogen *atoms* will desorb into the gas phase[2,3] while at lower temperatures hydrogen atoms will adsorb at gold surfaces and recombine to form gaseous molecular hydrogen.[2] What is known of the chemisorption of molecular hydrogen has been reviewed in Section 5.5.1; the situation with massive gold is briefly as follows. There is ample evidence for the existence of a weakly held molecular state at low temperatures, particularly on rough surfaces, but its desorption is complete well below room temperature, so that it is very unlikely to have any role in catalytic reactions. Hydrogen *atoms* certainly chemisorb at low temperature,[2] but they also quickly desorb on warming;[4] this suggests that reluctance to form the chemisorbed state lies not only with the activation energy, but also with the strength of the Au–H bond. Theoretical studies serve mainly to confirm what is already known experimentally; interesting questions such as whether there exists a ligand effect involving a neighbouring Group 10 metal, able to activate a gold atom somewhat, have not yet been addressed.

The present position regarding the dissociative chemisorption of hydrogen on small supported gold particles has also been reviewed in Section 5.5.1, so only a very brief summary is necessary now. For many years it has been felt that the qualitative dependence of activity on particle size for reactions in which hydrogen is a partner must imply that the molecule can dissociate if the particle is sufficiently small.[5–8] This suspicion has only recently been confirmed by direct experimentation with Au/Al_2O_3 catalysts having particles between 1 and 5 nm in size;[9] volumetrically measured isotherms were reported and new features in XANES observed, but it was concluded that adsorption was limited to atoms at edges and corners, which naturally proliferate on very small particles. The mystery of the apparent inactivity of gold in hydrogenations has therefore been dispelled, as being caused by the use of particles that were too large. This constraint on the locus of adsorption means that gold catalysts however carefully made are unlikely ever to rival the platinum group metals in terms of activity, but as we shall see this limitation also brings a benefit, in that incomplete coverage by hydrogen atoms allows *selective* hydrogenation of multifunctional molecules, to a degree that is impossible with platinum, unless selectively poisoned. Hydrogen therefore joins carbon monoxide in requiring very specific locations on gold particles in order to chemisorb.

The solubility of hydrogen atoms in gold is very small; at 273 K the H/Au ratio was 4.4×10^{-3}, and the activation energy for desorption between 250 and 380 K was $57 \pm 5 \, kJ \, mol^{-1}$; the heat of absorption was[1] $\sim 9 \, kJ \, mol^{-1}$. Hydrogen atoms are also able to diffuse easily through a gold layer when supplied *via* a palladium hydride phase.[10]

9.2. Reactions of Hydrogen[11–13]

The reactions with which we will be mainly concerned in this section are the hydrogen-deuterium equilibration

$$H_2 + D_2 \rightleftharpoons 2HD \qquad (9.1)$$

and the *para*-hydrogen conversion to *ortho*-hydrogen. This latter reaction, which requires the inversion of the nuclear spin of one of the atoms, can be catalysed by a strong paramagnetic centre without the need for dissociation, although dissociation followed by random recombination is an alternative route. This reaction has been little used for many decades, although formerly in vogue[12,14] because it is readily performed and its progress is easily followed by thermal conductivity. For the isotopic equilibration, dissociation and recombination are of course essential. It can also be monitored by thermal conductivity change, but mass spectrometry is more often used; low-temperature gas–solid chromatography (e.g. $MnCl_2/Al_2O_3$ at $77\,K$)[15] has also been used. The equilibrium constant is 3.2 at ordinary temperatures, rather than the statistically expected value of 4, due to the operation of the zero-point energy difference.

We may first consider work performed with massive gold, since no complication arising from a support can interfere. Both *para*-hydrogen conversion[16,17] and the isotopic equilibration[18,19] require somewhat high temperatures, and activation energies are also substantial (see Table 9.1). There are two possible explanations for these findings, which are surprising because the steady-state concentration of adsorbed atoms should be vanishingly small under these conditions: activity may be caused by (i) traces of other noble or base metals, or (ii) by thermal excitation of electrons from the filled $5d$ level to the incomplete $6s$ level, thereby creating d-band vacancies and a greater tendency towards chemisorption.[19] Perhaps less surprising are the observations that reaction is accelerated by pre-adsorption of hydrogen atoms, either at $78\,K$ using film or above about $345\,K$ on foil. Their presence allows the reaction to proceed by a Rideal–Eley mechanism[16]:

$$\underset{*}{H} + D_2 \rightarrow H \cdots \underset{*}{D} \cdots D \rightarrow HD + \underset{*}{D}. \qquad (9.2)$$

Table 9.1: Reactions of hydrogen on massive gold.

Reaction	Form	Temp. range (K)	Activation energy (kJ mol^{-1})	Reference
$H_2 + D_2$	Foil	603–673	115	19
$H_2 + D_2$	Foil	385–570	90	18
$H_2 + D_2$	Foil	345–417	46[a]	18
p-H_2 conv.	Wire	500–800	145[b]	16

[a]Surface precovered with H atoms. [b]Much lower values ($\sim25\,kJ\,mol^{-1}$) were reported for temperatures between 273 and 550 K.

This pathway has a lower activation energy $(46 \, \text{kJ mol}^{-1})$ than that before the deposition of the atoms (Table 9.1).

A number of attempts have been made to study the isotopic equilibrium on supported gold, but before reviewing these it is necessary to note a possible complication. It is now well established that under certain circumstances hydrogen atoms can move from a metal particle onto the support and back again;[13] these processes are termed, respectively, *spillover* and *reverse spillover*. One of the clearest ways of detecting them is by observing the exchange of hydrogen atoms in hydroxyl groups on the support with deuterium using infrared spectroscopy;[13] what happens is sketched in Scheme 9.1. In the case of gold, this exchange was first observed to occur at room temperature with Au/SiO_2 as long ago as 1969,[20] and again (above 368 K) in 1973.[5] More recently it has been seen with Au/Al_2O_3 $(d_{Au} = 3 \, \text{nm})$ at room temperature, but with larger (10.7 nm) particles only above 400 K.[21] It is therefore necessary to take care that equilibration is taking place only on the metal, and that spillover and its reverse are not contributing to the way it is proceeding, since only in the absence of spillover and its reverse will the H/D ratio remain constant.

Hydrogen–deuterium equilibration took place at around 473 K with Au/SiO_2, Au/MgO and Au/Al_2O_3 with particle sizes between 6 and 9 nm; rate constants and activation energies $(75\text{–}116 \, \text{kJ mol}^{-1})$ were independent of dispersion. Rates were slower with larger particles $(\geq 16.5 \, \text{nm})$ and the

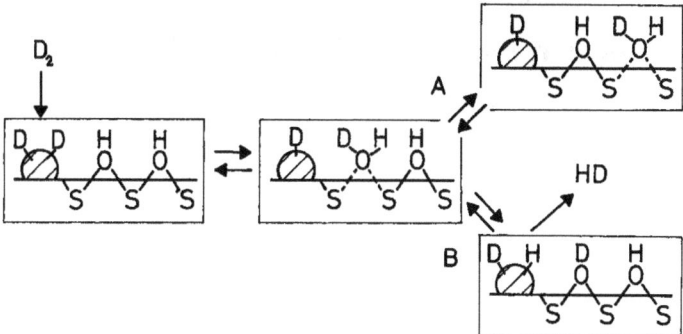

Scheme 9.1: A mechanism for metal-catalysed hydrogen spillover, shown by the exchange of support hydroxyls with deuterium. The process can extend to the whole surface (A), but HD is formed by reverse spillover (B), followed by desorption.

reaction was poisoned by nitric oxide.[22] Au/Al_2O_3 ($d_{Au} = 7\,m$) brought the reaction to equilibrium in less than $1\,h$, but the support by itself was only slightly less active; $\gamma\text{-}Al_2O_3$ was not effective below $500\,K$.[21] Slow reaction on $Au/\alpha\text{-}Al_2O_3$ has also recently been reported.[9] Au/MgO and Au/SiO_2 were also reported to be active at $423–453\,K$, the activation energy for the former being[23] $50\,kJ\,mol^{-1}$. On these catalysts the reaction was accelerated by adsorbed oxygen atoms,[24] perhaps by the reversible formation and decomposition of hydroxyl groups, although there was no simultaneous equilibration of oxygen isotopes. About 0.3% $Au/PTFE$ (polytetrafluoroethene) was active between 393 and $518\,K$, and a kinetic equation was derived that was consistent with bimolecular four-centre transition state;[15] the activation energy was $29\,kJ\,mol^{-1}$. Deuterium also exchanges with preadsorbed hydrogen on Au/TiO_2, Au/ZrO_2 and Au/ZnO above $523\,K$.[25]

Gold is often considered as the terminal member of a bimetallic series in which the other member is most frequently a metal of Group 10. The important work of Couper and Eley[16] using *para*-hydrogen conversion showed a significant *increase* (30%) in activity as gold was added to palladium, without change in the activation energy. A weak ligand effect might be suspected, but this should affect the activation energy, and the observation[26] of a similar enhancement in the rate of hydrogen atom recombination suggests an increase in surface mobility may be responsible.

9.3.   Hydrogenation of Unsaturated Molecules and Related Reactions

9.3.1.   The carbon–carbon double bond

The addition of hydrogen to the C=C bond in a simple alkene such as ethene is formally a very straightforward reaction that can yield only a single product, namely, the corresponding alkane, but the simplicity is more apparent than real, although the complexity is only revealed when the reaction is conducted either with deuterium rather than hydrogen or with a higher homologue. These ancillary reactions have been described in detail in other monographs,[11–13,27] and a brief summary will be adequate for present purposes.

When ethene reacts with deuterium on a metal catalyst, the products typically consist of (i) deuterated ethenes containing from one to four deuterium atoms, formed by *alkene exchange*, (ii) deuterated ethanes containing from zero to six deuterium atoms formed by *addition* and (iii) hydrogen

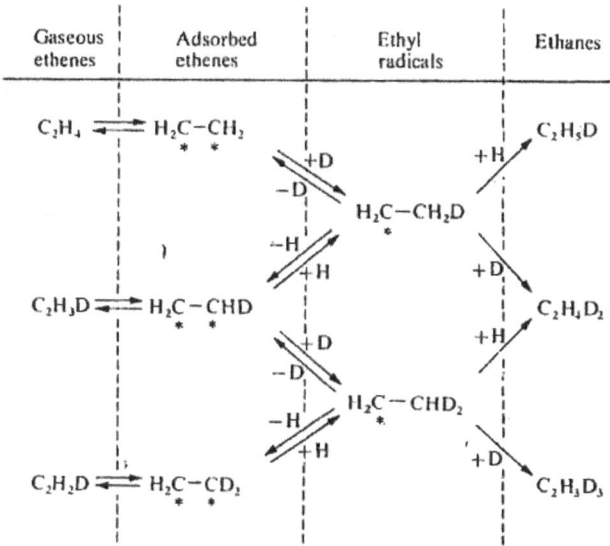

Scheme 9.2: Mechanism of the reaction of ethene with deuterium over a metal catalyst.

and hydrogen deuteride (Scheme 9.2). The extent of alkene exchange varies greatly from one metal to another, being large with palladium and nickel and small with platinum and iridium; the amount of hydrogen appearing balances the amount of deuterium in the alkene. These observations, which are of long standing, are explained by the Horiuti–Polanyi mechanism, in which hydrogen or deuterium atoms are added in sequence, forming first an adsorbed alkyl radical, which may then acquire a second atom to become an alkane or lose an atom by *alkyl reversal* (either the same or different from that just accepted) to re-form the alkene. Reiteration of these steps, with the recombination of adsorbed atoms, accounts for all the observed products.[13,27] Corresponding reactions take place with the butenes: hydrogenation of 1-butene can afford (besides *n*-butane) both *Z*- and *E*-2-butene via a 2-butyl radical by *double-bond migration* (Scheme 9.3), while the reaction of *Z*-2-butene will give (besides *n*-butane) *Z*-2-butene by *Z-E isomerisation*, as well as a little 1-butene.

While these reactions with the simple alkenes have little practical significance, they epitomise those experienced by more complex molecules, as well as being a source of entertainment for academic scientists, and informative of the basic properties of gold in catalysis.

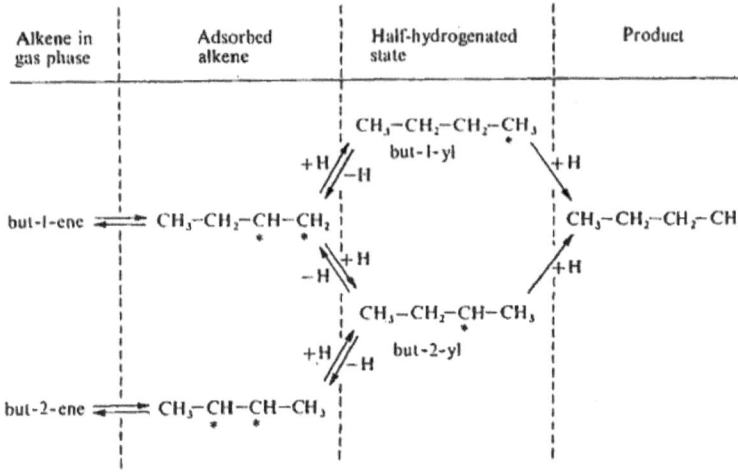

Scheme 9.3: Mechanism of double-bond migration and hydrogenation of butanes over a metal catalyst.

Turning now to what is observed with gold, we may as before note the little work that has been done with massive gold surfaces. On gold microspheres, between 573 and 673 K, 1-butene isomerised to 2-butenes in the absence of hydrogen by an intramolecular rearrangement, rather than suffering dehydrogenation to butadiene, which was the case with palladium and palladium-rich alloys.[28] Hydrogen atoms were found to diffuse through a gold film electroplated into a palladium–gold foil exposed to molecular hydrogen at 383 K, and at the gold surface they were collected by 1-butene or cyclohexene, which reacted with approximately equal rates.[10] Cyclohexene in the presence of hydrogen can either be hydrogenated or above about 473 K can dehydrogenate to form benzene. These reactions have been studied in detail using gold powder between 478 and 558 K; activation energies were, respectively,[29] 61 and 92 kJ mol^{-1}.

Results of somewhat greater interest have been obtained with supported gold catalysts, especially with Au/SiO$_2$ formed by impregnation with aqueous HAuCl$_4$ and its thermal decomposition in air.[5,8,23] Catalysts were active for 1-pentene hydrogenation at 373 K, but activity showed a strange dependence on gold loading; as this was decreased from 5% the specific activity per gold atom passed through a minimum at 1% and then increased progressively, so that it was 7000 times greater at 0.01% gold than at 1.25% gold. This increase in specific activity was linked with the appearance of a mauve colour in the catalyst, and the detection of an epr signal betokening the existence of very small particles. The reaction over 5% Au/SiO_2 gave 2-pentenes as well as *n*-pentane, but with low concentrations of gold there was no isomerisation; why alkyl reversal should not occur on small particles was not explained, nor has there been any subsequent work on this or related systems to throw further light on the matter. Au/Al_2O_3 preparations showed lower activity than the Au/SiO_2 materials, and the specific rate did not increase in the same way when the gold loading was lowered.

The reaction of ethene with hydrogen and with deuterium shed some further light; the rate at 453 K was first order in hydrogen and 0.4 order in ethane, and in the reaction with deuterium the 5% Au/SiO_2 gave deuterated ethenes and ethanes in much the same proportions as given by Pt/SiO_2.[23] The possibility that reaction was caused by traces of platinum in the gold was considered but ruled out.[5] The more likely explanation is that the active sites are atoms of low coordination number, and that these have some vacant *d*-orbitals and hence resemble platinum. Some confirmation of this view is provided by the observation that single gold atoms bonded to the surface of magnesia were active for ethene hydrogenation at 353 K.[30] The hydrogenation of ethene on 5% Au/SiO_2 was also accompanied by the formation of oligomers containing three to six carbon atoms, signifying some C–C bond breaking and re-forming. The significance of this earlier work[5,8,23] lies in the fact that it provided the first recognition of the importance of particle size, but it was not appreciated until re-discovered some 10 years later by Masatake Haruta.

A detailed study has been carried out of the interconversion of benzene and cyclohexane by the transfer of hydrogen atoms. This was detected by placing a radioactive label in one molecule and observing its appearance in the other; this reaction occurred with Au/MgO and Au/Al_2O_3 between 473 and 523 K.[31]

The conversion of cyclohexane to a mixture of (i) cyclohexene and (ii) benzene plus hydrogen occurred on gold film between 469 and 612 K, with activation energies of respectively,[32] 74 and 105 kJ mol^{-1}. Its exchange with deuterium took place stepwise between 423 and 513 K with an activation energy of 27 kJ mol^{-1}.

Other much less detailed results concerning gold-catalysed hydrogenation have been reported. Norbornene was hydrogenated on Au/MgO above 353 K,[33] and propenol (allyl alcohol) was reduced by very small dendrimer-encapsulated gold particles; PdAu particles worked better.[34] Diethylitaconate has been reduced by means of supported Au^{3+} ions, reinforcing the idea that electron-deficient species can also be active,[35] as indeed they seem to be in the water-gas shift (Chapter 10).

9.3.2. Alkynes and alkadienes

Hydrogenation of the C≡C bond and of pairs of C=C bonds, either conjugated or separated, introduces a new feature, because the reaction can proceed in two distinct stages, giving first an alkene and subsequently and alkane. These two processes can also occur at the same time, and the extent to which alkene appears in the products is termed the *degree of selectivity*. In disubstituted alkynes the product alkene adopts predominantly the *Z*-configuration although various amounts of the *E*-form may also appear. With conjugated alkadienes where the distinction is possible, either of the double bonds may be reduced first or by double-bond migration one of them may move: thus 1,3-butadiene, which is the only one so far examined with gold catalysts, the products consist of 1-butene and *Z*- and *E*-2-butenes.[13,36]

Both of these reactions have very important industrial uses (Section 14.3.9). In order to obtain alkene streams of sufficient purity for further use, the products of steam-cracking or catalytic cracking of naphtha fractions must be treated to lower the concentration of alkynes and alkadienes to very low levels (<5 ppm). For example, residual alkynes and dienes can reduce the effectiveness of alkene polymerisation catalysts, but the desired levels of impurities can be achieved by their *selective hydrogenation* (Scheme 9.4) with palladium catalysts, typically Pd/Al$_2$O$_3$ with a low palladium content. A great deal of literature exists,[13,37] particularly on the problem of hydrogenating ethyne in the presence of a large excess of

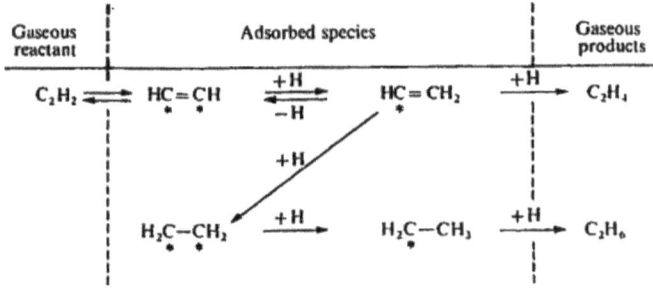

Scheme 9.4: Mechanism of hydrogenation of ethyne over a metal catalyst.

ethene. C_3 and C_4 alkynes and alkadienes must also be eliminated. A further important example of selective hydrogenation of dienes is to be found in *fat hardening* (see below).

A remarkable and potentially very useful property of gold catalysts is that they are essentially fully selective for alkene formation.[20] This is because alkynes and alkadienes are much more strongly adsorbed than the derived alkenes, which are therefore displaced from the surface and unable to return, at least until the first stage of the reaction is complete. The 2-butenes were however found to isomerise after butadiene had completely disappeared.[38]

The results of several studies of the hydrogenation of 1,3-butadiene on gold catalysts are shown in Table 9.2, and for comparison results[13,36] for Pt/Al_2O_3 and Ir/Al_2O_3 catalysts are also given. The butene analyses obtained with gold catalysts fall into two groups, one of which matches those for platinum, where

$$1\text{-butene} > E\text{-2 butene} > Z\text{-2 butene}$$

while the other agrees more closely with that found with Ir/Al_2O_3, where the amount of 1-butene is less and Z-2-butene $> E$-2-butene. There is no obvious explanation for this difference in behaviour; gold film, which must have large crystallites, behaves similarly to small supported particles. Detailed mechanisms to account for butene yields have been presented elsewhere,[13,36] and it is unnecessary to consider

Table 9.2: Hydrogenation of 1,3-butadiene on gold catalysts: selectivities and product distributions, with comparable results for other metals.

Metal	Support	T (K)	Conversion (%)	S	S_1	S_{2Z}	S_{2E}	Reference
Au	Al_2O_3	∼415	53	?	75	8	17	40
Au	SiO_2	∼415	6	?	72	9	19	40
Au	TiO_2	∼415	69	?	70	7	23	40
Au	Al_2O_3	443	?	100	53	30	15	41
Au[a]	Al_2O_3	513	15	99	57	24	19	39
Au	(film)	431	?	100	73	11	16	86
Pt	Al_2O_3	273	—	50	72	10	17	36
Ir	Al_2O_3	273	—	43	59	22	19	36
Pd	Al_2O_3	273	—	100	65	2	33	36

S: % selectivity to total butenes; S_1: %1-butene; S_{2Z}: %Z-2-butene; S_{2E}: %E-2-butene. [a]0.02% Au/Al_2O_3 in multichannel reactor.

them further. Supported gold catalysts lose activity with time,[39] and product distributions sometimes vary with temperature; with Au/TiO_2 and Au/Al_2O_3 the yield of 1-butene decreased with rise in temperature, those of the others increasing correspondingly.[40] Although rates differed considerably, turnover frequencies lay within a narrow band (0.5×10^{-3} to $2 \times 10^{-3} s^{-1}$ at 423 K);[40] the reaction therefore appears to be structure-insensitive. At 443 K the reaction was first order in hydrogen and zero in butadiene, with an activation energy[41] of 36.5 kJ mol^{-1}. In the reaction with deuterium, poor mass-balances were obtained due to exchange with support hydroxyls; their extended exchange with deuterium did not however cure the problem.[41]

Norbornadiene has been hydrogenated on Au/MgO above 353 K, but no details have been given.[33]

Quite the most industrially important application of the selective hydrogenation of dienic compounds is *fat-hardening*.[27] The principal component of vegetable and fish oils are long-chain (mainly C_{18}) unsaturated esters of glycerol, formed from C_{18} fatty acids containing two (linoleic) or three (linolenic) nonconjugated C=C bonds: they are liquid, unpalatable and unstable, and they are rendered usable, as in margarine, by partial hydrogenation, leaving only one double-bond. The glyceride ester of this oleic acid is termed triolein. All the double-bonds have the Z-configuration, and Z–E isomerisation is undesirable, as is full hydrogenation to the ester of stearic acid. These objectives are satisfactorily accomplished by hydrogenation over a nickel catalyst, although the merits of palladium are occasionally urged. Only one study[42] has been made of gold catalysts (1 and 5% Au/SiO_2 and Au/γ–Al_2O_3 made by impregnation); canola oil (a type of rape seed oil) was successfully hydrogenated between 423 and 523 K at 3.5–5.6 atm pressure, and higher maximum yields of the monoene were obtained than that given by a standard nickel catalyst. The product was colourless and devoid of dissolved gold, and complete reduction of the linolenic component was achieved with less formation of E-isomers than was possible with nickel. A number of potential uses for gold catalysts in fat hardening were mentioned, and it is a complete mystery why their use has not be exploited, especially since any gold residues in the product would be less harmful than nickel.

There have been two recent studies of the hydrogenation of ethyne on gold catalysts. With 1% Au/TiO_2 made by thermally decomposing the supported complex $[Au_2(PPh_3)_6](BF_4)_2$ and having particles of average size 4.6 nm, reaction occurred at 453 K with about 90% selectivity to ethene, but activity was lost quite rapidly.[43] However, the selectivity shown under the conditions used to Pd/TiO$_2$ and PdAu/TiO$_2$ catalysts was initially only 10–30%; normally such catalysts are expected to give very high selectivities, so like the clock that struck 13 this casts a shadow of suspicion on what has gone before. In the second study, activity with Au/Al$_2$O$_3$ was found between 313 and 523 K; at 360 K the ethene selectivity was 100%, the activation energy 34 kJ mol^{-1} and the orders of reaction in ethyne and hydrogen were respectively, 0.1 and 0.4.[44] Both reactants were therefore chemisorbed, the former more strongly. 2-Butene has been selectively reduced by Au/boehmite (one of the hydrated oxides of aluminium) at 400–490 K, giving 80% of Z-2-butene and equal amounts of the other isomers.[41] It is hard to understand why these very attractive properties have not been more fully exploited by the petrochemical industry.

9.3.3. Aromatic molecules

The aromatic nucleus is usually more difficult to hydrogenate than other forms of the C–C multiple bond, but when it occurs it normally proceeds directly to the cycloalkane, and only over certain ruthenium catalysts are significant amounts of cycloalkenes formed. The hydrogenation of naphthalene proceeds in stages *via* tetralin, then octalin and finally decalin (Scheme 9.5); to comply with new specifications for mid-distillate fuels intended for use in diesel engines, the naphthalene component needs to be fully hydrogenated and then further cracked to form alkylcyclohexanes. Au/γ–Al$_2$O$_3$ and Au/SiO$_2$ catalysts were both active for hydrogenation to decalin at 448 K, although both deactivate quickly; a bimetallic Pt$_1$–Au$_1$/SiO$_2$ was however somewhat more stable.[45] Strong adsorption of the intermediate tetralin by large gold particles favoured the use of more highly

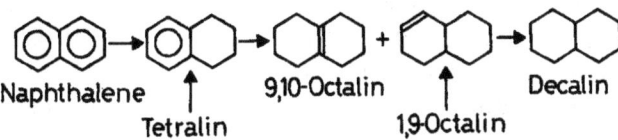

Scheme 9.5: Hydrogenation of naphthalene: note, the other possible octalin isomers are usually converted to the two shown by double-bond migration, since the C=C bond is most stable when most heavily substituted.

dispersed catalysts. Bimetallic Pd–Au/Al$_2$O$_3$ catalysts also performed better than Pd/Al$_2$O$_3$ for hydrogenating naphthalene and toluene in the presence of dibenzothiophene.[46] Benzene was slowly hydrogenated on gold powder at high temperature.[47]

9.3.4. Related reactions

Gold has only low activity for the hydrogenolysis of C–C bonds, which even on metals such as platinum or palladium needs quite high temperatures. *Neo*pentane was found to react over gold powder at 709–741 K, the activation energies for hydrogenolysis and for skeletal isomerisation to *iso*pentane being respectively[48] 213 and 201 kJ mol^{-1}; isomerisation was the main reaction. *n*-Dodecane reacted over Pt/HY zeolite and PtAu/HY between 543 and 623 K to give C$_4$–C$_7$ cracked alkanes and *iso*dodecane; at the same degree of conversion (43%) the Pt$_4$Au$_1$ composition gave maximum isomerisation, to the extent of 80% at 588 K.[49] The hydrocracking of polyethylene has also been effected by a gold catalyst.[50] Sulfur atoms formed a monolayer (S:Au = 1:2) on polycrystalline gold at 543–603 K, and were removed by hydrogen above 443 K.[51]

On gold film at 195 K a fast exchange was observed between CH$_3$SiH$_3$ and CH$_3$SiD$_3$, but no deuterium entered the methyl group.[52]

9.4. Chemoselective Hydrogenation

The term 'chemoselective' applies to the selective reduction of one reducible function in the presence of another type in the same molecule. The selective reduction of the C=O bond when conjugated to a C=C bond is a major challenge to the catalytic chemist; over metals of Group 10 the C=C bond is much the more reactive, and hydrogenation of the C=C–C=O group almost invariably leads first to the saturated aldehyde or ketone, together with the saturated alcohol (Scheme 9.6). There is keen industrial interest in the synthesis of unsaturated alcohols, such as would be produced by selective reduction of C=O function; they are required as flavours and fragrances, and as intermediates in the synthesis of pharmaceutical products.[7,53] Very considerable success has attended efforts to modify platinum catalysts by metal chloride promoters such as GeCl$_4$ and SnCl$_2$, and with certain types of molecule very high degrees of chemoselectivity have been reported. It was therefore natural to explore the behaviour of gold catalysts for this type of

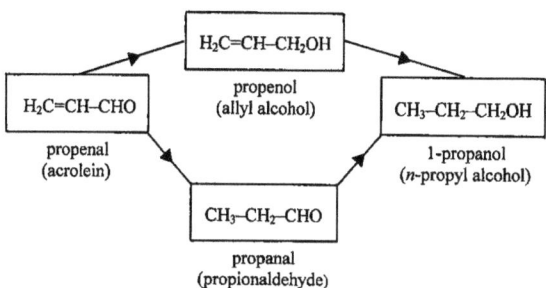

Scheme 9.6: Hydrogenation of unsaturated aldehydes and ketones, illustrated by propenal.

reaction, and some very positive results have been obtained.[6,7,25,54] A fair measure of understanding the locus of the desired reaction is also available.

Four main factors determine the degree of chemoselectivity that can be obtained: (i) the architecture of the reactant molecule, viz. the number and kind of substituents around the C=C–C=O group, (ii) the size, shape and morphology of the gold particles, which is partly determined by the choice of support,[54] (iii) the presence of modifiers and (iv) the operating conditions. Regarding (i), the substituents around the C=C group are important, since if alkyl groups are present they may provoke steric hindrance to the chemisorption of the double bond and thus increase selectivity to the unsaturated alcohol.[55–57] Even the methyl group can so act; it is always found that propenal (acrolein) affords lower selectivity than does 2-butenal (crotonaldehyde) (Table 9.3), and high selectivity at high conversion (>85% at 100% conversion) has been seen with 1-phenyl-2-butenal (cinnamaldehyde) using Au/Fe$_2$O$_3$ (see Scheme 9.7 for structures).[55] Au/Al$_2$O$_3$ proved to be less selective, however. Very favourable results were also obtained using Au/Fe$_2$O$_3$ with citral, the E-isomer giving geraniol and the Z-isomer nerol, with selectivity exceeding 96%, there being only trace amounts of citronellal and citronellol formed[55] (see Scheme 9.7 for structures). This splendid result must surely be due in part to the steric hindrance provided by the methyl groups about the C=C bonds. With unsaturated ketones,[55,56] one methyl group on the terminal carbon was not enough to produce high selectivity because the ketonic methyl group also discouraged adsorption of the C=O bond (Table 9.4), although two methyl groups or a phenyl group led to selectivities above 60%. When no methyl groups were in the terminal position, there was a catastrophic loss of selectivity[57] (see also Table 9.4).

Table 9.3: Hydrogenation of unsaturated aldehydes and ketones (S is the selectivity to unsaturated alcohol).

Reactant	Metal (promoter)	Support	Method	[Au] (%)	d_{Au} (nm)	T (K)	S	Reference
Propenal	Au	ZnO	IMP	1.6	9.0	593	34	7,25,62
	Au (In)	ZnO	IMP	1.6	10.1	593	63	7,25,62
	Au	TiO$_2$	DP	7.5	1.9	393	30	7
	Au	TiO$_2$	DP	1.7	5.3	513	23	39
	Au	ZrO$_2$	DP	1.2	2.1, 7.4	513	43	39
	Au	ZrO$_2$	DP	2.2	7.7	513	35	7
	Au	SiO$_2$	IMP	1.6	3.9	513	23	7
	Pt (Sn)	Nylon	—	~1	—	318	65	87
2-Butenal	Au (S)	ZnO	COPPT	5	~3	523	54	61
	Au (S)	ZnO	COPPT	5	~3	523	87	60
	Au	ZrO$_2$	COPPT	5	~3	523	51	61
	Au (S)	ZrO$_2$	COPPT	5	~3	523	48	61
	Au	TiO$_2$	DP urea	7.5	1.9	393	68	6
	Au	TiO$_2$	DP	~1	3.4	500	24	40
	Au	SiO$_2$	IMP	5	—	523	~0	61
	Au	SiO$_2$	IMP	1.6	3.9	513	39	7
	Au	SiO$_2$	GG^a	~1	5.3	500	9	40
	Au	Al$_2$O$_3$	DP	~1	4.6	500	3	40
1-Phenylpropenal	Pt (Ge)	Nylon	—	2	—	333	95	7
	Pt (Sn)	Nylon	—	~1	—	333	75	87
Benzalacetone	Au	Fe$_2$O$_3$	COPPT	3.1	—	333	57	55,56
	Au	Fe$_2$O$_3$	COPPT	16.6	6.6	333	67	55,56
	Au	Fe$_2$O$_3$	COPPT	5.4	39.4	333	61	55,56
	Au	Al$_2$O$_3$	DP	2.3	4.3	333	11	55,56

^{a}GG is the gas-phase grafting with Me$_2$Au(acac).

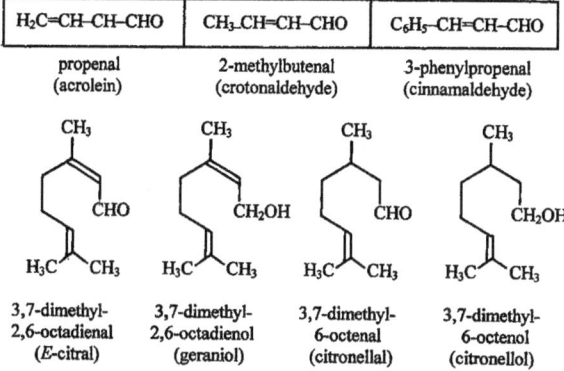

H₂C=CH–CH–CHO	CH₃–CH=CH–CHO	C₆H₅–CH=CH–CHO
propenal (acrolein)	2-methylbutenal (crotonaldehyde)	3-phenylpropenal (cinnamaldehyde)

3,7-dimethyl-2,6-octadienal (*E*-citral)   3,7-dimethyl-2,6-octadienol (geraniol)   3,7-dimethyl-6-octenal (citronellal)   3,7-dimethyl-6-octenol (citronellol)

Scheme 9.7: Structures of unsaturated aldehydes and reaction products of citral.

Table 9.4: Chemoselective hydrogenation of unsaturated ketones to secondary alcohols.

Catalyst	T (K)	R_1	R_2	R_3	S^a	Reference
5.4% Au/Fe₂O₃	333	H	C₆H₅	H	61	55,56
5.4% Au/Fe₂O₃	333	H	CH₃	H	16	55,56
5.4% Au/Fe₂O₃	333	CH₃	CH₃	H	65	55,56
25% Au/ZrO₂	~380	CH₃	H	H	58	57
25% Au/ZrO₂	~380	H	CH₃	H	45	57
25% Au/ZrO₂	~380	H	H	CH₃	9	57

[a] Selectivity to unsaturated alcohol.

The liquid-phase hydrogenation of *E*-4-phenyl-3-buten-2-one (benzalacetone) has been followed with gold supported on various forms of iron oxide, reduced at only 343 K; the sequence of selectivities was

FeO(OH)(goethite) > γ-Fe₂O₃(maghemite) > α-Fe₂O₃(haematite)

but on low-area haematite (6 m² g⁻¹) the particles were large (62 nm) and the selectivity was a mere 6%.[54]

The size and shape of the gold particles are clearly important parameters in determining activity and selectivity. The conclusions described in the literature are not always perfectly in accord,[6] but new insights are obtained when reasons for discrepancies are sought. We have to recognise, and try to rationalise, the following observations. (i) Activity and selectivity do not

necessarily vary with particle size in the same way for all catalysts.[6] (ii)
The shape of the particles depends on both the support and the conditions
of pretreatment.[6,7] (iii) Particles may be either single crystals or multiply-
twinned.[25] Failure to recognise the importance of (ii) and (iii) may go some
way to accounting for divergent experimental results.[9,58]

It is not unreasonable to expect that reaction characteristics will depend
on particle size, since this is very important with other reactions. It does
not however follow that the reductions of both the C=C and C=O functions
will respond to size variation in the same way. Thus the rate of 2-butenal
hydrogenation on Au/TiO$_2$ was seen to rise rapidly as particle size was
decreased;[6] in other work[59] a decreased activity at about 1 nm was ascribed
to a possible disappearance of small particles into the support. Despite the
variation in rate over Au/TiO$_2$, the selectivity remained about constant;[6]
this was explained by the suggestion that the *rate* depended on the availability of hydrogen atoms, since as we have seen it is now quite clear that small
gold particles can chemisorb some hydrogen,[6,9] while *selectivity* depended
on the mode of adsorption of the 2-butenal (see Table 9.3). We shall return
to this matter later. With another support (ZnO), the selectivity for 2-
butenol increased with the temperature used for reducing the catalyst,[60,61]
and thus with particle size; this may have been due to either improved
crystallinity of the gold particles or to some decoration by zinc atoms.

The shape adopted by gold particles depends on the interfacial chemistry; after hydrogen pretreatment at 573 K they have been seen as hemispherical on silica and titania, but more crystalline with extended facets
on other supports such as zinc oxide.[7] Increasing the temperature of pretreatment of Au/TiO$_2$ not only led to larger single-crystal particles, but
these crystals exhibited larger facets the structure of which could be identified by TEM;[6] the (111) plane was parallel to the (110) plane of the rutile
support. Quite small crystals can be multiply-twinned, and this turns out
to be important, since single crystals have shown better selectivities to
propenol than multiply-twinned assemblies of about same size.[7] The occurrence of multiple-twinning can be recognised not only by TEM,[7,25] but also
by disparities between estimates of size obtained by TEM and XRD or
XAFS, since the last two methods differentiate orientations of the crystal
lattice, while TEM (unless performed at high resolution) merely sees the
whole gathering. Quite large differences (to almost tenfold) are sometimes
shown, the XAFS value being of course the smaller.[9] These considerations
may go some way to explaining the failure to observe unequivocal size

effects in the reducing of unsaturated ketones,[56] as well as the reluctance of Au/Al_2O_3 and Au/SiO_2 to give significant selectivities to unsaturated alcohols (Table 9.3).

There is one more important factor to consider. There are several examples of effective modification of gold catalysts by selectivity promoters;[7,25,60,61] their beneficial effects on platinum catalysts have been known for some time (see Table 9.3 for an example). Partial sulfidation of Au/ZnO and Au/ZrO_2 led to much improved selectivity with 2-butenal, and usually this also led to an increase in rate.[60] When indium was applied to reduced Au/ZnO by impregnation with the nitrate, the rate of propenal hydrogenation was halved but the selectivity increased almost twofold[7,25,62] (see Table 9.3). Careful HRTEM image analysis showed that the indium covered the plane surfaces of the facetted crystallites, leaving the edges clear; these were thus thought to be where the selective reaction occurred. This is quite reasonable, because this is where hydrogen molecules can chemisorb, and their atoms apparently cannot migrate elsewhere. Now even the best catalysts will simultaneously also hydrogenate the C=C bond, giving either the saturated aldehyde or the saturated alcohol (Scheme 9.6) also at the edge sites, so that which of the two bonds is preferentially reduced will depend on which is preferentially engaged to the active centres. It may be that some polarisation of the Au–M bond (M = In, S) encourages chemisorption at the polar C=O; but detailed explanation of the role of modifiers and of the other factors (such as molecular structure) on selectivity is still awaited.

9.5. Hydrogenation of Carbon Dioxide

Methanol is one of the ten largest organic chemicals, manufactured worldwide on a scale of some 30×10^6 tonnes per year: it is made by hydrogenation of a mixture of carbon monoxide using a $Cu–ZnO/Al_2O_3$ developed by ICI in 1966.[63] It is now recognised that the reaction proceeds by first reducing the dioxide to the monoxide, the relevant equations being

$$H_2 + CO_2 \rightarrow CO + H_2O, \qquad \Delta H_{298} = +41.2\,\text{kJ mol}^{-1}, \qquad (9.3)$$

$$2H_2 + CO \rightarrow CH_3OH, \qquad \Delta H_{298} = -90.6\,\text{kJ mol}^{-1}, \qquad (9.4)$$

$$3H_2 + CO \rightarrow CH_3OH + H_2O, \quad \Delta H_{298} = -49.5\,\text{kJ mol}^{-1}. \qquad (9.5)$$

The equilibrium concentration of methanol therefore decreases as temperature rises and as pressure is lowered; the maximum conversion using carbon monoxide at 50 bar and 533 K would be about 50%. The first of these three reactions is the reverse water-gas shift. An outline reaction scheme is shown

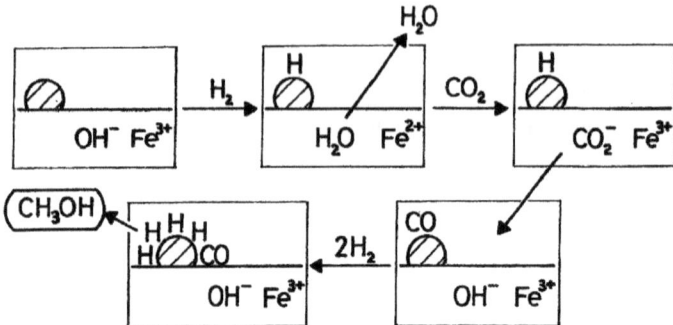

Scheme 9.8: A mechanism for the synthesis of methanol from carbon dioxide on Au/Fe_2O_3.

in Scheme 9.8; with copper catalysts, which have been intensively studied,[63] some methane can also be formed. Several studies using gold catalysts have been reported,[64-67] and although they are much less active than the Cu–ZnO/Al$_2$O$_3$ catalyst they have the advantage of being highly selective in converting carbon dioxide to methanol at least at low conversion,[64] but at higher conversions parasitic reactions giving methanal (formaldehyde) and carbon monoxide cause the selectivity to fall. No methane was formed over Au/ZrO_2,[65] but its formation occurred above 523 K with Au/TiO$_2$ and Au/Fe_2O_3.[67]

The principal features of the characteristics of gold catalysts in this reaction were laid down in early work by Masatake Haruta and his associates.[2,67,68] As with so many other reactions, success hinges on selecting the appropriate support and method of preparation. With 5% Au/MO_x at 523 K and 8 or 50 bar pressure, hydrogenation of carbon dioxide gave conversions that increased with the electronegativity of the cation M (i.e. with the acidity of the support), but the selectivity to methanol decreased, due to the onset of the reverse water-gas shift.[2,67,68] Use of a strongly basic support such as zinc oxide appears to be essential in achieving good methanol selectivity; the same is also reported for 33% Au/ZnO and $Au/ZnFe_2O_4$, which were actually better than Cu–ZnO/Al$_2$O$_3$.[2,67] The importance of the support acidity suggests that part of the reaction at least occurs on the support or perhaps at the perimeters of the particles (Scheme 9.8), and indeed the water-gas shift and its reverse involve the active participation of the support (see Chapter 10).

Various binary supports have been used, including $ZnFe_2O_4$ (see above), $ZnTiO_3$ (which gave a maximum methanol *yield* at [Zn] = 40%)[67] and $3ZnO \cdot ZrO_2$;[64] with this last, and a very high gold concentration (62.5%), Au_3Zn was formed together with Au^0 under reducing conditions. Exposing the amorphous alloy $Au_{25}Zr_{75}$ to reaction conditions at 493 K gave an Au/ZrO_2 catalyst with 8 nm particles,[66] and this was active in the range 413–533 K, but the maximum selectivity ($\sim$30%) decreased with increasing temperature.

On the basis of these results, it appears unlikely than any gold catalyst will supplant the widely used $Cu–ZnO/Al_2O_3$, unless the selectivity consideration becomes a critical factor.

9.6. Dehydrogenation

9.6.1. Introduction

This last section is concerned with reactions in which hydrogen is a product, often through the decomposition of an organic molecule. Dehydrogenation in the presence of oxygen, i.e. oxidative dehydrogenation, where water is the product, was covered in Chapter 8. Gold is particularly adept for dehydrogenation because of its ability to accept hydrogen as atoms from a chemisorbed molecule, especially from one with a functional group that ensures its attachment to the surface. The observation[69,70] that gold could catalyse the decomposition of ammonia was probably the first manifestation of its catalytic power, but a number of early patents also attest to its ability in this direction. A possible future application for catalytic dehydrogenation would develop if substances such as methanol or methanoic (formic) acid were to be adopted as hydrogen reservoirs. Methanol is available from a number of renewable sources as well as by synthesis as described in Section 9.5, and its ready decomposition to yield up its two molecules of hydrogen is a reaction of great interest. It is even better to *reform* methanol with water so as to utilise the hydrogen that it also contains:

$$CH_3OH + H_2O \rightarrow 3H_2 + CO_2. \tag{9.6}$$

9.6.2. Methanol

Methanol physically adsorbed on Au(310), a surface that contains both steps and kinks, suffered breaking of its O–H bond above 150 K, the methoxy group being stable to 500 K;[71] This did not happen with either the (110) surface, which is stepped, or with the flat (100) surface, highlighting once again the superior activity of gold atoms of low coordination number.

Au/TiO$_2$ catalysts made by impregnation were active at room temperature for the photocatalytic reforming of methanol in deionised water; rate maxima at about 0.2 and 2% gold were observed[72] (not easily explained). It was suggested that the reaction involved band-gap excitation of electrons in the titania, to produce O$^-$ ions that then converted methanol via methoxy species to hydrogen and carbon dioxide. Sites at the metal–support interface were thought to be responsible.

Decomposition and reforming of methanol on Au/ZnO and Au/TiO$_2$ catalysts have been intensively studied using FT-IR and quadrupole mass-spectrometry.[73] Unlike the Cu–ZnO system, no formate groups were observed on Au/ZnO; because of the larger size of the gold atom, the same kind of solid solution precursor is not formed, and gold is therefore unable to activate the support in the same way.

Addition of gold to Pd/CeO$_2$ increased activity for methanol decomposition at 453 K; it was suggested that this was due to formation of bimetallic palladium–gold ensembles.[58]

The oxidative dehydrogenation of methanol to methanal is mentioned in Chapter 8.

9.6.3. Methanoic acid

Methanoic (formic) acid decomposes on metal surfaces according to the equation

$$HCOOH \rightarrow H_2 + CO_2. \qquad (9.7)$$

This reaction has been described as 'belonging to the museum of catalysis', and indeed it has not been studied for a number of decades. This short section therefore relies entirely on work performed before 1970.

The reaction on gold was studied by Dutch scientists in the 1960s as part of their comprehensive examination of the process, using both Au/SiO$_2$ and gold powder.[74–76] As part of this study, the mechanism was unravelled by careful use of isotopically labelled reactants: at 423 K the decomposition of unlabelled methanoic acid in the presence of deuterium gave only light hydrogen (H$_2$), decomposition of the monodeutero-compound (HCOOD) gave an equilibrium mixture of the three hydrogen isotopes, as did the decomposition of a mixture (HCOOH + DCOOD). It was also observed that although the hydrogen + oxygen reaction was slow at 393 K, methanoic acid was quickly oxidised at this temperature, and oxidation of its mixture with deuterium gave only light water (H_2O) and no HDO or D_2O. It appears that deuterium cannot access the surface in the presence of methanoic acid, although the atoms that it releases combine randomly.

Remarkably consistent activation energies of about $60 \, kJ \, mol^{-1}$ were found in this early work with film,[77] wire[78] and Au/SiO_2.[75] The nature of the active sites has also been considered;[79] using 'spongy crystalline cohesive pellets' the reaction slowed down over 24 h, and it was thought to occur at defects that might possibly be associated with impurities. Activation energies derived from a rate *constant* were about $50 \, kJ \, mol^{-1}$.

9.6.4. Other reactions producing hydrogen

It merely remains to note very briefly a number of other reactions catalysed by gold that produce hydrogen; they mainly derive from the older literature.

We have already noted the oldest demonstration of gold catalysis, namely, the decomposition of ammonia;[69,70] this work has not apparently been repeated. Hydrogen peroxide decomposition took place on gold wire at moderate temperature with an activation energy[80] of $61.5 \, kJ \, mol^{-1}$, the near-identity of this value with those for methanoic acid decomposition suggests that recombinative desorption of hydrogen atoms is the slow step,[81] and it may therefore equate to the dissociation energy of the Au–H bond. Photocatalytic dissociation of the water molecule was catalysed by gold supported on the perovskite $K_2La_2Ti_3O_{10}$ in aqueous KOH.[82] Hydrogen iodide decomposition was followed on gold wire between 803 and 1090 K by C.N. Hinshelwood in the very early days of quantitative catalytic work; the reaction followed zero order kinetics with an activation energy of $105 \, kJ \, mol^{-1}$. We have already noted a number of patents dating from before World War II claiming dehydrogenation of ethane-1,2-diol (ethylene glycol),[83] cyclohexanone,[84] and ethylpyridine.[85]

References

1. L. Stobiński and R. Duś, *Appl. Surf. Sci.* **62** (1992) 77.
2. G.K. Boreskov, V.I. Savchenko and V.V. Gorodetskii, *Dokl. Akad. Nauk SSSR* **189** (1969) 537.
3. M.V. Kislyuk and I.I. Tret'yakov, *Kinet. Katal.* **17** (1976) 1515; *Chem. Abstr.* **86** (1977) 79319.

4. A.G. Sault, R.J. Madix and C.T. Campbell, *Surf. Sci.* **169** (1986) 347.
5. G.C. Bond, P.A. Sermon, G. Webb, D.A. Buchanan and P.B. Wells, *J. Chem. Soc. Chem. Commun.* (1973) 444.
6. A. Zanella, C. Louis, S. Giorgio and R. Touroude, *J. Catal.* **223** (2004) 328.
7. P. Claus, *Appl. Catal. A: Gen.* **291** (2005) 222.
8. G.C. Bond and P.A. Sermon, *Gold Bull.* **6** (1973) 102.
9. E. Bus, J.T. Miller and J.A. van Bokhoven, *J. Phys. Chem. B* **109** (2005) 14581.
10. R.S. Yolles, B.J. Wood and H. Wise, *J. Catal.* **21** (1971) 66 and references therein.
11. G.C. Bond and V. Ponec, *Catalysis by Metals and Alloys*, Elsevier, Amsterdam, 1996, Ch. 10.
12. G.C. Bond, *Catalysis by Metals*, Academic Press, London 1962, Ch. 8.
13. G.C. Bond, *Metal-Catalysed Reactions of Hydrocarbons*, Springer, New York, 2005, Ch. 3.
14. D.D. Eley and J.R. Rossington, in *Chemisorption*, W.E. Garner, (ed.), Butterworths, London, 1957, p. 137.
15. I. Iida, *Bull. Chem. Soc. Jpn.* **52** (1979) 2858.
16. A. Couper and D.D. Eley, *Discuss. Faraday Soc.* **8** (1950) 172.
17. S.J. Holden and D.R. Rossington, *J. Phys. Chem.* **68** (1964) 1061.
18. H. Wise and K.M. Sancier, *J. Catal.* **2** (1963) 149.
19. R.J. Mikovsky, M. Boudart and H.S. Taylor, *J. Am. Chem. Soc.* **76** (1954) 3814.
20. D.J.C. Yates, *J. Coll. Interface. Sci.* **29** (1969) 194.
21. A.C. Gluhoi, H.S. Vreeburg, J.W. Bakker and B.E. Nieuwenhuys, *Appl. Catal. A: Gen.* **291** (2005) 93.
22. S. Galvagno and G. Parravano, *J. Catal.* **55** (1978) 139.
23. P.A. Sermon, G.C. Bond and P.B. Wells, *J. Chem. Soc. Faraday Trans.* **75** (1979) 385.
24. S. Naito and M. Tanimoto, *J. Chem. Soc. Chem. Commun.* (1988) 832.
25. P. Claus, H. Hofmeister, Ch. Mohr and J. Radnik, EUROPACAT VI (2003); P. Claus, H. Hofmeister and Ch. Mohr, *Gold Bull.* **37** (2004) 181.
26. P.G. Dickens, J.W. Linnett and W. Palczewska, *J. Catal.* **4** (1965) 140.
27. G.C. Bond, *Heterogeneous Catalysis: Principles and Applications*, 2nd edn., Oxford Univ. Press, 1987.
28. S.H. Inami, B.J. Wood and H. Wise, *J. Catal.* **13** (1969) 397.
29. R.P. Chambers and M. Boudart, *J. Catal.* **5** (1966) 517.
30. J. Guzman and B.C. Gates, *Angew. Chem. Int. Ed.* **42** (2003) 690; *J. Catal.* **226** (2004) 111.
31. G. Parravano, *J. Catal.* **11** (1968) 269; **18** (1970) 320.
32. J. Erkelens, C. Kemball and K. Galwey, *Trans. Faraday Soc.* **59** (1963) 1181.
33. V. Amir-Ebrahimi and J.J. Rooney, *J. Molec. Catal.* **67** (1991) 339.
34. R.W.J. Scott, O.M. Wilson, S.-K. Oh, E.A. Kenik and R.M. Crooks, *J. Am. Chem. Soc.* **126** (2004) 15583.
35. S. Carrettin, A. Corma, M. Iglesias and F. Sánchez, *Appl. Catal. A: Gen.* **291** (2005) 247.
36. P.B. Wells, in *Surface Chemistry and Catalysis*, A.F. Carley, P.R. Davis, G.J. Hutchings and M.S. Spencer, (eds.), Kluwer, Dordrecht, 2003.
37. A. Borodziński and G.C. Bond, *Catal. Rev. Sci. Eng.*, accepted.
38. Ho-Geun Ahn and H. Niiyama, *Kongop Hwahak* **12** (2001) 920.
39. S. Schimpf, M. Lucas, C. Mohr, U. Rodemerck, A. Brückner, J. Radnik, H. Hofmeister and P. Claus, *Catal. Today* **72** (2002) 63.
40. M. Okumura, T. Akita and M. Haruta, *Catal. Today* **74** (2002) 265.

41. D.A. Buchanan and G. Webb, *J. Chem. Soc. Faraday Trans. I* **70** (1978) 134.
42. L. Caceres, L.L. Diosady, W.F. Graydon and L.J. Rubin, *J. Am. Oil Colour Chem.* **62** (1985) 906.
43. T.V. Choudary, C. Sivadinarayana, A.K. Datye, D. Kumar and D.W. Goodman, *Catal. Lett.* **86** (2003) 1.
44. J. Tia, K. Haraki, J.N. Kondo, K. Domen and K. Tamaru, *J. Phys. Chem. B* **104** (2000) 11153.
45. B. Pawelec, A.M. Venezia, V. La Parola, S. Thomas and J.L.G. Fierro, *Appl. Catal. A: Gen.* **283** (2005) 165.
46. B. Pawelec, E. Cano-Serrano, J.M. Campos-Martin, R.M. Navarro, S. Thomas and J.L.G. Fierro, *Appl. Catal. A: Gen.* **275** (2004) 127.
47. D.A. Cadenhead and N.G. Masse, *J. Phys. Chem.* **70** (1966) 3558.
48. M. Boudart and L.D. Ptak, *J. Catal.* **16** (1970) 90.
49. G. Riahi, M. Gąsior, B. Grzybowska, J. Haber, M. Polisset-Thfoin and J. Fraissard, *Proc. 13th Internat. Congr. Catal.*, Paris, 2004, pp. 1–130.
50. M. Legate and P.A. Sermon, in *Proc. 6th Internat. Congr. Catal.*, G.C. Bond, P.B. Wells and F.C. Tompkins, (eds.), Chem. Soc. London, 1976, p. 603.
51. P. Leitgeb and B. Bechtold, *Z. Phys. Chem. (Frankfurt)* **95** (1975) 165.
52. D.I. Bradshaw, R.B. Moyes and P.B. Wells, *J. Chem. Soc. Chem. Commun.* (1975) 137; *Proc. 6th Internat. Congr. Catal.*, G.C. Bond, P.B. Wells and F.C. Tompkins, (eds.), Chem. Soc. London, 1976, p. 1042.
53. B. Chen, U. Dingerdissen, J.G.E. Krauter, H.G.J. Lansink Rotgerink, K. Möbus, D.J. Ostgard, P. Panster, T.H. Rietmeier, S. Seebald, T. Tacke and H. Trauthwein, *Appl. Catal. A: Gen.* **280** (2005) 17.
54. C. Milone, R. Ingoglia, L. Schipilliti, G. Neri and S. Galvagno, *J. Catal.* **236** (2005) 80.
55. C. Milone, M.L. Tropeano, G. Gulino, G. Neri, R. Ingoglia and S. Galvagno, *Chem. Commun.* (2003) 868.
56. C. Milone, R. Ingoglia, A. Pistone, G. Neri, F. Frusteri and S. Galvagno, *J. Catal.* **222** (2004) 348.
57. M. Shibata, N. Kawata, T. Masumoto and H. Kimura, *J. Chem. Soc. Chem. Commun.* (1988) 154.
58. M.P. Kapoor, Y. Ichiyashi, T. Nakamori and Y. Matsuura, *J. Molec. Catal. A: Chem.* **213** (2004) 251.
59. P. Claus, A. Brückner, C. Mohr and H. Hofmeister, *J. Am. Chem. Soc.* **122** (2000) 11430.
60. J.J. Baillie, H.A. Abdullah, J.A. Anderson, C.H. Rochester, N.V. Richardson, N. Hodge, J.-G. Zhang, A. Burrows, C.J. Kiely and G.J. Hutchings, *Phys. Chem. Chem. Phys.* **3** (2001) 4113.
61. J.J. Baillie and G.J. Hutchings, *Catal. Commun.* (1991) 2151.
62. C. Mohr, H. Hofmeister, J. Radnik and P. Claus, *J. Am. Chem. Soc.* **125** (2003) 1905.
63. R.J. Farrauto and C.H. Batholomew, *Fundamentals of Industrial Catalytic Processes*, Chapman and Hall, London, 1997.
64. J. Stoczyński, R. Grabowski, A. Kozłowska, P. Olszewski, J. Stoch, J. Skrzypek and M. Lachowska, *Appl. Catal. A: Gen.* **278** (2004) 11.
65. A. Baiker, M. Kilo, M. Maciejewski, S. Menzi and A. Wokaun, in *Proc. 10th Internat. Congr. Catal.*, L. Guczi, F. Solymosi and P. Tétényi, (eds.), Elsevier, Amsterdam, 1993, Vol. B, p. 1257.

66. R.A. Koeppel, A. Baiker, C. Schild and A. Wokaun, *J. Chem. Soc. Faraday Trans.* **87** (1991) 2821.
67. H. Sakurai and M. Haruta, *Catal. Today* **29** (1996) 361; *Appl. Catal. A: Gen.* **127** (1995) 93; H. Sakurai, S. Tsubota and M. Haruta, *Appl. Catal. A: Gen.* **102** (1993) 125.
68. M. Haruta, *Cat. Surveys Japan* **1** (1997) 61.
69. P.L. Dulong and L.G. Thenard, *Ann. Chim. Phys.* **23** (1823) 440.
70. A.J.B. Robertson, *Catalysis of Gas Reactions by Metals*, Logos, London, 1970.
71. C.P. Vinod, J.W. Niemantsverdriet and B.E. Nieuwenhuys, *Appl. Catal. A: Gen.* **291** (2005) 93.
72. M. Bowker, L. Millard, J. Greaves, D. James and J. Soares, *Gold Bull.* **37** (2004) 170.
73. M. Manzoli, A. Chiorino and F. Boccuzzi, *Appl. Catal. B: Env.* **57** (2004) 201.
74. G.C. Bond and D.T. Thompson, *Catal. Rev.-Sci. Eng.* **41** (1999) 319.
75. J. Fahrenfort, L.L. van Reijen and W.M.H. Sachtler, in *The Mechanism of Heterogeneous Catalysis* J.H. de Boer, (ed.), Elsevier, Amsterdam, 1960, p. 23.
76. W.M.H. Sachtler and N.H. de Boer, *J. Phys. Chem.* **64** (1960) 1579; P. Mars, J.J.F. Scholten and P. Zwietering, *Adv. Catal.* **14** (1963) 35.
77. J.K.A. Clarke and E.A. Rafter, *Z. Physikal. Chem. NF* **67** (1969) 169.
78. D.D. Eley and P. Luetic, *Trans. Faraday Soc.* **53** (1957) 1483.
79. M.A. Bhata and H.A. Taylor, *J. Chem. Phys.* **44** (1966) 1264.
80. D.D. Eley and D.M. MacMahon, *J. Coll. Interface Sci.* **38** (1972) 502.
81. V.M. Berenblit and G.L. Pavlova, *Zhur. Priklad. Khim.* **43** (1970) 1057; *Chem. Abstr.* **73** (1970) 48908.
82. Y.-W. Tai, J.-S. Chen, C.-C. Yang and B.-Z. Wan, *Catal. Today* **97** (2004) 95.
83. French Patent 2,007,925.
84. C.N. Hinshelwood and C.R. Pritchard, *J. Chem. Soc.* **127** (1925) 1552.
85. British Patent 1,152,817; U.S. Patent 3,476,808.
86. U.S. Patent 3,553,220.
87. R.B. Moyes, P.B. Wells, J. Grant and N.Y. Salman, *Appl. Catal. A: Gen.* **229** (2002) 251.
88. Z. Poltarzewski, S. Galvagno, R. Pietropaolo and P. Staiti, *J. Catal.* **102** (1986) 190.

CHAPTER 10

The Water-Gas Shift

10.1. Introduction[1-4]

The development of the industrial economy in the 19th century required a source of cheap and easily transportable form of energy for space-heating, lighting and industrial processes. The carbonisation of coal provided *coal-gas*, consisting of hydrogen, carbon monoxide, methane and ethane, as well as coal tar and a residue termed *coke*. The passage of steam through a bed of coke made red-hot by a flow of air produced *water-gas* by the reactions:

$$C + H_2O \rightarrow CO + H_2, \qquad \Delta H^0 = 131.2\,kJ\,mol^{-1}, \qquad (10.1)$$
$$C + 2H_2O \rightarrow CO_2 + 2H_2, \qquad \Delta H^0 = 90.0\,kJ\,mol^{-1}. \qquad (10.2)$$

Since these reactions are endothermic, the coke was cooled, and so air had to be passed intermittently in order to raise the bed temperature again. The oxidation of carbon by the process

$$2C + O_2 \rightarrow 2CO, \qquad \Delta H^0 = -220.8\,kJ\,mol^{-1} \qquad (10.3)$$

is extremely exothermic. The combination of coal-gas and water-gas (with the former predominating) gave *town gas,* which until replaced by *natural gas* (largely methane), was widely used for domestic and industrial purposes.

Some chemical operations, however, demand a supply of pure hydrogen; these include ammonia synthesis and fat-hardening, and so it became necessary to find a way of altering the composition of water-gas to achieve this. Its gaseous components can be brought into equilibrium by the *water-gas shift*

$$CO + H_2O \rightarrow CO_2 + H_2, \qquad \Delta H^0 = -41.2\,kJ\,mol^{-1} \qquad (10.4)$$

using suitable catalysts;[5-7] traditionally this has been done in two stages, using first an iron–chromia catalyst (Fe_3O_4–Cr_2O_3) at 'high temperature' ($\sim$900 K), followed by a copper–zinc catalyst (Cu/ZnO–Al_2O_3) at 'low temperature' ($\sim$600 K) to complete the task. The forward reaction being exothermic, the equilibrium moves to the right as temperature is

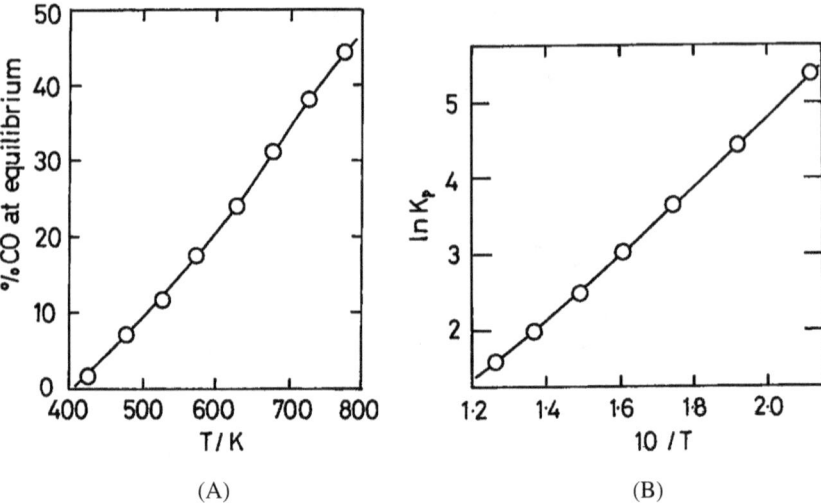

Figure 10.1: (A) Dependence of the equilibrium concentration of carbon monoxide on temperature with a 1:1 H_2O:CO ratio. (B) Dependence of ln K_p on reciprocal temperature.

lowered, and so highly active catalysts are wanted for operation at the lowest convenient temperature. Figure 10.1(A) shows how with an equimolar ratio of reactants the equilibrium amount of carbon monoxide falls with decreasing temperature, and Figure 10.1(B) shows the dependence of ln K_p on reciprocal temperature. The position of equilibrium can of course be altered favourably by increasing the water/carbon monoxide ratio in the feed (Figure 10.2). The carbon dioxide can easily be removed by absorption in water, so quite pure hydrogen would be available if the forward reaction could be used at room temperature. Careful choice of catalyst is necessary, because the components of water-gas are also capable of forming other products such as methane and, by the Fischer–Tropsch process, higher hydrocarbons and oxygenated molecules. Indeed methanation has been used to remove residual carbon monoxide for hydrogenations such as ammonia formation, where catalysts are fatally poisoned by it. Final purification is also possible if small amounts of carbon monoxide in the hydrogen can be selectively oxidised to carbon dioxide. Efforts are in hand to develop gold catalysts for this purpose (see Chapter 7).

From the middle of the last century, interest developed in the use of hydrocarbons in place of coal, as they already contained much hydrogen: the

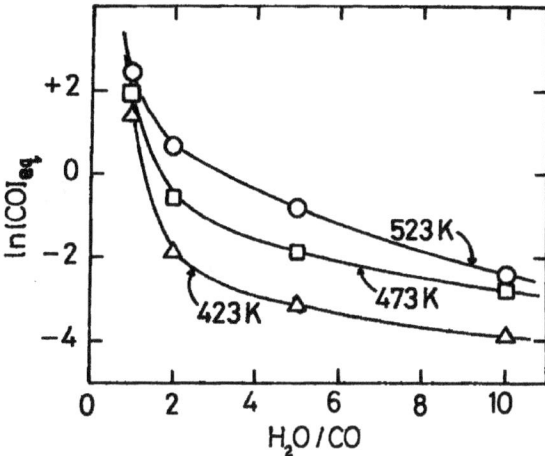

Figure 10.2: Dependence of the equilibrium concentration of carbon monoxide on the H_2O:CO ratio at various temperatures.

steam-reforming of hydrocarbons using nickel catalysts therefore became a major process,[3] but the technology of water-gas shift was still needed to make pure hydrogen. A possible important future use for pure hydrogen is in *fuel cells,* for which the basic source of hydrogen would be either a hydrocarbon or an alcohol, which can also be steam-reformed (see next section).

The water-gas shift is also one of the key steps occurring in vehicle exhaust treatment; taking the equilibrium to the right is desirable not only because it lowers the concentration of the toxic carbon monoxide, but also because hydrogen is a more effective reductant for nitrogen oxides.[8,9] The role of gold catalysts in these processes is also being explored (see Chapter 11).

The activities of a large number of alumina-supported metals were examined some years ago,[10] to make sure that nothing more active than copper for the low-temperature reaction had been overlooked. The Arrhenius parameters are shown as a compensation plot in Figure 10.3. Based on their turnover frequencies at 573 K, the metals fall into four groups: (i) extremely active (Cu); (ii) high activity (Co, Ru and Re); (iii) moderate activity (Ni, Os and Pt) and (iv) low activity (Fe, Rh, Pd, Ir and Au). This only goes to show how important it is to compose the catalyst in the best way, since as we shall see gold can be elevated from the bottom to the top of the list. A recent theoretical analysis[11] fails to recognise its outstanding potential, but the method employed assumes (incorrectly) that the reaction takes place

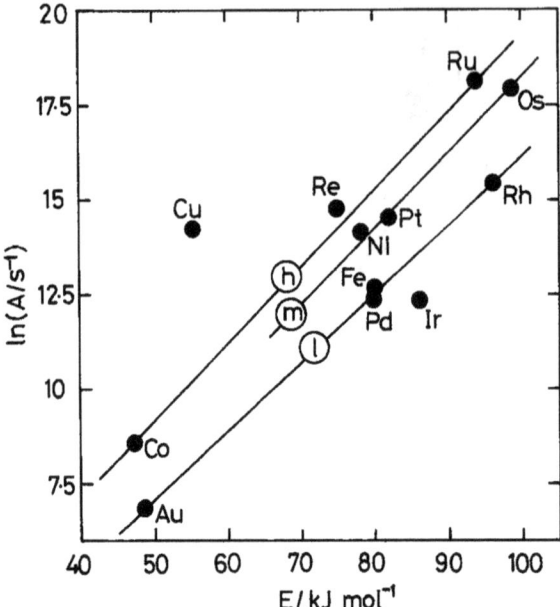

Figure 10.3: Activities of alumina-supported metals for water-gas shift: Arrhenius parameters shown as a compensation plot (l = low, TOF_{573} < $0.4\,s^{-1}$; m = medium, TOF_{573} = 0.06–$0.10\,s^{-1}$; h = high, TOF_{573} = 0.2–$0.4\,s^{-1}$).

only on the metal. Another use of DFT does however embrace the role of ceria as support[12] (see Section 10.3.1).

10.2. Gold as a Catalyst for the Water-Gas Shift[4,8,9,13–15]

10.2.1. Introduction

An important possible future use for pure hydrogen is in *proton-exchange-membrane fuel cells* (PEMFCs); the basic source for the hydrogen could be either a hydrocarbon or an alcohol, either of which can be steam-reformed to produce water-gas.[16,17] As explained above, the equilibrium concentration of carbon monoxide decreases as the temperature falls (Figure 10.1), but as little as 1% is detrimental to the operation of platinum-based catalysts in a fuel cell. Excess water, which is commonly used,[18] serves to move the equilibrium to the right, but there is still every incentive to find highly active catalysts that will perform the reaction at the lowest possible temperature. A comparison of titania-supported Group 11 metals showed the order of activity to be Au > Cu ≫ Ag, the latter being quite inactive. The greater activity of the gold compared to copper (and platinum) is largely due to its weaker chemisorption of carbon monoxide. The use of a suitable gold catalyst would therefore be advantageous, perhaps combined with a similar catalyst to remove the last traces of carbon monoxide form hydrogen by selective oxidation (see Chapter 7). Very pure hydrogen would thus become a weight- and cost-effective energy source for fuel cells.[19–27]

10.2.2. Titania, ferric oxide and similar supports

The earliest work[20,21,23] showed that Au/Fe_2O_3 catalysts performed well in the water-gas shift; Au/TiO_2 catalysts also worked satisfactorily, and both have been subjected to detailed studies. Au/Fe_2O_3 catalysts prepared either by coprecipitation (COPPT) using Na_2CO_3 at 333 K[20,21,23,24] or by deposition–precipitation (DP) of the gold precursor onto Fe_2O_3 or freshly-precipitated $Fe(OH)_3$[24,28–30] gave after calcination high activity that was detectable as low as 393 K (Figure 10.4). Comparison of a 3% Au/Fe_2O_3

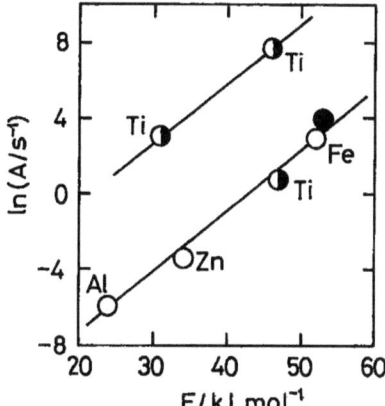

Figure 10.4: Compensation plot for various supported gold catalysts; the cation of the support is shown. The filled point is for a $Cu/ZnO–Al_2O_3$ low-temperature shift catalyst; the points for TiO_2-supported catalysts on the upper line were made by deposition-participation; all those on the lower line were made by coprecipitation. See Table 10.2 for further details.

COPPT catalyst with a conventional $CuO/ZnO–Al_2O_3$ catalyst showed that, although rates were similar above $513\,K$, below this temperature the gold catalyst was much superior. Strangely enough, an Au/Al_2O_3 catalyst prepared in the same way had virtually no activity at all,[20] suggesting that it was cooperation between the gold and the Fe_2O_3 support that was responsible for the high activity (Figure 10.4). Temperatures for 50% conversion were much lower when the mean particle size was $4\,nm$ than when it was greater than $10\,nm$, but particles smaller than $2.5\,nm$ were not highly active. Slightly lower rates were found when the DP method was applied to $Fe(OH)_3$ than to Fe_2O_3[24] (Table 10.1). A detailed study[31] of 3% Au/Fe_2O_3 has shown that using various chemical reducing agents led to better activities, the best being given by sodium tetrahydridoborate ($NaBH_4$). Addition of 2,6-dimethylpyridine resulted in loss of activity, from which it was concluded that acidic hydroxyl groups participated in the reaction, along with oxidic gold species.[32] The mean particle size increased from 15 to $32\,nm$ during the first $6\,h$ of use at $393\,K$. A role for cationic gold has been indicated by a correlation between activity and hydrogen consumption in temperature-programmed reduction.[33]

It might therefore be expected that gold deposited on other readily reducible oxides (e.g. Co_3O_4) would be active, while gold on less easily reduced oxides (e.g. TiO_2 and ZrO_2) would not. The order of activity shown in Table 10.1 contains some surprises; in particular the high activities shown by Au/TiO_2 and Au/ZrO_2[24,28,34] were quite unexpected. The good activity of the former has been confirmed,[25,35] but Au/Fe_2O_3, Au/Al_2O_3 and Au/ZnO made by coprecipitation, as well as the commercial $CuO–ZnO/Al_2O_3$, all had lower but comparable activities (see Table 10.2 and

Table 10.1: Activities of supported gold catalysts prepared by different methods (3% Au) for the water-gas shift; temperatures for 50% conversion.[24,28]

Support	COPPT		DP		ModDP[a]	
	d (nm)	$T_{50\%}$ (K)	d (nm)	$T_{50\%}$ (K)	d (nm)	$T_{50\%}$ (K)
Fe_2O_3	4	433	—	—	3.5	423
TiO_2	—	—	<5	403	<3	573
ZrO_2	—	—	<5	423	<3	573
Co_3O_4	8	508	—	—	7	473
CeO_2	—	—	4.5	473[b]	—	—

[a]Using freshly precipitated hydroxide as the support. [b]After 4 weeks of operation.

Table 10.2: Activities of supported gold catalysts for the water-gas shift; turnover frequencies at 373 K.[25,35]

Support	Method	%Au	d (nm)	TOF $(s^{-1} \times 10^4)$
TiO_2	DP	3.4	4.4	7.9
TiO_2	DP	10	4.4	9.2
Fe_2O_3	COPPT	5	3.3	0.9
Al_2O_3	COPPT	5	3.7	1.1
ZnO	COPPT	5	4.9	0.6

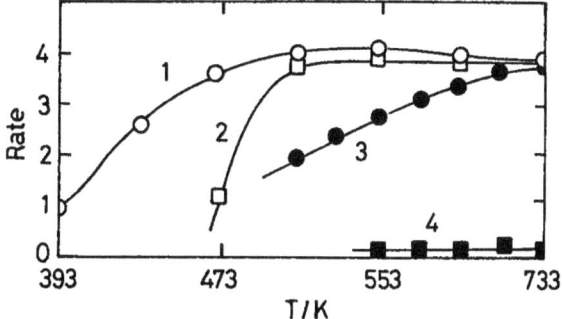

Figure 10.5: Rate as a function of temperature for various supported gold catalysts: (1) $Au/\alpha\text{-}Fe_2O_3$; (2) $CuO/ZnO\text{-}Al_2O_3$; (3) $\alpha\text{-}Fe_2O_3$; (4) Au/Al_2O_3 ((1) and (4) made by COPPT). Feed, 4.9 vol % CO in air; $P_{H2O} = 223$ Torr; SV = 4000 h^{-1}; 1 atm; rate in mol m^{-2} min^{-1} × 10^{-2}.

Figure 10.5). Gaps in the information available in these tables prevent the drawing of positive conclusions concerning activity, but Au/TiO_2 made by DP would seem to be best. In the case of Au/Al_2O_3 made by COPPT, however, very different activities have been reported (see Table 10.2 and Ref. 20). Au/Co_3O_4 was found to deactivate rapidly, caused by reduction of the support to metallic cobalt.[24]

The nature of the support and the preparation technique are clearly of great importance. The mixed oxides $Fe_2O_3\text{-}ZnO$ and $Fe_2O_3\text{-}ZrO_2^{36}$ turned out to be less effective than Fe_2O_3 itself,[30] $T_{50\%}$ values being, respectively, about 520 and 500 K. The usefulness of ferric oxide as a support may be limited by its inability to stabilise gold particles against sintering.[37] Use of freshly precipitated hydroxides or hydrous oxides of titanium and zirconium gave less active catalysts than the preformed oxides,[24,30] probably because

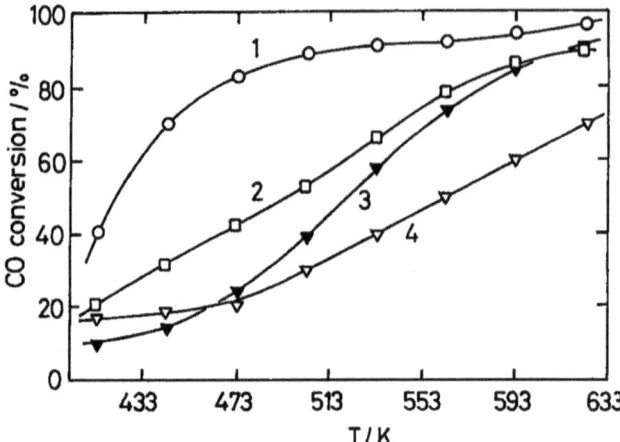

Figure 10.6: Conversion as a function of temperature for various supported gold catalysts: (1) Au/Fe$_2$O$_3$; (2) Au/Fe$_2$O$_3$–ZnO; (3) Au/Fe$_2$O$_3$–ZrO$_2$; (4) Au/ZrO$_2$; all were made by DP onto freshly precipitated hydroxides. Conditions similar to those of Figure 10.5.

some of the metal became encapsulated during calcination (Table 10.1 and Figure 10.6). In other work, however, zirconia was reported to improve the activity of Au/Fe$_2$O$_3$.[36] Gold on sulfated zirconia was highly effective for producing hydrogen suitable for fuel-cell use in both high and low temperatures regimes (623 and 473 K, respectively).[16] The nonpyrophoric catalyst was neither affected by the water produced nor by sulfur (at least in the S^{VI} state).

10.2.3. Ceria and ceria–zirconia as supports

Ceria has emerged as a very useful support; in particular it does not experience the deactivation that bedevils the other supports, the activity value in Table 10.1 being observed after four weeks of continuous operation.[24,38] This observation is of especial interest since ceria is extensively used as a component of automotive three-way emission-control catalysts, mainly due to its ability to undergo a relatively rapid change in oxidation state in consequence of changes in the redox potential of the exhaust gas.[39] This gives a high oxygen storage capacity, so that it can absorb oxygen when it is in excess and release it when it is needed. The activity of Au/CeO$_2$ catalysts may therefore entail the utilisation of the support's lattice ions and their regeneration by water[18,19,24,25,40,41] (see later). Only in one publication,[42] in which a 10% Au/CeO_2 catalyst was used, is there mention of rapid deactivation. When prepared by DP, Au/CeO_2 was better than Au/TiO_2, and even superior to the $Cu/ZnO-Al_2O_3$ catalyst, giving about 90% conversion at 500 K, and very little methane.[43]

About 3%Au/hydroxyapatite (a type of calcium phosphate) gave >90% conversion at 573 K, with little deactivation, and no formation of methane up to 653 K.[44] This result is of interest because creation of anion vacancies in the support is very unlikely to occur.

More recently it has been shown[37,39,45-47] that using CeO_2-ZrO_2 mixed oxide as support is also effective; this combination, also a support of choice in vehicle exhaust control,[39] affords greater stability, because the incessant fluctuation of oxidation state of the cerium ions can occur without the associated structural deterioration that ultimately afflicts the pure oxide. Gold ions also dissolve in this mixed oxide, as they are thought to do in ceria[48] (Section 3.5.3), and the solid solution $Ce_{1-x}Au_xZrO_y$ has been identified by XRD.[46] The DP method was preferred for depositing the gold, and the use of lattice oxygen to oxidise hydrogen and carbon monoxide was catalysed much more effectively by the small particles thus formed than by the larger particles formed by impregnation.

In the extensive work on carbon monoxide oxidation reported in Chapter 6, one of the matters for debate concerned the oxidation state of the gold in the working catalyst. In the studies on the water-gas shift summarised above, it has been reasonably supposed that the gold component of the catalysts was in the metallic state, and indeed spectroscopic evidence has shown that in the reverse reaction on Au/Fe_2O_3 and Au/TiO_2 the carbon monoxide that was formed resided only on Au^0 particles and not in the oxidic forms that were simultaneously present. It therefore came as a surprise to read that Au/CeO_2 catalysts from which metallic gold had been removed by treatment with sodium cyanide solution possessed activity for water-gas shift.[48-50] The observations were briefly as follows: the ceria support, which contained 10% La_2O_3, was in the form of very small particles ($d_{av} \sim 5$ nm, area ~ 150 m^2 g^{-1}), and catalysts made with ~ 3–5% Au either by DP or COPPT and calcined at 673 K were tested before and after the sodium cyanide treatment. It was found that this did not significantly change either the rate or the activation energy (48 kJ mol^{-1} for DP materials, 37 kJ mol^{-1} for COPPT); rates, which were expressed as m^{-2} of total surface, were in fact about 10 times lower than those for the commercial $CuO/ZnO-Al_2O_3$. Residual gold contents were 0.2–0.7% so that the reported rates based on gold content are comparable with those found by others. It was also shown by XPS that most if not all the Au^0 had in fact been leached out by the sodium cyanide treatment, and by TPR with carbon monoxide that oxygen species associated with Au^0 particles were absent.

Metallic gold has also been dissolved from pure ceria,[51] from gadolinium-doped ceria,[51] and from catalysts made by a 'urea gelation/coprecipitation method',[52] without significant effect on their activities for water-gas shift. The only problem that has been identified is the formation of a cerium hydroxycarbonate, which occurs during shut-down when water is present. The very extensive work performed on this system[48–52] ensures that its conclusions are given full consideration.

The implication that cationic gold is alone capable of effecting the water-gas shift is not however supported by recent work[37,47] on Au/CeO_2–ZrO_2 employing XAFS and XANES, aided by DFT calculations. Noting that on introducing the reactants to the catalyst precursor there was a rapid change in its colour and an exotherm caused by the partial reduction of the ceria lattice, it was concluded that the reducing power of the reactants converted all or almost all the gold to Au^0, the stability of the cationic species being lowered by the support's reduction. About 15% of the gold was however re-oxidised by air at 423 K, the support's re-oxidation then stabilising cationic gold species, although these could only be formed by gold atoms actually in contact with the support. This model suggests that Au^0 particles may be in the form of hemispheres 1.3 nm in diameter and containing about 50 atoms. The interfacial gold atoms, occupying cation vacancies on the support, would stabilise gold particles, and would tend to be positively charged; they might participate in the active centre of the reaction. This idea bears some relationship to the model developed on the basis of observations made on the oxidation of carbon monoxide.[53]

A XANES study of the Au/Al_2O_3 system has shown that Au^{3+} was reduced to Au^0 in helium at 623 K, and that sodium cyanide treatment dissolved some 80% of the gold; the remaining Au^{x+} was however reduced by a second helium treatment.[54]

10.3. Mechanism of the Gold-Catalysed Water-Gas Shift

If the trumpet sound an uncertain note, who shall obey the call?

10.3.1. Gold on ceria and ceria–zirconia

Consideration of the mechanism of this reaction on gold catalysts is made difficult by the variety of interpretations placed upon the experimental results. There are essentially three conflicting proposals: (1) reaction on or close to Au^0 particles, with adjacent oxide ions, formed from water molecules, acting as oxidant;[9,24,28] (2) reaction involving Au^{x+} cations and redox processes on the support[49] and (3) reaction mediated by formate ions on the support.[27] The problem is highlighted by two spectroscopic studies of the reaction on Au/CeO_2. One of them[9,24] focuses on the role of Au^0 particles, while recognising the existence of Au^{x+} and of carbonate, bicarbonate and formate species that show IR absorption between 1200 and 1800 cm^{-1} (see Section 5.3.4), but these are regarded as spectators and not reaction intermediates. A full reaction scheme has been developed on this basis by Donka Andreeva and her associates[24,28] (Scheme 10.1). After initial creation of anion defects by reaction of oxide ions close to Au^0 particles with carbon monoxide, reaction proceeds by their re-oxidation by water with the liberation of hydrogen: further oxidation utilises oxygen ions originating in the water molecule. In this mechanism, no role is assigned to formate ions or lattice oxygen other than that coming from the water; it accords with mechanism (1) above. A mechanism awarding a major role to Au^0 particles, perhaps in association with $Au^{\delta+}$ species, is clearly indicated

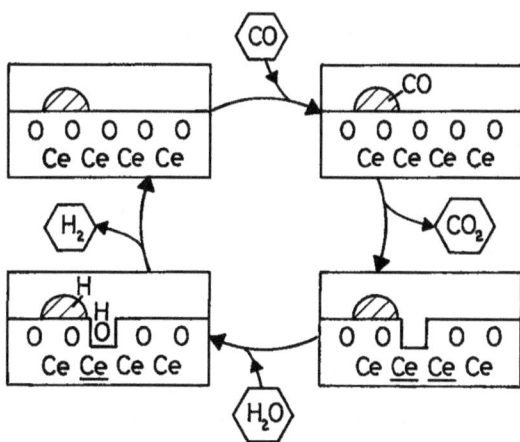

Scheme 10.1: Mechanism (1) for water-gas shift on Au/CeO_2; $O = O^{2-}$; $Ce = Ce^{4+}$; $\underline{Ce} = Ce^{3+}$.

by the XAFS/XANES results obtained with $Au/CeO_2–ZrO_2$ mentioned above.[37,47]

The other study[27] notes the formation of geminal hydroxyl groups on the cerium ions and proposes their reaction with carbon monoxide to give formate ion, which then decomposes into products. The exact way in which this happens was not however made clear, and the role of the gold was not emphasised. This mechanism, corresponding to (3) above, is supported by the results of transient experiments showing that hydrogen and carbon dioxide were formed simultaneously,[43] and not sequentially as would be required by the ceria redox mechanism (2). The service provided by the gold may then be to make $Au^{\delta-}$ species (or possible $Au^{\delta+}$ species as in mechanism (2) above), on which carbon monoxide adsorbs before reacting with a Ce^{3+}–OH to give the formate ion. The hydroxyl groups may have been created either by hydrogen spillover or by reaction of water with a surface anion vacancy made by loss of a lattice oxide ion.[55]

Clearly a definitive mechanism cannot be established on the basis of the available evidence. It would be incorrect to conclude that one mechanism was right and the other wrong, and in particular it is risky to rely solely on spectroscopic measurements, which by their very nature show only snap-shots of parts of the process, however informative they may be. The lack of any detailed kinetic analysis or reaction modelling is keenly felt, as is the absence of isotopic tracing. The only kinetic expression published so far[45] has been obtained using an $Au/CeO_2–ZrO_2$ catalyst: it shows

$$r = P_{CO}^{0.75} P_{H_2O}^{0.57} P_{CO_2}^{-0.27} P_{H_2}^{-0.99}. \qquad (10.5)$$

This does not immediately suggest a mechanism or the location of the active centres, but the strong inhibition by hydrogen might indicate its slow reversible release from a water molecule or from hydroxyl groups that it has formed (see Scheme 10.1). This cycle can of course work in either direction.

The reducibility of the ceria support when gold is present also occurs at lower temperatures than in its absence, but only the surface layer is reduced.[9,27,49] Redox processes are thought to occur with Au/CeO_2 even at room temperature;[24] water appears to re-oxidise anion vacancies close to the metal, and the failure of XANES measurements to show any change in the Ce^{3+}/Ce^{4+} ratio on exposure of a reduced catalyst to water vapour even at $623\,K$[27] may be due to the limited number of sites participating in redox reactions. Re-oxidation of reduced ceria surfaces by water in the absence of metal has long been known.[56]

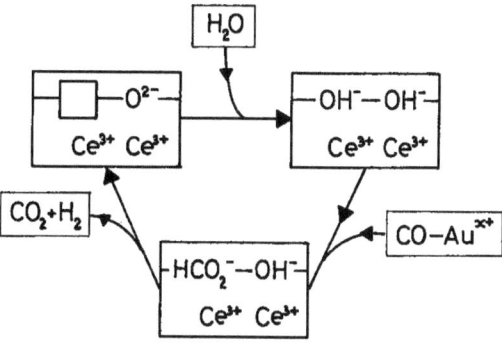

Scheme 10.2: A mechanism involving formate species proceeding only on the support.

One could also envisage a mechanism proceeding entirely on the ceria support, making use of anion defects created by carbon monoxide migrating from Au^0 particles or perhaps even from Au^{x+} ions or clusters[12] (Scheme 10.2). This also provides a role for geminal hydroxyl groups and formate ions. Any more specific formulation of mechanism also needs to incorporate the detailed structure of the ceria surface, which contains two types of oxygen.[39,49] It has also been suggested that gold particles nucleate preferentially on pre-existing defects.[12,57] Recent DFT calculations[12] led to a mechanism that is certainly plausible and that incorporates features of the other proposals. It was concluded that gold atoms can be oxidised by ceria, and that only $Au^{\delta+}$ could adsorb carbon monoxide strongly enough for subsequent catalysis, which proceeded through a formate species. The active sites were neither single atoms nor large particles, but 'ultra-small gold clusters' of four gold atoms that were positively charged and anchored on an anion defect. Importance was attached to the empty nonbonding f states of ceria that act as an electron buffer in reactions, rather as the delocalised Fermi-level states in metals do.

10.4. Gold on other Oxides

Oxides other than ceria and ceria-zirconia are very effective supports for gold in the water-gas shift, perhaps even better, but there is no agreement as to the comparative merits of ceria and titania (compare Table 10.1 and Ref. 43). We need not suppose that the same mechanism will operate in all cases, and indeed it seems unlikely that Au^{x+} ions will actually dissolve in the lattices of titania and of ferric oxide; they are much more easily reduced on these supports than on ceria. However, the idea that lattice oxide ions and the anion vacancies created by their removal are important suggests that the reducibility of the support may be a useful parameter for assessing or even predicting catalytic activity for the water-gas shift.[58] Temperature-programmed reduction (TPR) of Au/Fe_2O_3 with hydrogen occurred at a lower temperature (by 140 K) than that of Fe_2O_3 by itself, but only the first stage ($Fe_2O_3 \rightarrow Fe_3O_4$) was affected.[24] It appeared that reduction took place by the spillover of hydrogen atoms from the gold particles to the support, followed by their reaction with oxide ions:

$$Fe^{3+}O^{2-} \xrightarrow{H^\bullet} Fe^{2+}OH^- \tag{10.6}$$

and subsequent dehydration of the hydroxyl-covered surface. This is a well-established mechanism for the metal-catalysed reduction of oxides.[59] In this case further reduction is made difficult by the inaccessibility of the Fe^I oxidation state. The question of hydrogen chemisorption on small gold particles has been discussed in Chapter 5; FT-IR evidence for it has been obtained with Au/TiO_2 and Au/Fe_2O_3, and in the former case Ti^{3+} ions formed by hydrogen spillover have been seen by XPS and EPR.[24] Zr^{3+} paramagnetic centres have also been observed in Au/ZrO_2 treated with hydrogen.

A mechanistic cycle has been proposed[21] for the reaction on Au/Fe_2O_3 (Scheme 10.3) that bears some resemblance to that suggested for Au/CeO_2 (Scheme 10.2) in that an anion vacancy near to a gold particle is required.

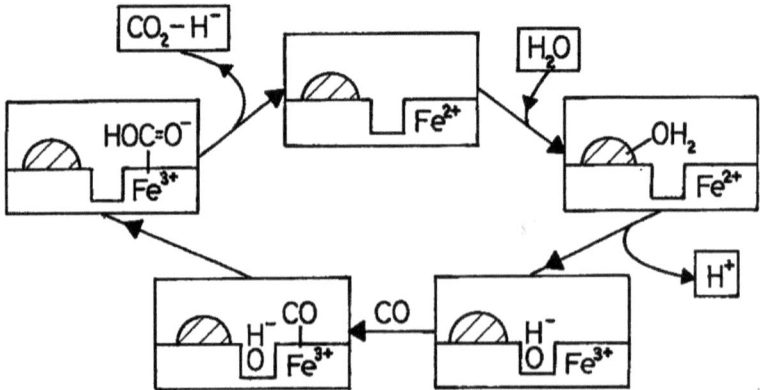

Scheme 10.3: A mechanism proposed for water-gas shift on Au/Fe_2O_3.

There is no specific provision for the chemisorption of carbon monoxide on the gold, although this could occur before it reacts with an adjacent hydroxyl ion. The way in which hydrogen ions recombine to form molecules also remains unclear.

Some studies have employed the reverse reaction, and the formation of carbon monoxide on Au/TiO_2 and on Au/Fe_2O_3 has been observed to take place specifically on the Au^0 particles, and not on Au^{x+} ions.[28] Application of the Principle of Microscopic Reversibility therefore demands that the forward reaction should also start in this way.

10.5. Gold-Containing Bimetallic Catalysts

Studies have been reported of the effect of combining gold with other metals that are active for the water-gas shift. Ruthenium[60,61] and nickel[61] when added to Au/Fe_2O_3 produced modest increases in rate ($\sim$x2 at 373 K, x1.5 at 513 K), the effect with ruthenium being particularly marked below 420 K where the activation energy was very low.[60] Above this temperature, comparison of the activation energies shown by the separate and combined metals implied that the reaction on the mixture had the character of ruthenium rather than gold, but despite extensive characterisation it was impossible to tell which was doing what to the other. A number of other metals gave small increases in rate at higher temperatures.[60]

The effect of adding gold to the $Cu/ZnO–Al_2O_3$ catalyst has not yet been investigated.

References

1. *Catalyst Handbook* (M. Twigg, ed.), 2nd edn., Wolfe Publ. Ltd., London, 1989.
2. K. Kochloefl, in *Handbook of Heterogeneous Catalysis*, G. Ertl, H. Knozinger and J. Weitkamp, (eds.), VCH, Weinheim, 1997, Vol. 4, p. 1831.
3. G.C. Bond, *Heterogeneous Catalysis: Principles and Applications*, Clarendon Press, Oxford, 1987.
4. D.L. Trimm, *Appl. Catal. A: Gen.* **296** (2005) 1.
5. G.C. Bond and D.T. Thompson, *Catal. Rev.-Sci. Eng.* **41** (1999) 319.
6. M.B. Cortie and E. van der Lingen, *Mater. Forum* **26** (2002) 1.
7. R.J. Farrauto and C.H. Bartholomew, *Fundamentals of Catalytic Processes*, Chapman and Hall, London, 1997, Ch. 6.
8. D. Andreeva, V. Idakiev, T. Tabakova, L.I. Ilieva, P. Falaras, A. Bourlinos, A. Travlos, *Catal. Today* **72** (2002) 51.
9. T. Tabakova, F. Boccuzzi, M. Manzoli and D. Andreeva, *Appl. Catal. A: Gen.* **252** (2003) 385.
10. D.C. Grenoble, M.M. Estadt and D.F. Ollis, *J. Catal.* **67** (1981) 90.

11. N. Schumacher, A. Boisen, S. Dahl, A.A. Gokhale, S. Kandoi, L.C. Grabow, J.A. Dumesic, A. Mavrikakis and I. Chorkendorff, *J. Catal.* **229** (2005) 265.

12. Z.-P. Liu, S.J. Jenkins and D.A. King, *Phys. Rev. Lett.* **94** (2005) 196102.

13. D.T. Thompson, *Appl. Catal. A: Gen.* **243** (2003) 201.

14. M. Kinne, R. Leppelt, V. Plzak and R.J. Behm, *Proc. CatGold 2003*, Vancouver, Canada (2003).

15. R.J.H. Grisel, Ph.D. Thesis, Leiden University (2001).

16. A. Kuperman and M.E. Moir, WO Patent 2005500532 A2 to Chevron Inc. (2005).

17. D. Cameron, R. Holliday and D. Thompson, *J. Power Sources* **118** (2003) 298.

18. Q. Fu, S. Kudriavtseva, H. Saltsburg and M. Flytzani-Stephanopoulos, *Chem. Eng. J.* **93** (2003) 41.

19. G. Pattrick, E. van der Lingen, C.W. Corti, R.J. Holliday, and D.T. Thompson, *Proc. CAPoC-6 Conference*, Brussels, Belgium (2003); *Topics Catal.* **30/31**, 273.

20. D. Andreeva, V. Idakiev, T. Tabakova, and A. Andreev, *J. Catal.* **158** (1996) 354.

21. D. Andreeva, V. Idakiev, T. Tabakova, A. Andreev and R. Giovanoli, *Appl. Catal. A: Gen.* **134** (1996) 275.

22. L.I. Ilieva, D.H. Andreeva and A.A. Andreev, *Thermochim. Acta* **292** (1997) 169.

23. D. Andreeva, T. Tabakova, V. Idakiev, P. Christov and R. Giovanoli, *Appl. Catal. A: Gen.* **169** (1998) 9.

24. D. Andreeva, *Gold Bull.* **35** (2002) 82.

25. H. Sakurai, A. Ueda, T. Kobayashi and M. Haruta, *Chem. Commun.* (1997) 271.

26. A. Venugopal, J. Aluha, D. Mogano and M.S. Scurrell, *Appl. Catal. A: Gen.* **245** (2003) 149.

27. G. Jacobs, P.M. Patterson, L. Williams, E. Chenu, G. Sparks, G. Thomas and B.H. Davis, *Appl. Catal. A: Gen.* **262** (2004) 177.

28. F. Boccuzzi, A. Chiorino, M. Manzoli, D. Andreeva and T. Tabakova, *J. Catal.* **188** (1999) 176.

29. T. Tabakova, F. Boccuzzi, M. Manzoli, J.W. Sobczak, V. Idakiev and D. Andreeva, *Appl. Catal. B: Env.* **49** (2004) 73.

30. T. Tabakova, V. Idakiev, D. Andreeva and I. Mitov, *Appl. Catal. A: Gen.* **202** (2000) 91.

31. A. Venugopal and M.S. Scurrell, *Appl. Catal. A: Gen.* **258** (2004) 241.

32. D. Boyd, S. Golunski, G.H.R. Hearne, T. Magadzu, K. Mallick, M.C. Raphulu, A. Vengopal and M.S. Scurrell, *Appl. Catal. A: Gen.* **292** (2005) 76.

33. D. Boyd, S. Golunski, G.H.R. Hearne, T. Magadzu, K. Mallick, M.C. Raphulu and M.S. Scurrell, *Appl. Catal. A: Gen.* **292** (2005) 76.

34. F. Boccuzzi, A. Chiorino, M. Manzoli, D. Andreeva, T. Tabakova, L. Ilieva and V. Iadakiev, *Catal. Today* **75** (2002) 169.

35. M. Haruta, *CATTECH* **6** (2002) 102.

36. J.-M. Hua, Q. Zheng, Y.-H. Zheng, K.-M. Wei and X.-Y Lin, *Catal. Lett.* **102** (2005) 99.

37. D. Tibiletti, A. Amieiro-Fonseca, R. Burch, Y. Chen, J.M. Fisher, A. Goguet, C. Hardacre, P. Hu and D. Thompsett, *J. Phys. Chem. B* **109** (2005) 22553.

38. M. Kinne, R. Leppelt, V. Plzak and R.J. Behm, *Proc. CatGold 2003*, Vancouver, Canada (2003).

39. *Catalysis by Ceria and Related Materials*, A. Trovarelli, (ed.), Imperial College Press, London, 2002.

40. T. Tabakova, F. Boccuzzi, M. Manzoli and D. Andreeva, *Appl. Catal. A: Gen.* **252** (2003) 38.

41. Y. Li, Q. Fu and M. Flytzani-Stephanopoulos, *Appl. Catal. B: Env.* **27** (2000) 17.
42. C.H. Kim and L. Thompson, Preprints of Symposia — American Chemical Society, Division of Fuel Chemistry, **48** (2003) 233.
43. H. Sakurai, T. Akita, S. Tsubota, M. Kiuchi and M. Haruta, *Appl. Catal. A: Gen.* **291** (2005) 179.
44. A. Venugopal and M.S. Scurrell, *Appl. Catal. A: Gen.* **245** (2003) 137.
45. J.P. Breen, R. Burch, J. Gomez-Lopez, A. Ameiro, J.M. Fisher, D. Thompsett, R.J. Holliday and D.T. Thompson, *Proc. Fuel Cell Seminar*, San Antonio, Texas, USA, November 2004.
46. M. Vicario, C. de Leitenberg, J. Llorca, G. Dolcetti and A. Trovarelli, EUROCAT VI (2003).
47. A. Amieiro-Fonseca, J.M. Fisher and D. Thompsett, Internat. Patent Appln. WO 2005/087656 A1.
48. A. Venugopal and M.S. Scurrell, *Appl. Catal. A: Gen.* **245** (2003) 137.
49. Q. Fu, H. Saltsburg and M. Flytzani-Stephanopoulos, *Science* **301** (2003) 935.
50. Q. Fu, J. DeJesus, W. Deng, H. Saltsburg and M. Flytzani-Stephanopoulos, *Proc. 13th Internat. Congr. Catal.*, Paris, July 2004, O4-025.
51. W.-L. Deng, J. de Jesus, H. Saltsburg and M. Flytzani-Stephanopoulos, *Appl. Chem. A: Gen.* **291** (2005) 126.
52. Q. Fu, A. Weber and M. Flytzani-Stephanopoulos, *Catal. Lett.* **77** (2001) 87.
53. G.C. Bond and D.T. Thompson, *Gold Bull.* **33** (2000) 41.
54. J.T. Calla and R.J. Davis, *Catal. Lett.* **99** (2005) 21.
55. G. Jacobs, S. Ricote, P.M. Patterson, U.M. Graham, A. Dozier, S. Khalid, E. Rhodus and B.H. Dacvis, *Appl. Catal. A: Gen.* **292** (2005) 229.
56. C. Padeste, N. Cant and D.L. Trimm, *Catal. Lett.* **18** (1993) 305.
57. V. Idakiev, Z.-Y. Yuan, T. Tabakova and B.-L. Su, *Appl. Catal. A: Gen.* **281** (2005) 149.
58. G. Munteanu, L.Ilieva, R. Nedyalkova and D. Andreeva, *Appl. Catal. A: Gen.* **277** (2004) 31.
59. G.C. Bond and J.B.P. Tripathi, *Trans. Faraday Soc.* **72** (1976) 933.
60. A. Venugopal, J. Aluha, D. Mogano and M.S. Scurrell, *Appl. Catal. A: Gen.* **245** (2003) 149.
61. A.K. Venugopal, J. Aluha and M.S. Scurrell, *Catal. Lett.* **90** (2003) 1.

CHAPTER 11

Reactions of Environmental Importance

11.1. Introduction

There is an increasing awareness of the need to preserve the quality of the earth's atmosphere and water. The introduction of legislation worldwide both to reduce pollution levels for gaseous emissions from road vehicles, to limit release of volatile organic compounds (VOCs) from industrial operations, and also to control impurities in aqueous effluents, is stimulating efforts to design catalytic processes to meet these requirements. In this chapter, we discuss complete oxidation processes, reduction of nitrogen oxides, ozone decomposition and removal of halocarbons, sulfur dioxide, dioxins and volatile organic compounds, as well as catalytic wet air oxidation (CWAO) systems for oxidizing organic compounds in water. The oxidation of saturated and unsaturated hydrocarbons, which has been extensively studied, is described in Section 11.3.2 (oxidation of volatile organic compounds); methane however merits a separate section (11.3.1).

11.2. Catalytic Treatment of Vehicle Exhaust

11.2.1. Introduction[1–3]

Before the introduction in the United States of legislation to limit emissions from vehicles, their pollution together with hydrocarbon vapour from oil refineries plus an adequate supply of sunlight resulted in the frequent occurrence of a *photochemical smog* over major cities, as well as unacceptable levels of carbon monoxide. The first legislative requirements entailed the lowering of carbon monoxide and hydrocarbon emissions, because this could be easily effected by catalytic oxidation using a platinum catalyst; the more stringent controls imposed later demanded the elimination of nitrogen oxides, which could only be secured by reaction with the reducing components of the exhaust and the addition of rhodium to the catalyst

Table 11.1: Reactions occurring in the treatment of vehicle exhaust and their products.

Reaction	Products				
	CO_2	N_2	H_2	H_2O	NH_3
1 $CO + O_2$	+				
2 $CO + NO$	+	+			
3 $CO + H_2O$	+		+		
4 $H_2 + NO$		+		+	+
5 $C_nH_m + O_2$	+			+	
6 $C_nH_m + NO$	+	+		+	+
7 $C_nH_m + H_2O$	+		+	+	

(Section 14.2). Table 11.1 lists the most important reactions that need to be catalysed in order to ensure that only harmless molecules emerge from a vehicle fuelled by an internal combustion engine. Now the relative amounts of the components leaving the engine vary with the air:fuel ratio being used, as this determines the temperature of the combustion in the cylinder (Figure 11.1(A)). When this ratio is below the stiochiometric value of about 14.5 (i.e. that which would theoretically give complete combustion to carbon dioxide and water), the chief pollutants are carbon monoxide and unburnt hydrocarbons; in this region most of the nitrogen oxides (NO_x) can be removed by reactions 2 and 6 (Table 11.1), any remaining reductants being removed later by reactions 1 and 5, for which additional air is needed. Above the stoichiometric ratio, the reductants are readily treated by oxidation, but the nitrogen oxides are unaffected. The optimum solution is obtained by ensuring that the engine always works very close to the air:fuel ratio of 14.5, since here there is a narrow 'window' in which a high degree of removal of all pollutants is possible (Figure 11.1(B)). This is made easier by incorporating in the catalyst an 'oxygen storage' capability, such as ceria, which mops up oxygen when in excess, and releases it when it is needed. A 'NO_x storage' feature, such as a barium compound, may also be used; this absorbs NO_x as nitrate ion under oxidising conditions, and releases it for reduction by temporarily running the engine under reducing conditions.

As a result of a major research effort lasting many years, catalyst systems employing metals of the platinum group have been successfully developed to solve most of the problems of vehicle exhaust treatment; they

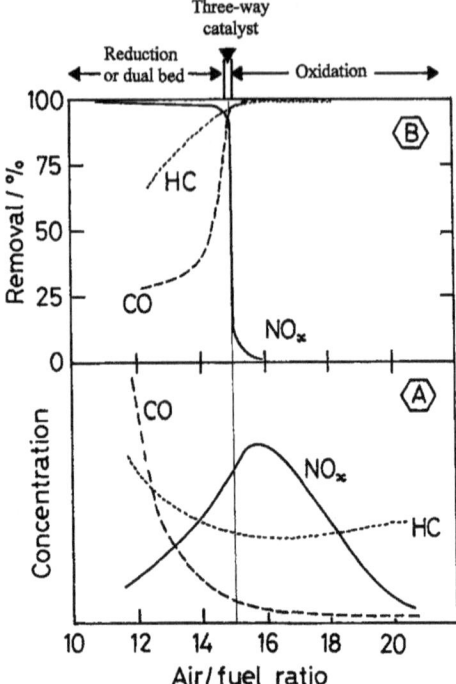

Figure 11.1: (A) Dependence upon air/fuel ratio of concentration (arbitrary units) of pollutants formed in an internal combusion engine (HC = unburnt hydrocarbons). (B) Extent of removal of pollutants with the catalyst system named above the diagram.

oxidise carbon monoxide and hydrocarbons and reduce the NO_x under near-stoichiometric conditions, hence they are known as 'three-way catalysts' (TWC). Rhodium, palladium and platinum have mainly been used, because they are durable under both reducing and oxidising conditions; they are supported on alumina on which some ceria is deposited. The catalyst is then applied as a 'wash-coat' onto a ceramic monolith having a honeycomb-like structure.[1-3]

One remaining difficulty, however, is that during the first few seconds of engine running the temperature of the catalyst has not reached the point where it can work effectively. The discovery of the superlative ability of gold catalysts to oxidise carbon monoxide at low temperatures has therefore led to a number of studies of their effectiveness in the component reactions shown in Table 11.1. Reactions 1 and 3 have been dealt with extensively in Chapters 6 and 10, respectively; the remainder of this section treats the other reactions, together with those involving the minor but troublesome by-products nitrous oxide and ammonia. A final sub-section shows how far gold catalysts have been developed for exhaust treatment.

11.2.2. The problem of the nitrogen oxides

Nitrogen oxides (NO, NO_2 and N_2O, collectively known as NO_x) are produced from the high-temperature reaction between nitrogen and oxygen during power station, oil refinery and other industrial operations, and during the fuel combustion stage of internal combustion (gasoline-fuelled) and diesel engines; they are primary atmospheric pollutants.[4-7] They can dissolve in water to form nitrous and nitric acids, which are consequently secondary pollutants, which together with sulfurous and sulfuric acid are present in acid rain. Nitrogen oxides can also combine with unburnt hydrocarbons to produce photochemical smog, which can often be seen as a haze above cities during warm weather.[5] Nitric oxide constitutes about 90–95% of total NO_x emissions from combustion sources; it is a colourless, water-insoluble gas, with a strong characteristic smell.[5,6] Nitrogen dioxide is also formed but in smaller quantities: this is a brown, extremely poisonous water-soluble gas, which forms rapidly when nitric oxide enters the atmosphere:

Diesel engines and some gasoline-fuelled engines operate under 'lean-burn' conditions, where there is 10–15% more oxygen than is needed to burn all the fuel.[8,9] The most important strategies for NO_x removal under these conditions are to use either (i) NO_x storage-reduction catalysts or (ii) selective catalytic reduction (SCR) by an added reductant; the term 'selective' implies that the reductant attacks the NO_x in preference to the oxygen.

Studies with gold catalysts have focused on the selective reduction of nitric oxide by *propene*, *carbon monoxide* and *hydrogen*; urea,[10] methane[11] and other hydrocarbons[9,11,13] have also been used. NO_x removal using the first three of these will now be discussed.

11.2.3. Selective reduction of nitrogen oxides with propene[9,14-17]

Under typical test conditions (NO, 1000 ppm; C_3H_6, 500 ppm; 5 vol% O_2; 1.8 vol% H_2O; balance, He), nitric oxide is oxidised to the dioxide before

it reacts with *propene*; under these conditions, supported gold catalysts
are effective for its reduction,[17] Au/ZnO, Au/α-Fe$_2$O$_3$ and Au/ZrO$_2$ being
among the most active at lower temperatures ($\sim$623 K). This is especially
noticeable at 523 K, where Au/ZnO gave a maximum of 49% conversion to
nitrogen, together with 16% nitrous oxide. This matches that for typical
platinum group metal catalysts.[18] Au/MgO and Au/TiO$_2$ showed middle
temperature ($\sim$623 K) conversion, but on Au/Al$_2$O$_3$ at higher temperature ($\sim$700 K) (Figure 11.2)[12,19] there may be a gold particle size effect
with maximum extent of NO$_x$ activity in the 15–30 nm range.[12] Smaller
gold particles appeared to favour the combustion of propene, lowering the
extent of NO$_x$ removal,[19] and this may account for the importance of gold
loading. With Au/Al$_2$O$_3$ made by the sol–gel method, 0.17% Au was more
active than 0.82%,[9,17] while for a series of Au/Al$_2$O$_3$ catalysts made by
deposition–precipitation (DP) having 0.2–1.7% Au the most active had
0.8% Au.[20] As with other reactions, the method of preparation is very
important for nitric oxide reduction, and the size of the gold particles is
one of the controlling factors for catalytic activity.[9,17]

Reduction using *propene* over Au/Al$_2$O$_3$ is tolerant to the presence of
water, which may even promote reduction to a small extent.[9,17,21] The
enhancing effect of moisture has also been observed in carbon monoxide
oxidation over other supported gold catalysts (see Chapter 6), and this
provides them with a significant potential advantage over others in applications involving combustion of exhaust gases, since these usually contain

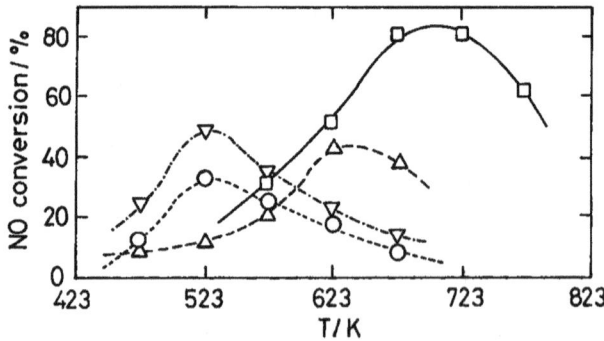

Figure 11.2: Typical results for the temperature-dependence of the reaction of nitric oxide with propene: 1.2% Au/ZnO (∇); 0.17% Au/Al$_2$O$_3$ ($\square$);
0.85% Au/MgO (Δ); 0.84% Au/ZrO$_2$ (O).[9,17]

more than 10% water by volume.[9] Catalysts containing oxides of transition metals, such as those of copper, silver and cobalt, are appreciably deactivated by moisture.[9,22–24] Platinum supported on metal oxides maintains nitric oxide conversion even in the presence of high concentrations of water, but this is accompanied by increased formation of nitrous oxide.[9,25,26] The presence of carbon monoxide had a positive effect on nitric oxide reduction by propene over Au/Al_2O_3, giving yields of nitrogen of 60–75%, at 673 K.[9] Nitric oxide conversion was very low ($\sim$8%) in the absence of oxygen,[9] but the addition of 4 vol% oxygen to the reactant stream increased conversion to 69%. Conversion to nitrogen remained almost constant for 6–20 vol% oxygen. Steady conversion of nitric oxide on Au/Al_2O_3 catalysts in the high oxygen concentrations is advantageous for applications to exhaust gases under lean burn conditions.[9]

Ethene, ethane and *propane* are also effective for NO_x reduction,[9,12,13] but more than 5% oxygen is necessary for maximum effect, while reduction of nitrogen dioxide does not require oxygen. The observation that gold catalyses oxidation of nitric oxide to nitrogen dioxide suggests that the latter is a reaction intermediate,[17] acting as an initiator or oxygen-transfer agent.[18] In line with this, nitric oxide reduction is faster when manganese oxide (Mn_2O_3) is mechanically mixed with Au/Al_2O_3.[9]

Another advantage of gold catalysts over those based on the platinum group metals for selective reduction with propene is their greater selectivity for nitrogen (S_{N_2}) than for nitrous oxide. On Au/Al_2O_3, values of S_{N_2} of $\sim$100% have been observed,[9,12,17,19,20] but at the lower temperatures needed over Au/ZnO, significant nitrous oxide formation has been observed.[9,17] The results obtained thus far therefore indicate that S_{N_2} values over gold catalysts might be better than those obtained over platinum group metal catalysts, which are typically $< 30\%$, although they are active at lower temperatures. Recently it has been shown that for $Pt–Au/SiO_2$ catalysts, if there was a strong Au–Pt interaction (as when a Pt_2Au_4 organometallic precursor was used in the preparation), the temperature for maximum NO_x reduction was significantly increased ($\sim$423 K), but there was a concomitant increase in S_{N_2} from $\sim$50 to 70–80%.[27,28]

A new class of spongy gold catalysts formed of unsupported powdered metal consisting of micrometer particles ($\sim$5 μ) that have nanometer-scale ($\sim$10 nm) internal skeletal structure were very active in the selective reduction of propene under lean-burn conditions;[29] alloying with palladium led to a significant widening of the temperature range for activity.

11.2.4. Reduction of nitrogen oxides with carbon monoxide

Reduction of nitric oxide by *carbon monoxide* was first reported to occur at low temperature (252–369 K) over gold powder (0.86 m^2 g^{-1});[30] the rate was maximal at 314 K, and nitric oxide inhibited the reaction. This paper also provides an early indication of the activity of gold for the oxidation of carbon monoxide near ambient temperature. Nitric oxide reduction occurs in this way in catalytic converters for the treatment of engine exhaust gases:[9]

$$2NO + 2CO \rightarrow N_2 + 2CO_2 \qquad (11.1)$$

Although there are some differences in the literature concerning the relative activities of the platinum group metals for this reaction, the most likely sequence is Rh > Pd > Pt.[9] Over Au/Al$_2$O$_3$, the reduction of nitric oxide in the absence of oxygen occurs as low as 323 K, producing nitrous oxide:

$$2NO + CO \rightarrow N_2O + CO_2 \qquad (11.2)$$

Over Rh/Al$_2$O$_3$ it takes place at temperatures only above 400 K,[31,32] showing that gold catalysts easily outperform those based on the platinum group metals for this reaction under these conditions. The activity of supported gold catalysts depends strongly on the metal oxide support employed. Supports containing iron provide the highest catalytic activity: for catalysts such as Au/NiFe$_2$O$_4$ and Au/MnFe$_2$O$_4$, the reduction to nitrous oxide (at ∼298 K) or nitrogen (at 423 K) is nearly complete.[9] Gold catalysts are therefore superior in their low-temperature catalytic activity to rhodium,[32] platinum[33] and palladium[34,35] catalysts. This extraordinarily high activity of gold catalysts could be maintained in the presence of moisture.[9]

However, as expected, reduction of nitric oxide to dinitrogen under the more commercially interesting *lean-burn conditions* is more difficult with gold catalysts, which are not as active as those based on platinum group metals at lower temperatures.[36] Over Au/Al$_2$O$_3$, the reaction between nitric oxide and carbon monoxide was strongly inhibited by oxygen, giving a maximum conversion of nitric oxide of only 5%,[9] because oxidation of carbon monoxide took place in preference to nitric oxide reduction. On the other hand, a mechanical mixture of Mn$_2$O$_3$ with Au/Al$_2$O$_3$ gave a higher conversion to dinitrogen (21%) at 623 K; Mn$_2$O$_3$ probably catalysed the oxidation of nitric oxide to nitrogen dioxide, which then reacts with carbon monoxide adsorbed on the Au/Al$_2$O$_3$:

$$2NO_2 + 4CO \rightarrow N_2 + 4CO_2 \qquad (11.3)$$

11.2.5. Reduction of nitrogen oxides by hydrogen

Reduction of nitrogen oxides can also be achieved by *hydrogen:*

$$2NO + H_2 \rightarrow N_2O + H_2O \tag{11.4}$$

$$2NO + 2H_2 \rightarrow N_2 + 2H_2O \tag{11.5}$$

$$2NO + 5H_2 \rightarrow 2NH_3 + 2H_2O \tag{11.6}$$

$$N_2O + H_2 \rightarrow N_2 + H_2O \tag{11.7}$$

Hydrogen is generated in the exhaust gases from combustion of hydrocarbons either by the water-gas shift (see Chapter 10) or by steam reforming above 770 K:

$$C_3H_8 + 6H_2O \rightarrow 3CO_2 + 10H_2 \tag{11.8}$$

For the reduction of both nitric and nitrous oxides, the activity of Au/Al_2O_3 was greatly improved by the addition of transition metal oxides and ceria.[37]

The activities for reduction of nitric oxide by hydrogen fall in the sequence Pt > Rh > Pd > Au, Ru > Ir;[9] gold gives an intermediate activity amongst the platinum group metals for both nitric oxide decomposition and hydrogen oxidation. The combination of nitric oxide oxidation and nitrogen dioxide reduction by hydrogen may also give further improvement of NO_x conversion to dinitrogen.

Au/Al_2O_3 and Au/TiO_2 were active for the reduction of nitric oxide by hydrogen;[38] the first showed activity even at room temperature. Au/SiO_2 was inactive unless pre-oxidised and excess hydrogen was present; then it was moderately active above 523 K.[38] Au/MO_x catalysts (M = Co, La, Ce) were also active for this reaction, but this type of catalyst supported on silica was less active than when supported on alumina due to easier sintering of the gold particles. Nevertheless, the presence of an additional metal oxide was beneficial to the activity of Au/SiO_2. Addition of CoO_x and LaO_x improved the selectivity towards dinitrogen formation.[38,39]

11.2.6. Removal of nitrous oxide

Nitrous oxide (N_2O, also known as dinitrogen oxide) is one of the main products of nitric oxide reduction. It is regarded as an undesirable and harmful component of automotive exhaust gases,[40] and it is a more harmful greenhouse gas than carbon dioxide, possibly contributing to ozone depletion in the upper atmosphere.[6,13] In spite of this, there are still only a few papers dealing with nitrous oxide decomposition and reduction, compared with the number dealing with nitric oxide. Au/TiO_2 and Au/CoO_x–TiO_2 are promising catalysts for its reduction by carbon monoxide.[41] Studies on the nitrous oxide/hydrogen and nitrous oxide/carbon monoxide reactions[38,40] using Au/TiO_2 and Au/Al_2O_3 and mixed oxides indicated a synergistic effect for mixtures of metal oxide supports, e.g. Au/CeO_x–Al_2O_3 and Au/Li_2O–CeO_x–Al_2O_3 where T_{50} values of $\sim$325 K were obtained for the reaction with hydrogen. The partly reducible metal oxide additive may contribute to the formation of new active sites and increase dissociation of the nitrous oxide, but in addition alkali and alkaline-earth metal oxides stabilize gold particles against sintering.

11.2.7. Gold-containing catalysts for treating vehicle exhaust

Over Au/Al_2O_3 the reaction between carbon monoxide and nitric oxide is severely inhibited by oxygen, and this unfortunately militates against the promise of its use for three-way conversion at low temperature. Nevertheless the effectiveness of a gold-containing catalyst developed at the Anglo-American Research Laboratories in South Africa has been demonstrated (Figure 11.3).[42] It consisted of 1% Au/CoO_x in admixture with

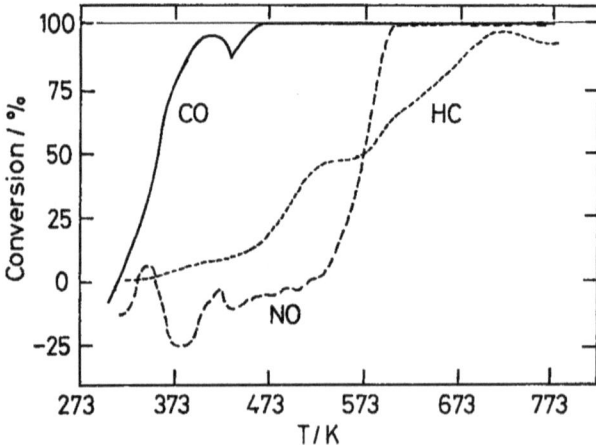

Figure 11.3: Temperature-dependence of conversion of pollutants effected by the Anglo-American Research Laboratories' catalyst under reducing gasoline conditions (see text for details).[42]

zironia-ceria, zirconia and titania, with promoters (0.1% Rh, 2% ZnO, 2% BaO); gold particles were large in size (40–100 nm). The catalyst was evaluated under simulated gasoline engine emissions for oxidizing (0.9% O_2) and reducing (0.6% O_2) conditions close to stoichiometric. Under oxidising conditions the catalyst had a T_{50} for carbon monoxide of 348 K, a T_{50} for hydrocarbons of 513 K, and as expected no NO_x reduction. The T_{50} for carbon monoxide removal stayed almost the same under reducing conditions (358 K), whereas the T_{50} for hydrocarbon oxidation increased to 583 K, complete conversion of NO_x occurring at approximately 603 K, with a T_{50} of 573 K. This catalyst under simulated diesel engine conditions with 7.5% oxygen showed significant activity for NO_x reduction.

A Toyota patent[43] claims that a catalyst containing cationic gold stabilised by dispersion within the lattice of an oxide in Group 2, having for example the formula $Au_2Sr_5O_8$, is effective in treating vehicle exhaust, giving a T_{50} of about 623 K (see Section 14.2.2).

11.3. Destruction of Other Atmospheric Pollutants

11.3.1. Methane[44]

The complete catalytic oxidation of hydrocarbons to carbon dioxide and water has been intensively studied in both industrial and academic research laboratories throughout the world, in order to reduce their emission from motor vehicles and processing plants; the removal of hydrocarbon pollutants is one of the major global environmental challenges.[45–49] Methane is the hydrocarbon most difficult to oxidise, and is often used as a model for activity tests; in addition, it is itself a powerful greenhouse gas.[45] Supported gold catalysts have been examined for this reaction; Au/Co_3O_4 showed activity at 473–523 K,[50–52] and this was enhanced by adding 0.2% platinum.[51] 5% Au/CeO_2 ($d_{Au}{\sim}8$ nm) was also active, as was Au/Al_2O_3, but mixed oxide supports (MO_x–Al_2O_3) showed improved activity in the following order:[53–55]

$$CuO_x > MnO_x > CrO_x > FeO_x > CoO_x > NiO_x > ZnO_x$$

Oxide ions of the support were shown to participate in the reaction through a Mars-van Krevelen mechanism (Section 1.4), and the supports also acted as structural promoters to stabilise the small gold particles.[55]

11.3.2. Removal of volatile organic compounds (VOCs)[44]

Emission into the atmosphere of gaseous compounds arising from domestic or industrial activities[56–61] leads to pollution that is harmful to public health; this may be caused directly by those compounds or by others formed in the atmosphere by chemical reactions brought about by sunlight or ozone. These organic compounds have saturated vapour pressures above 10 Pa at 293 K; more than 1500 of them have been identified, including aromatic solvents, hydrocarbons, oxygenated compounds and chlorocarbons. Their toxic or malodorous nature and their contribution to ozone and smog formation[56] demand their elimination. Gas streams requiring treatment often contain only low concentrations of pollutant (100–1000 ppm), however, so that the heat of reaction is rarely sufficient to raise the catalyst's temperature to its operating value; if this is so, external heating has to be supplied.

Catalytic combustion is likely to prove the best technology for the destruction of VOCs.[56–61] It is preferred to a thermal method due to the lower temperature required, which implies a considerable saving of energy, and to its greater selectivity to complete oxidation. Another advantage is that it can operate with dilute effluent streams (<1% VOCs). For this reason, supported noble metals (Pt, Pd and Rh) or copper, chromium or manganese oxides have been the conventional catalysts used,[62–68] but recently Au/Fe_2O_3,[57–59,69] Au/Al_2O_3[56] and Au/CeO_2[68] have been found to have high activities.[56,68] This has been explained by the capacity of small gold particles to increase the mobility of the lattice oxygen in the case of ferric oxide,[58,59] and with Au/CeO_2 to weakening of the surface Ce–O bonds adjacent to gold atoms, thus activating the surface capping oxygen,[68] which is involved in a Mars-van Krevelen reaction mechanism.[58,59,60]

Gold is generally inferior to palladium and platinum catalysts for the combustion of *unsaturated hydrocarbons*, but it is at least as active for *saturated hydrocarbons* such as propane. This is probably because alkenes can adsorb on defect sites of small gold particles a little too strongly, and so retard the migration of oxygen adsorbed at the interface with the support. Palladium and platinum adsorb alkenes much more strongly than does gold, however, and this aids the dissociation of C–C bonds, making this oxidation faster than that of alkanes.[70]

Gold catalysts have nevertheless been used for the oxidation of various saturated and unsaturated hydrocarbons;[71–74] Au/Co_3O_4 exhibits the highest catalytic activity, perhaps because the support is the most active of the base metal oxides.[50–52,71,72] Based on the temperature for 50% conversion for the oxidation of propane and propene, the activity for 10% Au catalysts decreased in the following order:[50]

$$Au/Co_3O_4 > Au/NiFe_2O_4 > Au/ZnFe_2O_4 > Au/Fe_2O_3$$

with mixed oxide supports MO_x/Al_2O_3 (M = Ce, Mn, Co and Fe) the most active catalyst for propene oxidation was the Au/CeO_2–Al_2O_3, giving a T_{50} of 463 K.[75,76]

Complete oxidation of propene by Pt/Al_2O_3 in the presence of carbon monoxide is made easier by admixture with Au/TiO_2, which oxidises the carbon monoxide and prevents its toxic effect on the platinum.[77]

Au/V_2O_5 supported on titania and zirconia has been evaluated for the complete oxidation of *benzene*;[46–49,78] a strong synergistic effect was observed between gold and vanadia when molecular oxygen was used as oxidant, this effect being more pronounced with titania than with zirconia.[46,48,79] Activation of oxygen was thought to take place on the gold particles, while the vanadium oxide surface was responsible for activating benzene. This reaction goes faster when ozone is used as the oxidant, as it decomposes to give highly reactive atomic oxygen on the gold surface.[80]

Au/V_2O_5 supported on ceria also has a high and stable activity for this reaction.[78,79,81,82] The presence of gold raised the reducibility and reactivity of vanadia and ceria surfaces, and this resulted in a lowering of the reaction temperature.[78,81,82] The strong synergistic effect observed between gold and vanadia when they are present simultaneously on ceria could be caused by a specific interaction between them, and to the possibility of obtaining nanosize gold and ceria particles, producing more active oxygen species.[82] In experiments with Au/V_2O_5 supported on ceria-alumina, a 1:1 ceria:alumina ratio had higher activity than a 1:4 ratio; the alumina in the mixed support increased the stability.[83] Au/V_2O_5 supported on ceria (3% Au, 4% V_2O_5) gave a T_{50} value below 423 K for this reaction.[84]

Oxygenated VOCs include methanol, ethanol, 2-propanol and acetone, and in general the comparative ease of destruction is:

$$alcohols > aldehydes > ketones > esters$$

For the oxidative decomposition of nitrogen-containing organic molecules, ferric oxides and nickel ferrites ($NiFe_2O_4$) have the highest level of catalytic activity owing to their strong affinities for nitrogen.[85] The oxidative decomposition of trimethylamine, which is a typical odour-producing compound,

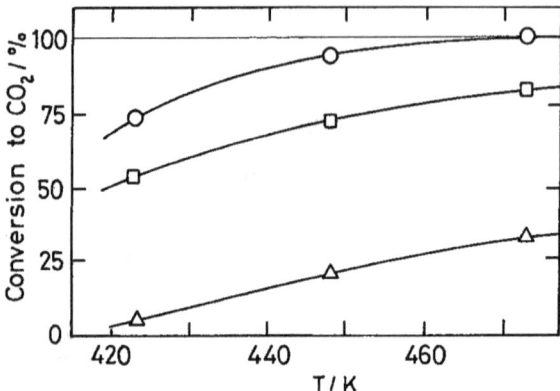

Figure 11.4: Temperature-dependence of formation of carbon dioxide in the oxidation of trimethylamine: 2 at% Ir/La_2O_3 ($\triangle$); 2 at% Au/Fe_2O_3 ($\square$); combination of the two ($\circ$).[86]

proceeds over $Au/NiFe_2O_4$ below 373 K (Figure 11.4), yielding mostly dinitrogen and carbon dioxide, while over palladium and platinum catalysts nitrous oxide is mainly produced even at higher temperatures.[70,85] The activity of Au/Fe_2O_3 was greatly improved by the addition of Ir/La_2O_3.[86]

About 1% $Au/\alpha\text{-}Fe_2O_3$ is active for the combustion of methanol and its decomposition products (methanoic acid and methanal) at temperatures below 373 K, at a space velocity of $2000\,h^{-1}$. Comparison of T_{50} values for Au/Fe_2O_3 and conventional combustion catalysts Pd/Al_2O_3 and Pt/Al_2O_3 gave the order of activity:[87]

$$Pd/Al_2O_3 > Au/Fe_2O_3 > Pt/Al_2O_3,$$

which is in contrast to that determined in earlier work,[88,89] in which gold was found to have low activity. In all cases, the only product observed was carbon dioxide. In addition to the obvious advantages of a catalyst with comparable activity to platinum and palladium for low temperature oxidation, the Au/Fe_2O_3 activity is improved by the presence of moisture. The reason for this is not clear, although FTIR studies have indicated that with the Au/Fe_2O_3 catalyst the adsorption of carbon monoxide increases.[90] Methanol has also been oxidised on Au/TiO_2 and Au/CeO_2 using non-calcined supports of high area, giving complete conversion at, respectively,

503 and 353 K.[91] Ethanol was successfully oxidised on Pd–Au/Al$_2$O$_3$ at 573 K, the presence of the gold stabilising the palladium against oxidation.[92]

11.3.3. Oxidative decomposition of dioxins

Dioxins are a family of heterocyclic compounds that are poisonous by-products of the manufacture of some herbicides and bactericides. The most toxic are polychlorinated dibenzo-*p*-dioxins (PCDDs) and dibenzo-furans (PCDFs, also called dioxins).[93] Amongst the most dangerous are the isomers of 2,3,7,8-tetrachlorodibenzo-*p*-dioxin (TCDD) and 2,3,7,8-tetrachlorodibenzofuran (TCDF). The former occurs in small amounts in some herbicides and defoliants, including the so-called 'Agent Orange' (a highly toxic herbicide sprayed as a defoliant in chemical warfare).[5] Dioxins also occur in the gaseous discharges from the combustion of domestic refuse, so that control of their emission from incinerators is one of the most urgent needs in environmental protection today.[5,86] There are several treatments currently in use, including raising the waste gas temperature to cause non-catalytic oxidative decomposition.[86] However, when the initial investment and mechanical complexity are taken into account, this is not practicable for small-scale incinerators. A useful solution would be to decompose dioxins and their derivatives at the dust filter using catalytic oxidation, preferably below 473 K, the temperature to which the waste gases are usually lowered after efficient heat recovery.

2,3,7,8 tetrachlorodibenzo-*para*-dioxin   2,3,7,8 tetrachlorodibenzofuran

Since gold is stable in the presence of halogens, some supported gold catalysts are more active and stable than other noble metal catalysts for reactions involving them, such as the oxidative decomposition of chlorofluoro-rocarbons and dioxins.[85,94−96] Au/Fe$_2$O$_3$–La$_2$O$_3$ had very good activity for the decomposition of several PCDDs and PCDFs at 413 K,[97] while others have used platinum, palladium and iridium catalysts supported on silica–boria–alumina oxides and zeolites as a first catalyst, and supported gold and some other metals (such as Ag, Cu, Fe, Sb, Se, Te and Ta and their oxides) as a second catalyst.[98−100] These eliminated dioxins and furans by oxidative decomposition, while suppressing the formation of secondary pollutants from gaseous precursors.[98−100]

Multi-component catalysts containing various supported noble metals have been developed;[86] these were found to act synergistically for several catalytic reactions. Although Ir/La_2O_3 exhibited low activity in the decomposition of dioxin derivatives, its presence, together with Pt/SnO_2 enhanced the activity of Au/α-Fe_2O_3. This helped to achieve 98% decomposition of dioxins from the outlet gases of incinerators even at 423 K.[86,101] This multi-component catalyst is therefore a promising material for purifying dioxin-containing gases emanating from incinerators.

11.3.4.   Hydrodechlorination of chlorofluorocarbons

Chlorofluorocarbons (CFCs) have been used as non-inflammable refrigerants, blowing agents in foams, aerosol propellants and solvents,[102−109] although their use is now forbidden. They are extremely chemically and thermally stable, and therefore when emitted into the atmosphere do not react before reaching the stratosphere. Once there, they photodissociate, producing chlorine radicals which participate in ozone depletion.[95−102] They also make a relatively large contribution to the greenhouse effect, estimated to be up to 25%. The hydrodechlorination of CFCs is therefore is very desirable from an environmental standpoint.

The development of CFC alternatives has focused on non-chlorinated compounds, such as hydrofluorocarbons (HFCs), which have similar physical properties, but break down in the lower atmosphere.[110−111] Consequently, hydrodechlorination is a simple method for the production of HFCs and also for destruction of CFCs[108,109,112−114]. Palladium is unique among the catalytic metals for its activity and selectivity in this process,[102−108,115,116] but using Pd–Au/SiO_2 (Pd:Au = 60:40) increased the selectivity for difluoromethane from difluorodichloromethane (CFC-12) shown by Pd/SiO_2 at 453 K from 40 to 95%.[109,112−114]

The decomposition of chlorodifluoromethane (HCFC-22) using gold supported on sulfated TiO_2/ZrO_2 proceeded as follows:[117]

$$CHClF_2 + H_2O \rightarrow CO + HCl + 2HF \qquad (11.8)$$

the carbon monoxide then being oxidised to the dioxide. A side-reaction with hydrogen fluoride led to the formation of trifluoromethane and hydrogen chloride:

$$CHClF_2 + HF \rightarrow CHF_3 + HCl \qquad (11.9)$$

The catalyst was deactivated by reaction with hydrogen chloride and hydrogen fluoride, but some of the activity could be recovered by exposing the deactivated catalyst to carbon monoxide at high temperatures.

11.3.5. Hydrodechlorination of 2,4-dichlorophenol[118]

2,4-Dichlorophenol has a large-scale use in pharmaceutical and herbicide synthesis, but is classed as a priority recalcitrant environmental pollutant; conversion to 2-chlorophenol, which can be recycled, is therefore a desirable reaction. Ni/SiO_2, Au/SiO_2 and $NiAu/SiO_2$ have been prepared from diaminoethane complexes and tested for this reaction; their activities rising in the sequence

$$Au/SiO_2 < Ni/SiO_2 < Ni/SiO_2 + Au/SiO_2 < NiAu/SiO_2$$

Selectivity to phenol with the $NiAu/SiO_2$ was 100%, the other catalysts forming 2-chlorophenol as well. Reactivation by hydrogen after the first 8 h of use gave improved activities, especially for the bimetallic catalyst. This was due to a narrowing of the particle size range, initially 10–150 nm, to 2–60 nm, and to homogenisation of the initially largely separate nickel and gold particles, to give a genuinely bimetallic catalyst. The presence of the gold influenced the temperature-programmed reduction of the precursors in a peculiar way: the hydrogen uptake for the Ni/SiO_2 was very large, due to hydrogenolysis of the ligands, but the $NiAu/SiO_2$ precursor only showed two small peaks (at 536 and 552 K); no explanation for this difference was suggested.

11.3.6. Removal of chlorinated hydrocarbons[119]

Chlorinated hydrocarbons (CHCs) are widely used in industry but bring both environmental and health risks;[5,120] catalytic oxidation is a low cost method for their destruction. The most active catalysts are the platinum group metals supported on alumina, but high temperature is needed to obtain a satisfactory rate and to overcome chloride poisoning,[121] but hydrogen chloride attacks the alumina support, so the use of other supports that are themselves active at lower temperature, such as Co_3O_4 and Cr_2O_3–Al_2O_3,[120] is advantageous. The rate of decomposition of dichloromethane over 5% Au/Co_3O_4 at 573 K was ten times higher than over Cr_2O_3–Al_2O_3, 70 times higher than over 0.5% Pt/Al_2O_3, and 560 times higher than over 0.5% Pd/Al_2O_3. Moreover, Au/Co_3O_4, catalysts were stable and selectively converted dichloromethane to carbon monoxide and hydrogen chloride. No detectable by-products (i.e. other CHCs) or partial oxidation products (chlorine, phosgene or methanal) were formed.

The activity of Pd/Al_2O_3 catalysts for hydrodechlorination of trichloroethene at room temperature is promoted by the presence of gold[122] (see Section 14.2.3).

CHCs have also been destroyed by hydrolysis with steam over acidic solids in the absence of any metal.[123]

11.3.7. Ozone decomposition

Ozone (O_3) is useful as a powerful oxidizing agent, but the molecule itself is toxic to both animal and plant life, and its release into the environment must be avoided; promoting ozone decomposition is therefore important. Its presence in the upper atmosphere is beneficial since it absorbs ultra-violet radiation, but close to the ground it is harmful, as it causes respiratory illness and encourages photochemical pollution. The maximum allowed ozone concentration in a working environment and the regulation threshold level for allowable exposure is 0.1 ppm.[124] For example, ozone needs to be removed from aircraft cabins, submarines, and office environments. Discharges from sterilisation, odour removal and wastewater treatment units must also have their ozone levels lowered.[5,125]

Although the decomposition of ozone to dioxygen is a thermodynamically favoured process,[126] it is thermally stable up to 523 K and catalysts are needed to decompose it at ambient temperature in ventilation systems, in the presence of water vapour and at high space velocity. A limited number of catalysts have been evaluated and active components are mainly metals such as platinum, palladium and rhodium, and metal oxides including those of manganese, cobalt, copper, iron, nickel and silver. Supports that have been used include γ-alumina, silica, zirconia, titania and activated carbon.[125,170]

Activated carbon and zeolites have been used in absorbent filters and need to be regenerated but catalytic decomposition has no such disadvantages, and therefore has attracted great interest. Most of the available information is to be found in the patent literature.[128–130] Supported silver and silver-containing catalysts have excellent activity for reducing ozone at high space velocity and in the presence of moisture, but they have low stability under severe reaction conditions. Gold is similar to silver as far as its mechanism of action is concerned,[131] but has superior performance, including resistance to moisture. About 1% Au/Fe_2O_3 and 1% Au/NiO are highly effective at 273 K, giving almost complete conversion at a space velocity of $10\,000\,h^{-1}$, being superior to both silver and nickel catalysts in both activity and stability. Gold catalysts were also used for the simultaneous removal of ozone and carbon monoxide[124] and Au/CeO_2 and Au/TiO_2 have achieved 100% ozone decomposition.[132]

About 3% Au/V_2O_5 supported on titania and zirconia catalysts ($d_{Au} = 5\,nm$) were very active for the decomposition of ozone at 293 and 303 K, respectively.[84,126]

Engelhard (USA) market a base metal catalyst system that converts ozone into oxygen when coated onto a car radiator at the moderate temperatures it generates. The inclusion of gold could make this catalyst even more effective.[124]

11.3.8. Reduction of nitrogen oxides with ammonia

Ammonia is sometimes used as a reducing agent for the selective reduction of NO_x in emissions from industrial installations, but unreacted ammonia creates a secondary air pollution problem because it is itself hazardous. Consequently, selective catalytic oxidation (SCO) is required to convert traces of ammonia to nitrogen downstream of the reactor:[134]

$$4NH_3 + 3O_2 \rightarrow 2N_2 + 6H_2O \qquad (11.10)$$

Cu/Al_2O_3 was found to be an active catalyst and very selective to nitrogen, and the addition of gold resulted in a significant increase in activity and a nitrogen selectivity of above 90%. The addition of gold raised the intensity of an absorption band at $1460\,cm^{-1}$, tentatively assigned to a surface imido-species, suggesting that the adsorbed ammonia molecule suffered dehydrogenation before its oxidation.

11.4. Removal of Sulfur Dioxide

Very large quantities of poisonous sulfur dioxide are formed by the combustion of fossil-derived fuels in factories, power plants, houses and automobiles, by the incineration of solid waste, and by volcanic activity.[5,6,135–148] Sulfur oxides (SO_2 and SO_3) are, together with nitrogen oxides, a primary and substantial contributor to acid rain; this is the main cause of damage to plants, fish and all biological systems by reducing the pH of the ground waters, streams, rivers and lakes, and accounts for the impoverishment of agricultural soils due to lixiviation of plant nutrients. Sulfur dioxide is also responsible for the depletion of the atmospheric ozone layer; it has harmful effects on human health, and acid rain causes the degradation of buildings and monuments in urban areas. The destruction of sulfur dioxide ($DeSO_x$) is consequently a very important requirement for improving the environment,[6,136,144–152] and new regulations emphasize the need for more efficient technologies to prevent the emission of sulfur dioxide formed in combustion processes.

The bond between adsorbed sulfur dioxide and bulk metallic gold is very weak (bonding energy $<42\,kJ\,mol^{-1}$) and the molecule does not dissociate;[146] this can be contrasted with its behaviour on most transition metal surfaces,[136,140,144,150] where the molecule adsorbs strongly and readily decomposes below 300 K. The only exception to this trend is silver, which adsorbs the molecule reversibly.[136,153] Titania is the catalyst most often used in the chemical industry and in oil refineries for removing sulfur dioxide by the Claus reaction[144,146,147] and by reducing it with carbon monoxide:[144,146]

$$2H_2S + SO_2 \rightarrow 2H_2O + 3S_{solid} \qquad (11.11)$$

$$SO_2 + 2CO \rightarrow 2CO_2 + S_{solid} \qquad (11.12)$$

The main products of the adsorption of sulfur dioxide on titania are sulfur trioxide and sulfate species.[146,147,154] In contrast to the behaviour of bulk gold, Au/TiO_2 catalyses desulfurisation with high efficiency;[146,154] the extraordinary ability shown by gold nanoparticles to adsorb and dissociate sulfur dioxide is comparable to that shown by the platinum group metals.[136,140,141,143,150] The interactions between the gold and titania are complex, each apparently assisting the activity of the other; for example, the presence of gold increases the rate at which oxygen vacancies move within the titania. Magnesia, frequently used in scrubbers for the absorption of sulfur dioxide,[140,155] also forms adsorbed sulfur trioxide and sulfate groups, without dissociation of the molecule. Au/MgO is more reactive towards sulfur dioxide than is bulk gold, but the magnesia seems to play only a minor role, and dissociation of sulfur dioxide on the catalyst is very limited because the Au...MgO interaction is very weak. The reactivities of

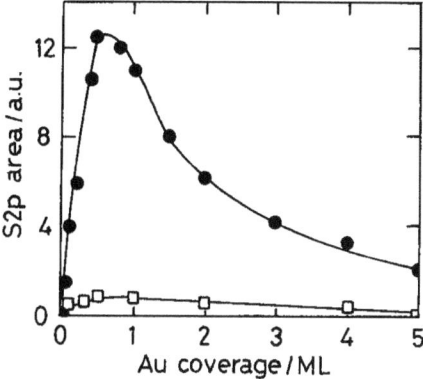

Figure 11.5: Amounts of sulfur formed by dissociation of 5 L of sulfur dioxide on Au/MgO(100) ($\square$) and Au/TiO$_2$(110) ($\bullet$) at 300 K as a function of Au coverage.[141]

model Au/MgO(100) and Au/TiO$_2$(110) surfaces towards sulfur dioxide at 300 K are compared in Figure 11.5.

11.5. Catalytic Wet Air Oxidation (CWAO)

Waste water is made environmentally acceptable by oxidizing the organic compounds it contains using oxygen or air at 453–588 K and 2–15 MPa.[156,157] The organic material present is first converted into simpler organic compounds, which are then further oxidized to carbon dioxide and water (Section 14.2.3). The last residues are organic acids, especially acetic acid. Organic nitrogen compounds are easily transformed into ammonia. Catalysts provide the possibility of using milder conditions; and CWAO processes studied to date have been based on platinum and palladium catalysts deposited on titania or titania–zirconia.[156]

Preliminary results have been reported[158] for the oxidation by Au/TiO$_2$ catalysts of an aqueous solution of succinic acid (5 g l^{-1}) at 463 K and 50 MPa in air; the reaction was structure-sensitive, small particles (2 nm) performing best, the reaction then being complete in 7 h. However, particle growth occurred during use, and re-used catalyst was less effective. AuRu/TiO$_2$ catalyst worked somewhat better, and several groups have demonstrated that platinum, palladium and ruthenium on titania or carbon are active catalysts for CWAO.[159–161]

References

1. G.C. Bond, *Heterogeneous Catalysis: Principles and Applications*, 2nd edn., Oxford University Press, Oxford, 1987, p. 160.
2. J.M. Thomas and W.J. Thomas, *Principles and Practice of Heterogeneous Catalysis*, VCH, Weinheim, 1997, p. 577.
3. E.S.J. Lox and B.H. Engler in *Handbook of Heterogeneous Catalysis*, G. Ertl, H. Knözinger and J. Weitkamp, (eds.), VCH, Weinheim, 1997, Vol. 4, p. 1559.
4. M. Bowker and R.W. Joyner, in *Insights into Speciality Inorganic Chemicals* D.T. Thompson, (ed.), Royal Society of Chemistry, Cambridge, UK, 1995.
5. M. Freemantle, *Chemistry in Action*, MacMillan Education, 1987.
6. S.A.C. Carabineiro, Ph.D. Thesis, New University of Lisbon, 2000.
7. P. Forzatti, *Appl. Catal. A: Gen.* **222** (2001) 221.
8. M.B. Cortie and E. van der Lingen, *Mater. Forum* **26** (2002) 1.
9. A. Ueda and M. Haruta, *Gold Bull.* **32** (1999) 3.
10. E. Seker, E. Gulari, R.H. Hammerle, C. Lambert, J. Leerat and S. Osuwan, *Appl. Catal. A: Chem.* **226** (2002) 183.
11. H. Ohtsuka, *Appl. Catal. B: Env.* **33** (2001) 325.
12. M.C. Kung, K.A. Bethke, J. Yan, J.-H. Lee and H.H. Kung, *Appl. Surf. Sci.* **121–122** (1997) 261.
13. B.W.-L. Jang, J.J. Spivey, M.C. Kung and H.H. Kung, *Energy Fuels* **11** (1997) 299.
14. B.E. Nieuwenhuys, *Adv. Catal.* **44** (1999) 259.
15. C. Mihut, C. Descorme, D. Duprez and M.D. Amiridis, *J. Catal.* **212** (2002) 125.
16. E. Seker, J. Cavataio, E. Gulari, P. Lorpongpaiboon and S. Osuwan, *Appl. Catal. A: Gen.* **183** (1999) 121.
17. A. Ueda, T. Oshima and M. Haruta, *Appl. Catal. B: Env.* **12** (1997) 81.
18. R. Burch and T.C. Watling, *Appl. Catal. B: Env.* **11** (1997) 207.
19. M.C. Kung, J.-H. Lee, A. Chu-Kung and H.H. Kung, *Stud. Surf. Sci. Catal.* **101** (1996) 701.
20. E. Seker and E. Gulari, *Appl. Catal. A: Gen.* **232** (2002) 203.
21. G. Pattrick, E. van der Lingen, C.W. Corti, R.J. Holliday and D.T. Thompson, Proc. CAPoC-6 Conference, Brussels, Belgium, 2003.
22. T. Miyadera, *Appl. Catal. B: Env.* **2** (1993) 199.
23. Y. Li and J.N. Armor, *J. Catal.* **145** (1994) 1.
24. T. Tabata, M. Kokitsu, O. Okada, T. Nakayama, T. Yasumatsu and H. Sakane, *Stud. Surf. Sci. Catal.* **88** (1994) 409.
25. A. Obuchi, A. Ohi, M. Nakamura, A. Ogata, K. Mizuno and H. Ohuchi, *Appl. Catal. B: Env.* **2** (1993) 71.
26. J.N. Armor, *Catal. Today* **26** (1995) 147.
27. G.R. Bamwenda, A. Obuchi, A. Ogata, J. Oi, S. Kushiyama, H. Yagita and K. Mizuno, *Stud. Surf. Sci. Catal.* **121** (1999) 263.
28. C. Mihut, B.D. Chandler and M.D. Amiridis, *Catal. Commun.* **3** (2002) 91.
29. G. Pattrick, E. van der Lingen, H. Schwarzer and S.J. Roberts, *Proc. CatGold 2003*, Vancouver, Canada, 2003.
30. N.W. Cant and P.W. Frederickson, *J. Catal.* **37** (1975) 53.
31. T.P. Kobylinsky and B.W. Taylor, *J. Catal.* **33** (1974) 376.
32. H. Murakami and Y. Fujitani, *Ind. Eng. Chem. Prod. Res. Dev.* **25** (1986) 414.
33. J.M. Schwartz and L.D. Schmidt, *J. Catal.* **148** (1994) 22.
34. A. El Hamdaoui, G. Bergeret, J. Massardier, M. Primet and A. Renouprez, *J. Catal.* **148** (1994) 47.

35. J.F. Trillat, J. Massardier, B. Moraweck, H. Praliaud, A.J. Renouprez, *Stud. Surf. Sci. Catal.* **116** (1998) 103.
36. M.B. Cortie and E. van der Lingen, *Mater. Forum* **26** (2002) 1.
37. A.C. Gluhoi, S.D. Lin and B.E. Nieuwenhuys, *Catal. Today* **90** (2004) 175.
38. M.A.P. Dekkers, Ph.D. Thesis, Leiden University, 2000.
39. M.A.P. Dekkers, M.J. Lippits and B.E. Niewenhuys, *Catal. Today* **54** (1999) 381.
40. A.C. Gluhoi, M.A.P. Dekkers and B.E. Nieuwenhuys, *J. Catal.* **219** (2003) 197.
41. N.W. Cant and N.J. Ossipoff, *Catal. Today* **36** (1997) 125.
42. J.R. Mellor, A.N. Palazov, B.S. Grigorova, J.F. Greyling, K. Reddy, M.P. Letsoalo and J.H. Marsh, *Catal. Today* **72** (2002) 145.
43. Y. Miyake and S. Tsuji, Eur. Pat. application 1 043 059 to Toyota.
44. R.L. Garten, R.A. Dalla Betta and J.C. Schlatter, in *Handbook of Heterogeneous Catalysis*, G. Ertl, H. Knözinger and J. Weitkamp, (eds.), VCH, Weinheim, 1997, Vol. 4, p. 1668.
45. W. Liu and M. Flytzani-Stephanopoulos, *J. Catal.* **153** (1995) 304.
46. D. Andreeva, T. Tabakova, L. Ilieva, A. Naydenov, D. Mehanjiev and M.V. Abrashev, *Appl. Catal. A: Gen.* **209** (2001) 291.
47. M.A. Centeno, M. Paulis, M. Montes and J.A. Odriozola, *Appl. Catal. A: Gen.* **234** (2002) 65.
48. V. Idakiev, L. Ilieva, D. Andreeva, J.L. Blin, L. Gigot and B.L. Su, *Appl. Catal. A: Gen.* **243** (2003) 25.
49. D. Andreeva, T. Tabakova, V. Idakiev, A. Naydenov, *Gold Bull.* **31** (1998) 105.
50. M. Haruta, *Now and Future* **7** (1992) 13.
51. S. Miao and Y. Deng, *Appl. Catal. B: Env.* **31** (2001) L1.
52. R.D. Waters, J.J. Weimer and J.E. Smith, *Catal. Lett.* **30** (1995) 181.
53. R.J.H. Grisel, Ph.D. Thesis, Leiden University, 2001.
54. R.J.H. Grisel, J.J. Slyconish and B.E. Nieuwenhuys, *Topics Catal.* **16–17** (2001) 425.
55. R.J.H. Grisel and B.E. Nieuwenhuys, *Catal. Today* **64** (2001) 69.
56. M.A. Centeno, M. Paulis, M. Montes and J.A. Odriozola, *Appl. Catal. A: Gen.* **234** (2002) 65.
57. S. Minicò, S. Scirè, C. Crisafulli and S. Galvagno, *Appl. Catal. B: Env.* **34** (2001) 277.
58. S. Minicò, S. Scirè, C. Crisafulli, R. Maggiore and S. Galvagno, *Appl. Catal. B: Env.* **28** (2000) 245.
59. S. Scirè, S. Minicò, C. Crisafulli and S. Galvagno, *Catal. Commun.* **2** (2001) 229.
60. S. Scirè, S. Minicò, C. Crisafulli, C. Satriano and A. Pistone, *Appl. Catal. B: Env.* **40** (2003) 43.
61. M. McGraph, *Appl. Catal. B: Env.* **5** (1995) N25.
62. E.M. Cordi and J.L. Falconer, *J. Catal.* **162** (1996) 104.
63. E.M. Cordi, P.J. O'Neill and J.L. Falconer, *Appl. Catal. B: Env.* **14** (1997) 23.
64. N.E. Quaranta, J. Soria, V. Cortes Corberan and J.L.G. Fierro, *J. Catal.* **171** (1997) 1.
65. T.A. Nijhuis, A.E.W. Beers, T. Vergunst, I. Hoek, F. Kapteijn and J.A. Moulijin, *Catal. Rev.-Sci. Eng.* **43** (2005) 45.
66. P. Papaefthimiou, T. Ioannides and X.E. Verykios, *Appl. Catal. B: Env.* **13** (1997) 175.
67. J.M. Gallardo-Amores, T. Armaroli, G. Ramis, E. Finocchio and G. Busca, *Appl. Catal. B: Env.* **22** (1998) 249.
68. M. Baldi, E. Finocchio, F. Milella and G. Busca, *Appl. Catal. B: Env.* **16** (1998) 43.

69. T. Mallat and A. Baiker, *Catal. Today* **19** (1994) 247.
70. A. Veda, T. Oshima and M. Haruta, *Appl. Catal. B: Env.* **18** (1998) 453.
71. M. Haruta, *Chem. Record* **3** (2003) 75.
72. A. Veda, T. Oshima and M. Haruta, *Appl. Catal. B: Env.* **18** (1998) 453.
73. R. Cousin, S. Ivanova, F. Ammari, C. Petit and V. Pitchon, *Proc. CatGold 2003*, Vancouver, Canada, 2003.
74. J.-N. Lin and B.-Z. Wan, *J. Chin. Inst. Chem. Eng.* **35** (2004) 149.
75. A.C. Gluhoi, N. Bogdanchikova and B.E. Nieuwenhuys, *J. Catal.* **229** (2005) 154.
76. A.C. Gluhoi, N. Bogdanchikova and B.E. Nieuwenhuys, *J. Catal.* **232** (2005) 96.
77. D. H. Kim, M. C. Kung, A. Kozlova, S. D. Yuan and H. H. Kung, *Catal. Lett.* **98** (2004) 11.
78. D. Andreeva, *Proc. CatGold 2003*, Vancouver, Canada (2003).
79. D. Andreeva, *Gold Bull.* **35/3** (2002) 82.
80. A. Naydenov, D. Mehanjiev, D. Andreeva and T. Tabakova, *Oxid. Commun.* **26** (2003) 492.
81. L.I. Ilieva, R.N. Nedyalkova and D.H. Andreeva, *Bulgarian Chem. Commun.* **34** (2002) 289.
82. D. Andreeva, R. Nedyalkova, L. Ilieva and M.V. Abrashev, *Appl. Catal. A: Gen.* **246** (2003) 29.
83. D. Andreeva, R. Nedyalkova, L. Ilieva and M.V. Abrashev, *Appl. Catal. B: Env.* **52** (2004) 157.
84. D. Andreeva, T. Tabakova, L. Ilieva, A. Naydenov, D. Mehanjiev and M.V. Abrashev, *Appl. Catal. A: Gen.* **209** (2001) 291.
85. A. Veda and M. Haruta, *Appl. Catal. B: Env.* **18** (1998) 115.
86. M. Okumura, T. Akita, M. Haruta, X. Wang, O. Kajikawa and O. Okada, *Appl. Catal. B: Env.* **41** (2003) 43.
87. M. Haruta, A. Ueda, S. Tsubota and R.M. Torres-Sanchez, *Catal. Today* **29** (1996) 443.
88. W.R. Patterson and C. Kemball, *J. Catal.* **2** (1963) 465.
89. C.S. Heneghan, G.J. Hutchings, I.D. Hudson and S.H. Taylor, *Nature* **384** (1996), 405.
90. F. Boccuzzi, A. Chiorino, S. Tsubota and M. Haruta, *Catal. Lett.* **29** (1994) 225.
91. M.-L. Jia, Y.-N. Shen, C.-Y. Li, Z.-R. Bao and S.-S. Sheng, *Catal. Lett.* **99** (2005) 235.
92. R. Brayner, D. Dos Santos Cunha and F. Bozon-Verduraz, *Catal. Today* **78** (2003) 419.
93. J.R. Hart, *Chemosphere* **54** (2004) 1539.
94. M. Haruta, *CATTECH* **6** (2002) 102.
95. G.J. Hutchings, *Gold Bull.* **29** (1996) 123.
96. B. Nkosi, M.D. Adams, N.J. Coville, G.J. Hutchings, *J. Catal.* **128** (1991) 378.
97. O. Kajikawa, X.-S. Wang, T. Tabata and O. Okada, *Organohalogen Compounds* **40** (1999) 581.
98. T. Sakurai, T. Iwasaki and E. Shibuya, *Zeolites* **17** (1996) 321. ·
99. Eur. Pat. Appl. 645172 (1995).
100. Jap. Appl. (1993) for N.E. Chemcat Corp., NKK Corp.
101. M. Haruta, *Proc. CatGold 2003*, Vancouver, Canada, 2003.
102. E.J.A.X. van de Sandt, A. Wiersma, M. Makkee, H. van Bekkum and J.A. Moulijn, *Catal. Today* **35** (1997) 163.
103. M. Makkee, E.J.A.X. van de Sandt, A. Wiersma and J.A. Moulijn, *J. Molec. Catal. A: Chem.* **134** (1998) 191.

104. J.A. Moulijn, M. Makkee, A. Wiersma and E.J.A.X. van de Sandt, *Catal. Today* **59** (2000) 221.

105. M. Makkee, A. Wiersma, E.J.A.X. van de Sandt, H. van Bekkum and J.A. Moulijn, *Catal. Today* **55** (2000)125.

106. E.J.A.X. van de Sandt, A. Wiersma, M. Makkee, H. van Bekkum and J.A. Moulijn, *Appl. Catal. A: Gen.* **173** (1998) 161.

107. A. Wiersma, E.J.A.X. van de Sandt, M.A. den Hollander, H. van Bekkum, M. Makkee and J.A. Moulijn, *J. Catal.* **177** (1998) 29.

108. Z. Karpiński, K. Early and J.L. d'Itri, *J. Catal.* **164** (1996) 378.

109. M. Bonarowska, B. Burda, W. Juszczyk, J. Pielaszek, Z. Kowalczyk and Z. Karpiński, *Appl. Catal. B: Env.* **35** (2001) 13.

110. L.E. Manzer, *Catal. Today* **13** (1992) 13.

111. L.E. Manzer and V.N.M. Rao, *Adv. Catal.* **39** (1993) 329.

112. M. Bonarowska, J. Pielaszek, W. Juszczyk and Z. Karpiński, *J. Catal.* **195** (2000) 304.

113. M. Bonarowska, A. Malinowski, W. Juszczyk and Z. Karpiński, *Appl. Catal. B: Env.* **30** (2001) 187.

114. M. Bonarowska, J. Pielaszek, V.A. Semikolenov and Z. Karpiński, *J. Catal.* **209** (2002) 528.

115. M. Ocal, M. Maciejewski and A. Baiker, *Appl. Catal. B: Env.* **21** (1999) 279.

116. A.L. Ramos, M. Schmal, D.A.G. Aranda and G.A. Somorjai, *J. Catal.* **192** (2000) 423.

117. S.Y. Lai, H. Zhang and C.F. Ng, *Catal. Lett.* **92** (2004) 107.

118. G. Yuan, J.L. Lopez, C. Louis, L. Delannoy and M.A. Keane, *Catal. Commun.* **6** (2005) 555.

119. Z. Ainbinder, L.E. Manzer and M.J. Nappa, in *Handbook of Heterogeneous Catalysis*, G. Ertl, H. Knözinger and J. Weitkamp, (eds.), VCH, Weinheim, 1997, Vol. 4, p. 1677.

120. B. Chen, C. Bai, R. Cook, J. Wright and C. Wang, *Catal. Today* **30** (1996) 15.

121. R.J. Farrauto, R.M. Heck and B.K. Speronello, *Chem. Eng. News* **7** (1992) 34.

122. M.O. Nutt, J.B. Hughes and M.S. Wong, *Env. Sci. Technol.* **39** (2005) 1346.

123. G.C. Bond and F. Rosa C., *Catal. Lett.* **39** (1996) 261.

124. Z. Hao, D. Cheng, Y. Guo and Y. Liang, *Appl. Catal. B: Env.* **33** (2001) 217.

125. B. Dhandapani and S.T. Oyama, *Appl. Catal. B: Env.* **11** (1997) 129.

126. R.H. Perry and D. Green, in *Perry's Chemical Engineer's Handbook*, McGraw-Hill, New York, 1989, p. 147.

127. P.M. Konova, A.I. Naydenov and D.R. Mehandjiev, *Bulgarian Chem. Commun.* **34** (2002) 437.

128. Y. Masafumi, N. Tadao, N. Kazuhiko and T. Tetsuo, *Jpn. Kokai Tokyo Koho.* **5** (1991).

129. M. Katsuhiko, H. Akira and I. Tetsuo, *Jpn. Kokai Tokyo Koho* **3** (1992).

130. Japanese Patent, 9150033 to Toyota, 1997.

131. Z. Hao, L. An and H. Wang, *React. Kinet. Catal. Lett.* **70** (2000) 153.

132. L.A. Petrov, *Stud. Surf. Sci. Catal.* **130C** (2000) 2345.

133. A. Naydenov, D. Mehanjiev, D. Andreeva and T. Tabakova, *Oxid. Commun.* **26** (2003) 492.

134. S.D. Lin, A.C. Gluhoi and B.E. Nieuwenhuys, *Catal. Today* **90** (2004) 3.

135. D.C. Anderson, P. Anderson and A.K. Galwey, *Fuel* **74** (1995) 1018.

136. J.J. Haase, *Phys. Condens. Matter* **9** (1997) 3647.

137. J. DeBarr and A.A. Lizzio, *Energy Fuels* **11** (1997) 267.

138. I. Mochida, K. Kuroda, S. Kawano, Y. Matsumura and M. Yoshikawa, *Fuel* **76** (1997) 533.
139. T. Zhu, L. Kundakovic, A. Dreher and M. Flytzani-Stephanopoulos, *Catal. Today* **50** (1999) 381.
140. J.A. Rodriguez, T. Jirsak, S. Chaturvedi and J.A. Hrbek, *J. Am. Chem. Soc.* **120** (1998) 11149.
141. T. Jirsak, J.A. Rodriguez, S. Chaturvedi and J. Hrbek, *Surf. Sci.* **418** (1998) 8.
142. J.A. Rodriguez and J. Hrbek, *Acc. Chem. Res.* **32** (1999) 719.
143. J.A. Rodriguez, T. Jirsak and S. Chaturvedi, *J. Chem. Phys.* **110** (1999) 3138.
144. J.A. Rodriguez, J.M. Ricart, A. Clotet and F. Illas, *J. Chem. Phys.* **115** (2001) 454.
145. J.A. Rodriguez, T. Jirsak, L. Gonzalez, J. Evans, M. Perez and A. Maiti, *J. Chem. Phys.* **115** (2001) 10914.
146. J.A. Rodriguez, G. Liu, T. Jirsak, J. Hrbek, Z. Chang, J. Dvorak and A. Maiti, *J. Am. Chem. Soc.* **124** (2002) 5242.
147. J.A. Rodriguez, M. Perez, T. Jirsak, J. Evans, J. Hrbek and L. Gonzalez, *Chem. Phys. Lett.* **378** (2003) 526.
148. S.A.C. Carabineiro, A.M. Ramos, J. Vital, J.M. Loureiro, J.J.M. Órfão and I.F. Silva, *Catal. Today* **78** (2003) 203.
149. S.C. Paik and J.S. Chung, *Appl. Catal. B: Env.* **5** (1995) 233.
150. T. Jirsak, J.A. Rodriguez and J. Hrbek, *Surf. Sci.* **426** (1999) 319.
151. M. Shelef and R.W. McCabe, *Catal. Today* **62** (2000) 35.
152. T.G. Kauffmann, A. Kaldor, G.F. Stuntz, M.C. Kerby and L.L. Ansell, *Catal. Today* **62** (2000) 77.
153. J.L. Solomon, R.J. Madix, W. Wurth and J. Stöhr, *J. Phys. Chem.* **95** (1991) 3687.
154. G. Liu, J.A. Rodriguez, T. Jirsak, J. Hrbek, Z. Chang, J. Dvorak and A. Maiti, *Proc. 18th Meeting of the North American Catalysis Society*, Cancun, Mexico, 2003.
155. J.A. Rodriguez, T. Jirsak, S. Chaturvedi and J.A. Hrbek, *J. Am. Chem. Soc.* **120** (1998) 428.
156. F. Luck, *Catal. Today* **27** (1996) 195.
157. F. Luck, *Catal. Today* **53** (1999) 81.
158. M. Besson, A. Kallel, P. Gallezot, R. Zanella and C. Louis, *Catal. Commun.* **4** (2003) 471.
159. S. Cao, G. Chen, X. Hu and P.L. Yue, *Catal. Today* **88** (2003) 37.
160. J. Trawczynski, *Carbon* **41** (2003) 1515.
161. J.-C. Beziat, M. Besson, P. Gallezot and S. Durecu, *J. Catal.* **182** (1999) 129.

CHAPTER 12

Catalysis by Soluble and Supported Gold Compounds

12.1. Overview of Homogeneous Catalysis by Gold

Until comparatively recently, i.e. 1998, significant applications-oriented homogeneous catalysis by gold complexes in solution was thought to be unlikely as the few examples in the literature had very small turnover numbers (TON = mole product per mole Au). This is in marked contrast to those for the platinum group metals, especially rhodium and palladium, which readily undergo catalytic oxidative-addition/reductive-elimination cycles and can have very high TONs. This has been rationalised by saying that this type of catalysis requires a very delicate balance between the stabilities of the two oxidation states involved, and this has not often been achieved for gold.[1] An additional factor was thought to be the reluctance of gold to form hydride complexes, so that the oxidation of Au^I by dihydrogen, or the formation of alkene complexes by β-elimination from Au^{III}-alkyl complexes was virtually unknown. Gold hydrides have however recently been postulated as intermediates[2] (see Section 12.3.4) and eventually a suitable choice of ligand could lead to the isolation of a stable gold hydride. For other precious metals, the relative stability of the two critical oxidation states has been successfully adjusted by appropriate choice of ligands, e.g. the inclusion of a good π-bonding ligand, such as carbonyl or a phosphine, increases the stability of the lower oxidation state. With gold, π-bonding has seemed to be of relatively little importance. This kind of thinking was supported experimentally to the extent that the few examples of homogeneous catalysis by gold reported in the literature were associated with very small turnover frequencies (TOF = mole product per mole Au per unit time) and yields of product per mole of catalyst. New thinking is, however, now required to rationalise the dramatically different results initially reported by Teles *et al.* (see Section 12.2) and now confirmed by a number of other groups of researchers. Some of the syntheses are uniquely promoted by gold catalysts and the reaction mechanisms could also have unique features. Homogeneous catalysis by gold is not new, however,[3] and even before 2004 there were about a hundred published papers on this topic, the first being the chlorination of naphthalene to octachloronaphthalene using AuCl or AuCl$_3$, reported in 1935:[4]

$$(12.1)$$

The gold compounds were amongst the most active catalysts but it is now known that other Lewis acid catalysts such as ferric chloride are superior for this reaction.

A significant advance was made in 2001 when the oxidation of sulfides to sulfoxides was achieved using HAuCl$_4$/AgNO$_3$ in acetonitrile as catalyst under 1 atm oxygen or air:[5]

$$R_2S + 0.5O_2 \rightarrow R_2SO. \qquad (12.2)$$

This catalyst had activity orders of magnitude higher than the previously used Ru^{II} or Ce^{IV} complexes, and this was a clear indication that sulfur does not poison the catalytic activity of gold. There is in fact clear evidence that sulfur can be a promoter for catalysis by gold.[6]

The oxidative carbonylation of amines to formamides was also reported in 2001 in up to 94% yield using Ph$_3$PAu^{I}Cl as catalyst and oxygen as oxidant in methanol:[7]

$$H_2N(CH_2)_6NH_2 + 2CO \rightarrow OCNH(CH_2)_6NHCO. \qquad (12.3)$$

(Ph$_3$P)Au^{I}Cl had also been used three years earlier for the dramatic breakthrough in gold catalysis made by Teles *et al.* at BASF (see Section 12.2). Wider use of liganded gold could have led to earlier realisation of the potential of homogeneous catalysis by gold, and investigation of a variety of ligands will undoubtedly increase its versatility. Until recently, however, only two intermediates of gold-catalysed reactions had been proven experimentally and these intermediates were both purely organic,[8] but an auraoxetane has now been isolated and characterized as a stable organometallic analogue of intermediates formed in gold-catalysed reactions[9] (Scheme 12.1):

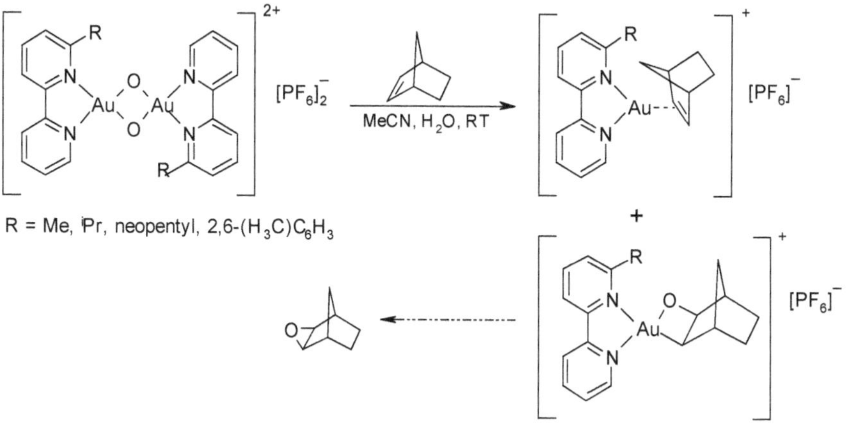

Scheme 12.1

The auraoxetane is depicted at the bottom right-hand side of Scheme 12.1. The reaction of the bridged μ-oxo dimer with norbornene led to a gold alkene complex (these are a rare species to date) and the unique auraoxetane. The epoxynorbornane was identified as the final product of the reaction.

The same reaction can often be catalysed by both Au^I and Au^{III}.[3] Since it is probable that the active species is formed *in situ* from the pre-catalyst by a change of oxidation state (e.g. either by the reduction of Au^{III} or by the disproportionation of Au^I to Au^{III} and Au^0), the same catalyst may be operating in both cases. In fact, the Au^{III} species most often used as starting compound for homogeneous catalysis has been $AuCl_3$.

An example of how gold oxidation state can control the structure of the product formed has been provided recently[3,10] when haloallenyl ketones were used in gold-catalysed cyclo-isomerisations to give furans (Scheme 12.2):

With $AuCl_3$ there was a preference for 2-bromofuran, obtained in up to 88% yield, possibly formed via the zwitterionic intermediate indicated in Scheme 12.2, whilst Et_3PAu^ICl exclusively produced 1-bromofuran, depicted at the bottom right-hand side of the scheme. This concept for the synthesis of halofurans has previously failed when using Pd(II) catalysts and a bromoallenyl compound; a yield of only 2% of the corresponding 2-chlorofuran was obtained.

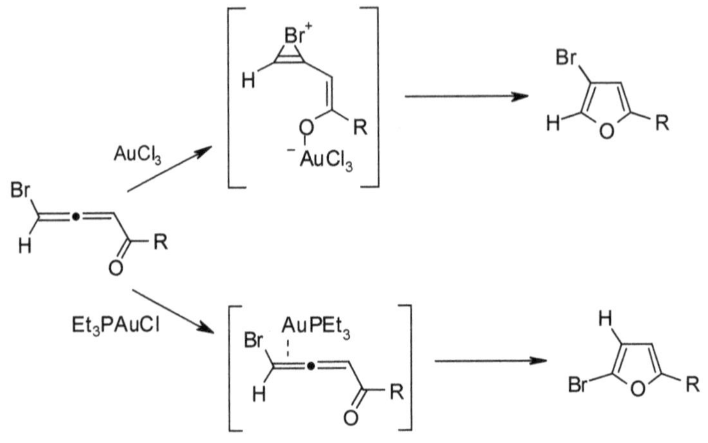

Scheme 12.2

The use of gold as a catalyst is desirable when it has a similar activity to that for a more expensive catalyst, or when it shows a higher activity or a higher selectivity than less expensive catalysts, and also when a completely new chemical transformation is possible using the gold catalyst. We will focus on examples of these three types in the following sections.

12.2.   Reactions with High Turnover Numbers and Frequencies

The status for homogeneous catalysis by gold in solution was dramatically transformed in 1998 by the results of Teles *et al.*[11] This BASF group described the use of cationic Au^I complexes of the type [L-Au$^+$] (where L is a phosphane, phosphite or arsine) for the addition of alcohols to alkynes in the presence of a Lewis or Brønsted acid as a co-catalyst in methanol at 313 K. An example of such a reaction is the conversion of 3-hexyne to 3,3-dimethoxyhexane:

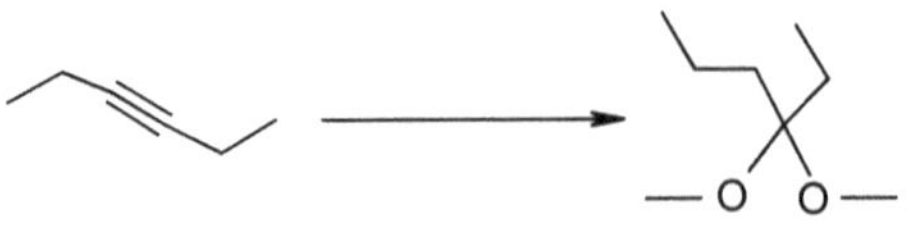

$$(12.4)$$

The turnover numbers for this type of reaction are up to 2×10^5 moles of product per mole of catalyst, with turnover frequencies of up to $5400\,\mathrm{h}^{-1}$.

These gold catalysts are a significant improvement on the mercury catalysts used previously and the reactions are conducted under mild conditions (293–323 K) in the presence of acid co-catalysts. 2-Propynol reacts
with excess methanol at 328 K in the presence of 0.01 mol% $CH_3Au^IPPh_3$
and sulfuric acid to give the following dioxane derivative, E-2,5-dimethyl-
2,5-dimethoxy-1,4-dioxane, in 93% yield after 20 h at 328 K:

$$(12.5)$$

Subsequently, TOFs of $3900\,h^{-1}$ have been reported[12] for the hydration
of 3-hexyne at 343 K in aqueous methanol using $Ph_3PAu^IOC(O)C_2F_5$ as
catalyst with $BF_3 \cdot Et_2O$ as co-catalyst to give 3-hydroxy-3-hexene and 3-
hexanone:

$$(12.6)$$

The catalyst was successfully recycled and reused five times without any
loss in activity. It was also demonstrated that (phosphane)Au^Icarboxylates
and sulfonates are highly active catalysts for hydration of non-activated
alkynes. Analogous Ag^I complexes are not active for these reactions due to
the fact that the Ag^I cations are much stronger acceptors for their ligands
and counterions compared with Au^I cations.

The research reported by Hayashi[13,14] indicated a highly efficient route
for the preparation of ketones from alkynes via hydration:

$$R^1-C{\equiv}C-R^2 + H_2O \rightarrow R^1-C{=}O-CH_2-R^2 + R^1-CH_2-C{=}O-R^2$$

$$(12.7)$$

The catalyst used was $Ph_3PAu^ICH_3$ (1 mol%) in sulfuric acid (50 mol% in
aqueous methanol) and high TOFs were obtained to give the Markovnikov
hydration product, 2-octanone, in high yield, without any anti-Markovnikov
hydration or methanol addition. Even higher yields and TOFs were
obtained using CF_3SO_3H (see Table 12.1). The reaction has also been
shown to be effective for a wide range of other substituted alkynes, e.g.
phenylacetylene.

Table 12.1: Effect of ligands on the hydration of alkynes.[a][13]

Substrate	Acid	Ligand	Yield of methylketone (TOF, h^{-1})
1-Octyne	H_2SO_4	None	35% (3500)
		CO (1 atm)	99% (9900)
		$(PhO)_3P$ (0.004 mmol)	90% (9000)
		$(EtO)PPh_2$ (0.01 mmol)	64% (6400)
	CF_3SO_3H[b]	None	70% (14000)
		CO (1 atm)	78% (15600)
Phenylacetylene	H_2SO_4	None	14% (1400)
		CO (1 atm)	33% (3300)

[a]$[(Ph_3P)Au^ICH_3]$ 0.002 mmol, acid 0.5 mmol, substrate 20 mmol, water 1 ml, methanol 10 ml, 343 K, 1 h.
[b]$[(Ph_3P)Au^ICH_3]$ 0.001 mmol.

Results varied with choice of additional phosphane ligands, and as can be seen from Table 12.1, the presence of CO had a significantly beneficial effect on both yield and TOF; indicating that coordination of the CO to the gold was probably involved in the mechanism.

Homogeneous catalytic hydrogenation has recently been achieved with high TOFs of up to ca 3900 h^{-1} and substrate: catalyst ratios of 1000:1 (see Section 12.3).

In summary, to date, all the most outstanding TONs/TOFs have been obtained with liganded Au^I complexes.

12.3. Gold Compounds as Catalysts for Organic Synthesis

The advances reported in the previous section have been paralleled by the work of several other research groups who have shown that soluble gold species can be used to catalyse the synthesis of cyclic organic molecules from alkynes to give products which have not previously had satisfactory synthetic routes, i.e. soluble gold catalysts have been shown to have unique advantages. Progress on these aspects of organic synthesis have been summarized by Stephen Hashmi.[3,8,15] Typical examples of the formation of products having new carbon–oxygen, carbon–nitrogen, and carbon–carbon bonds are given below. Reactions described here include some organic syntheses which have to date only been performed via use of gold catalysts, reactions where gold compounds are the best catalysts, or there is potential for development relevant to commercial applications.

12.3.1. Carbon–oxygen bond formation

The selective oxidation of methane to methanol has been reported[16] using homogeneous catalysis by gold and this is a very significant result. A 3 M solution of H_2SeO_4 in 96% sulfuric acid containing 27 mmol gold (added as 20 mesh gold powder to give clear yellow solution of cationic gold) led to the catalytic oxidation of methane (27 bar) to methanol with >90% selectivity at 453 K:

$$CH_4 + H_2SeO_4 \rightarrow CH_3OH + H_2SeO_3 \tag{12.8}$$

Both cationic gold and Se^{IV} need to be present, and TONs of up to 30 and TOFs of ca $10^{-3}\,s^{-1}$ were obtained in methanol concentrations of up to 0.6 M in sulfuric acid with >90% selectivity. The gold presumably has a catalytic role in promoting re-oxidation of the H_2SeO_3 to H_2SeO_4. Ways of increasing the performance of this catalyst could lead to industrial significance for homogeneous catalysis by gold. The reaction does not appear to proceed via free radicals and DFT calculations indicate that the active catalyst could be either Au^{III} or Au^I and they operate via mechanisms involving electrophilic C–H activation and oxidative functionalization. Au^{III} dissolved in sulfuric acid is also active for the carbonylation of alkenes to tertiary carboxylic acids, but the active intermediates are thought to be Au^I carbonyl species,[17] and the TONs are very small: this is an example of a Koch reaction, also known to be catalysed by Cu^I and Ag^I.

Propargyl and allyl ketones can be cyclised into furans using $AuCl_3$ (0.1 mol%) in acetonitrile:[18]

$$\tag{12.9}$$

Highly substituted furans play an important role in organic chemistry, both as key structural units in many natural products and important pharmaceuticals, and as useful building blocks in synthetic chemistry; examples of their synthesis using gold catalysis have recently been reported.[19]

Carbon–oxygen bonds can also be formed by the intramolecular reaction between an alkyne and an epoxide using 5 mol% AuCl$_3$ in acetonitrile:[20,21]

$$(12.10)$$

Thus, an oxygen nucleophile reacted intramolecularly with the alkynyl triple bond, and another example of this is provided by the synthesis of 2,5-disubstituted oxazoles from N-propargylcarboxamides under mild conditions using 5 mol% AuCl$_3$ in acetonitrile:[22]

$$(12.11)$$

While monitoring the conversion via ^{1}H NMR spectroscopy, an intermediate 5-methylene-4,5-dihydrooxazole could be observed and accumulated up to 95%; and this is the first direct and catalytic preparative access to such alkylidene oxazolines.

Superseding earlier negative predictions about the effectiveness of gold for reactions with alkenes, even unactivated alkenes have now been shown to react readily with weak nucleophiles such as phenols or carboxylic acids in the presence of 2 mol% of Ph$_3$PAu^{I}OTf (Tf = CF$_3$SO$_2$-) in toluene at 358 K to give up to 85 and 95% yields, respectively[23] (Scheme 12.3).

All these additions follow Markovnikov's rule. This regioselectivity is, in contrast to the additions to alkynes, connected with the formation of new stereogenic centres even when terminal alkenes are the starting compounds.

Gold-catalysed syntheses of heterobicyclic systems, where one of the rings contains an oxygen atom, have recently been reported.[24] Treatment of a 1,5-enyne alcohol with 5 mol% AuCl$_3$ in acetonitrile at 293 K gave a new product 6-oxabicyclo[3.2.1]octane in 89% yield (Scheme 12.4, structure on top right-hand side). The use of (Ph$_3$P)Au^{I}Cl/AgClO$_4$ as catalyst was equally effective. To rule out a possible involvement of the conjugate Brønsted acid in the alkyne activation, the same 1,5-enyne alcohol was

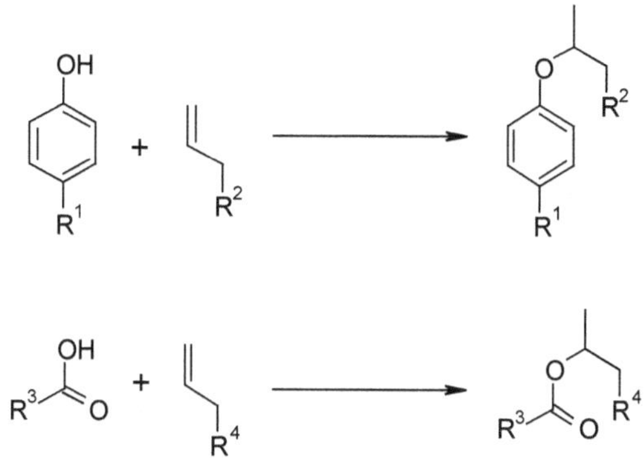

Scheme 12.3

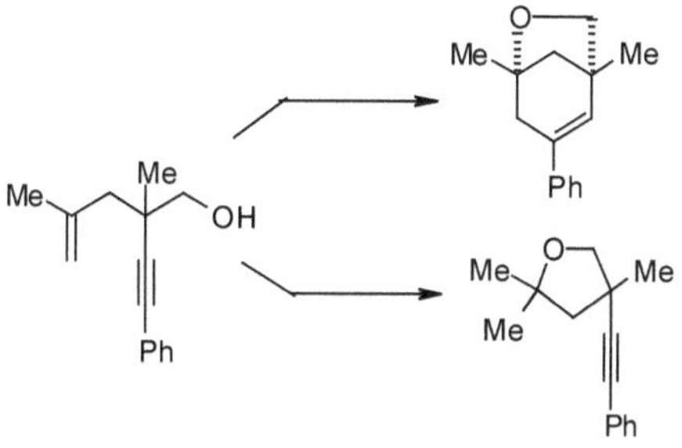

Scheme 12.4

treated with 50 mol% HCl in acetonitrile at 293 K instead of the gold catalyst: a much slower reaction ensued and the product was the substituted tetrahydrofuran depicted on the bottom right-hand side of Scheme 12.4 (70% yield).

12.3.2. Carbon–nitrogen bond formation

It is well known that late transition metal species such as Pd^{II} promote the addition of nucleophiles to carbon–carbon multiple bonds. Au^{III} has a similar d^8 configuration to Pd^{II} and has been found to catalyse the intramolecular addition of amines to give nitrogen heterocycles.[25,26] NaAuCl$_4$ catalyses the following reactions in acetonitrile or THF; 64–80% yields of the 2,3,4,5-tetrahydropyridine products being obtained after refluxing for 1 h:

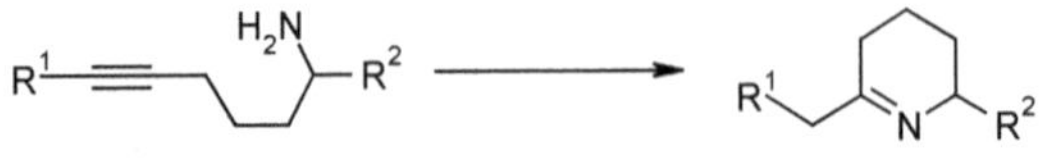

$$R^1 = H, Et, n\text{-}C_5H_{11}, n\text{-}C_6H_{13}, Ph$$
$$R^2 = H, Me, n\text{-}C_6H_{13}$$

(12.12)

Low concentrations of alkynylamines lead to good yields of cyclic product but high concentrations lead to precipitation of gold and low yields.

In a review of the use of gold catalysis in the synthesis of heterocyclic systems,[27] the following three-component coupling of aldehyde, alkynes and amines, illustrating highly efficient C–N bond formation, is highlighted.[28] Nearly quantitative yields of the propargyl piperidine is formed in most cases using water as solvent:

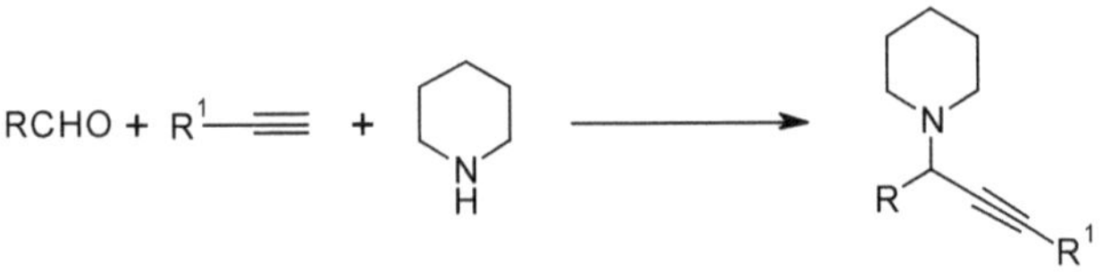

(12.13)

Less than 1 mol% of catalyst is required and the only by-product is water. Au^I and Au^{III} halides were active, with Au^{III} being slightly more active.

The use of $NaAuCl_4$ (4 mol%) as catalyst with the following substituted anilines in ethanol or aqueous ethanol at room temperature gives indoles in good to high yields:[29]

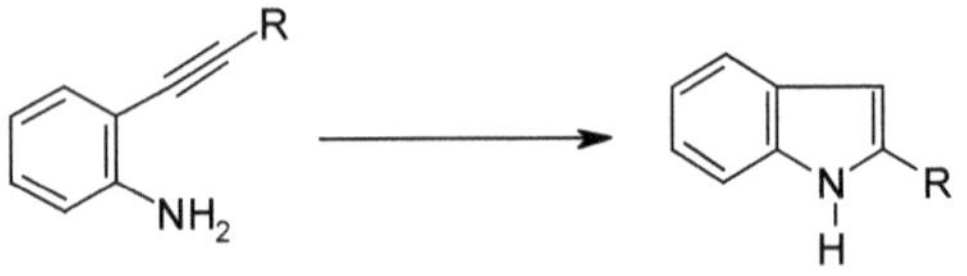

(12.14)

12.3.3. Carbon–carbon bond formation

The first example of a C–C bond-forming reaction catalysed by gold was the asymmetric aldol condensation developed in 1986.[30] The addition of an isocyano acetate to an aldehyde produces the *E*-oxazole as the major and *Z*-oxazole as the minor product in excellent *enantiomeric excess* (*ee*) in the presence of a cationic gold catalyst, $[Au(CyNC)_2]BF_4$, and a chiral diphosphanyl ferrocene ligand (see Scheme 12.5).

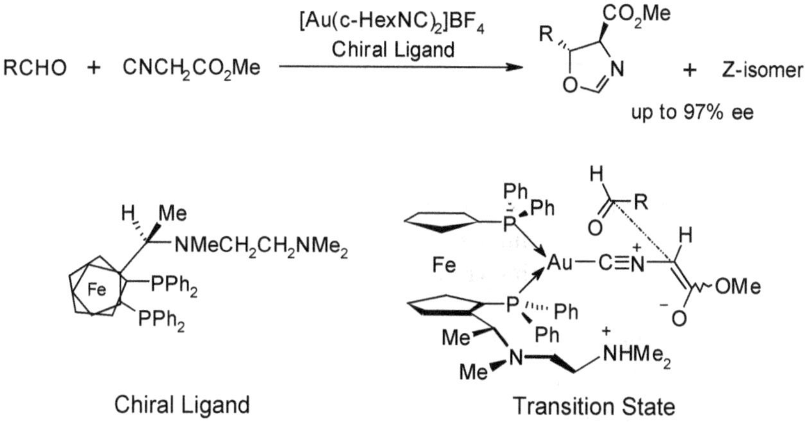

Scheme 12.5

Historically, it is interesting to note that this was the first example reported of a catalytic asymmetric aldol reaction.[31] The diastereomeric ratio of oxazole isomers was ca. 9:1 with an enantiomeric excess of up to 97%. This reaction has had a significant impact on organic chemistry, and already several summarising reviews have appeared on the reaction and its application in organic synthesis.[32,33]

Nucleophilic activation of propargylic alcohols by allylsilanes has provided another example of C–C bond formation. In the presence of 5 mol% $NaAuCl_4 \cdot 2H_2O$ in dichloromethane or dichloromethane/ethanol respectively at 298 K the allylsilanes give the direct nucleophilic substitution product depicted on the top right-hand side of Equation 12.15 in 82% yield, whereas ethanol gave the ketonic substitution/rearrangement product (bottom right-hand side) in 58% yield:[34]

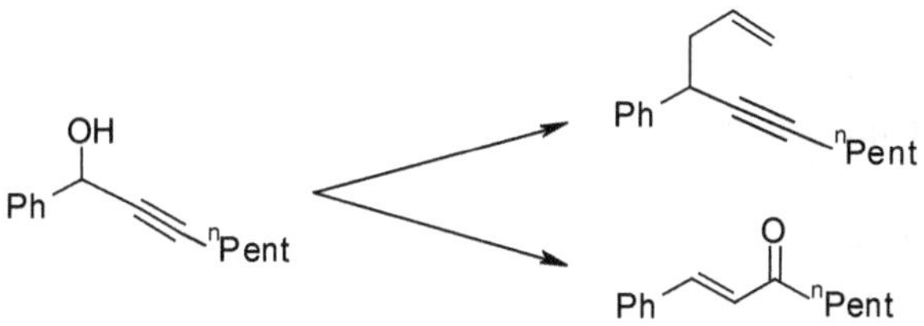

$$(12.15)$$

The following phenol synthesis from easily accessible furyl alkynyl starting material was the first gold-catalysed reaction to proceed via carbenes and arene oxides (Ts = *p*-toluene sulfonate; Cat = $AuCl_3$ in acetonitrile),[3,15,35,36] e.g.

$$(12.16)$$

Direct ^{1}H and ^{13}C NMR evidence was obtained for the formation of an epoxide intermediate. A number of other transition metals with a d^8 configuration also catalyse this transformation, but all are significantly less active than Au^{III}.

A gold-catalysed alkylation of arenes with epoxides has also recently been discovered: treating phenoxymethyloxiranes with AuCl$_3$/3AgOTf (Tf = CF$_3$SO$_2$-; triflate is trifluoromethanesulfonate i.e. OTf) (2.5 mol) in dichloromethane at 323 K yielded exclusively *endo* addition product 3-chromanols in good yield in 3 h:[37]

$$(12.17)$$

AuCl$_3$-catalysed benzannulation reactions have been reported.[38] Thus, naphthyl ketone derivatives were synthesized from *ortho*-alkynyl benzaldehydes in high yields:

$$(12.18)$$

Under similar conditions a copper(II)triflate/acid catalyst produces only unsubstituted naphthalenes.

The thermal cyclization of ketones onto alkynes, the Conia-ene reaction, can proceed thermally but only at inconveniently high temperatures.[39] Transition metals can catalyse it at lower temperatures but they require enolate generation, strong acid or photochemical activation. In the presence of phosphaneAu^I complexes, however, the reaction proceeds at ambient temperature under neutral conditions. For example, the following ketoester was converted into the cyclic product in 94% yield in 15 min in the presence of 1 mol% of gold catalyst $(Ph_3P)Au^ICl$ and 1 mol% AgOTf in dichloromethane at room temperature in 'open flask' conditions:

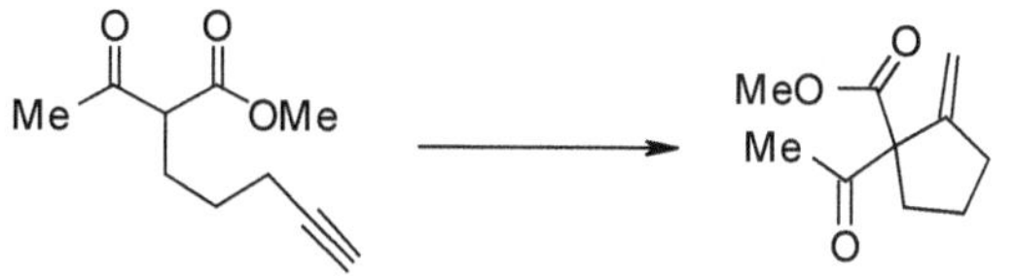

(12.19)

The high diastereoselectivities and mildness of these reaction conditions should make this reaction valuable for the synthesis of quaternary carbon centres and *exo*-methylenecycloalkanes.

The carbocyclization of acetylenic dicarbonyl compounds has also been reported.[40] Ph_3PAu^IOTf (1 mol%) catalysed the conversion of the following alkyne into the cyclic pentene in 93% yield in 10 min in CH_2Cl_2 at room temperature:

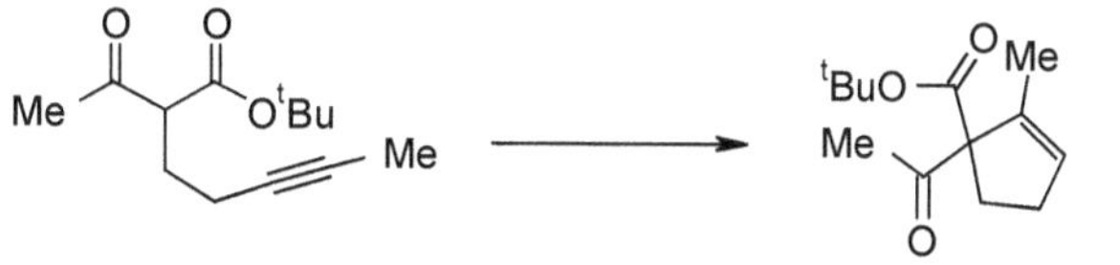

(12.20)

Under similar conditions, the use of 5 mol% AgOTf produced <5% conversion in 18 h, and the analogous copper catalyst did not catalyse the reaction at all.

$Ph_3PAuCl/AgSbF_6$ (1 mol%) was effective in catalysing the regioselective hydroheteroarylation of ethyne carboxylic ethyl ester in nitromethane at 313 K:[41]

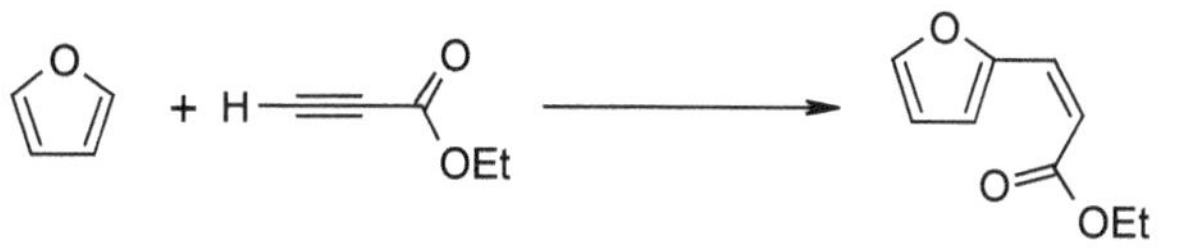

(12.21)

Thiophene reacted with the following imine in the presence of catalytic quantities of $AuCl_3/AgOTf$ in dichloromethane at 273 K:[42]

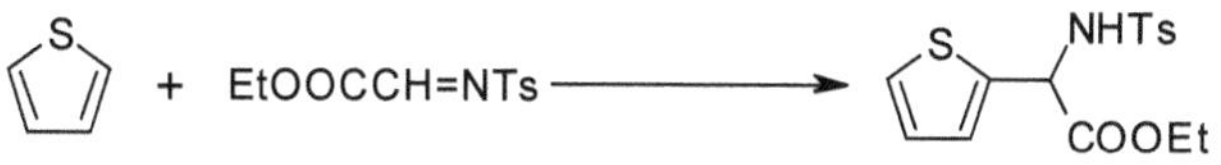

(12.22)

No reaction took place using Friedel–Crafts catalysts such as zinc and aluminium chlorides or $RuCl_3·3H_2O$, and the combined gold–silver catalyst gave higher yields than either of these compounds alone: the two examples of reactions given above are a portent that mixed metal systems should be investigated more generally in homogeneous catalysis as well as heterogeneous catalysis with gold (see also Sections 8.3–8.5).

The catalytic functionalization of aromatic C–H bonds to form C–C bonds under mild conditions is an attractive economic objective. An efficient hydroarylation reaction between alkynes and alkenes to form C–C bonds has been described.[43] The reactions are catalysed by Au^{III} complexes under mild and even solvent-free conditions at ambient temperature.

No excess of arene substrate is required for completion of the reaction. For example, reacting pentamethylbenzene with $HC{\equiv}CCO_2Et$ in dichloroethane or without any solvent gave a Z-arylalkene product in the presence of $AuCl_3$ pretreated with three equivalents of AgOTf at 296 K. The reaction was quantitative and the catalyst concentration could be lowered to 0.5 mol%. $AuCl_3$ alone gave a 25% yield, and in the absence of gold, no product at all was generated.

The direct functionalization of arenes by primary alcohol sulfonate esters has also been reported.[44] 5 mol% $AuCl_3$/AgOTf was again used as catalyst and high yields (up to 92%) of products obtained in some cases. Higher yields (e.g. 97, 93%) were obtained in intramolecular cycloalkylation reactions of a similar type.

The reaction of α, β-unsaturated ketones with electron-rich arenes is catalysed by $AuCl_3$, which was shown to be efficient under very moderate reaction conditions;[45] but in the case of sterically demanding products, HBF_4 was a better catalyst.

The addition of pentane-2,4-dione to styrene is catalysed by $AuCl_3(5\,mol\%)$/AgOTf(15 mol%) in CH_2Cl_2 or nitromethane:[46]

(12.23)

A reaction mechanism involving the formation of a gold hydride intermediate was tentatively suggested, in analogy with mechanisms known to operate for soluble platinum group metal catalysts such as those of palladium or rhodium phosphine hydride complexes.[47]

12.3.4. Catalytic hydrogenation

In heterogeneous catalysis, gold is much less active for hydrogenation than for oxidation, but this could make it have commercial potential for the selective hydrogenation of alkynes and dienes in olefin streams used for the production of polymers (see Section 9.2). An interesting example of homogeneous enantioselective hydrogenation has been described[48] using a new neutral dimeric Au^I complex {$(AuCl)_2([(R, R)MeDuphos]$ bearing the 1,2-bis[$(2R,5R)$-2,5-dimethylpholanebenzene] ($[(R, R)MeDuphos]$) ligand: this catalysed the asymmetric hydrogenation of alkenes and imines under mild reaction conditions with high TOFs (up to ca $3900\,h^{-1}$ and high enantiomeric excesses (up to 95% with a bulky substrate). The catalytic activities and selectivities were comparable with those of platinum and iridium complexes derived from the same ligand. A mechanism was proposed involving a gold hydride intermediate. This is a good illustration of the largely untapped potential of the effective use of phosphane ligands with gold as a catalyst.

12.3.5. Polymerisation of aniline by auric acid

Among conducting polymers, polyaniline has been extensively studied because of its high environmental stability, controllable electrical conductivity, and interesting redox properties. A new method for preparing polyaniline nanoballs, using $HAuCl_4$ in toluene as an oxidising agent, has now been reported,[49] in the presence of a phase-transfer catalyst at room temperature. During the reaction the aniline is oxidised and forms polyaniline whilst the $HAuCl_4$ is reduced and forms gold particles which decorate the nanoballs. This method is designed to result in intimate contact between the nanogold and the polyaniline. Among conducting polymers polyaniline has been extensively studied because of its high environmental stability, controllable electrical conductivity and interesting redox properties. Potential applications for polyaniline include organic lightweight batteries, microelectronics, electrochromic displays, electromagnetic shielding and sensors.

12.4. Supported Gold Complex Catalysts

Au^{III} supported on nanocrystalline ceria efficiently catalyses the homocoupling of boronic acids to give quantitative yields of biaryls in the presence of potassium carbonate:[50]

$$2ArB(OH)_2 + 2H_2O + Au^{3+} \rightarrow 2B(OH)_3 + ArAr + Au^+ + H_2 \quad (12.24)$$

Since the reactivity correlates with the amount of Au^{III} on the surface, it is assumed that the reaction is initiated by a twofold transmetallation from boron to Au^{III} followed by reductive elimination of the biaryl compound. The catalytic cycle is completed when Au^I is re-oxidized to Au^{III}. The reaction takes place in the absence of oxygen, and hydrogen can be detected by Raman spectroscopy. It also takes place in the absence of potassium carbonate but the catalyst is less stable. The TOF was at least 20 (calculated as the moles of boronic acid converted divided by two and by the moles of gold in the catalyst per hour).

Homogeneous and heterogenized Schiffs base complexes are catalysts for the self-coupling of aryl boronic acids.[51] Complexes such as the following were found to be catalysts in solution and when supported on MCM-41 and gave exclusively homocoupling product in high yields with TONs 53.3–66.0 mol converted per mol gold:

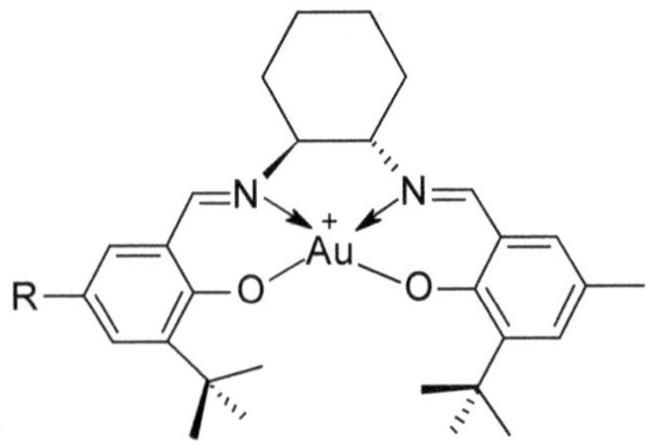

This is in contrast to the situation when palladium is used for the similar Suzuki reaction when the predominant products are from cross-coupling reactions.

12.5. Future Prospects

The recent surge of excitement in studying homogeneous catalysis by gold occurred a decade later than the similar movement in heterogeneous catalysis by gold and to date fewer researchers have become involved. The potential for activity and selectivity in homogeneous systems is at least as great, because many of the reactions occur under very mild conditions: and the present indications are that there is considerable potential for syntheses which are uniquely catalysed by gold in solution. Stereoselectivity has received little attention so far but is another valuable area for exploration. The limited versatility of gold compounds readily available has meant that much of the homogeneous investigations to date have been conducted with simple species such as $AuCl_3$, and there are now signs that the use of specially designed ligands will increase the potential for versatility in types of reactions which can be achieved. There are also some indications that investigation of gold-silver mixed-metal homogeneous systems give more interesting results than derivatives of either metal alone, and investigations of other mixed metal catalysts containing gold could bring further dividends. Speculative detailed mechanisms have been proposed in many of the papers we have quoted but there is little or no proof for these proposals and we have therefore omitted most of these schemes from this chapter. Since gold catalysts have enabled syntheses not catalysed by complexes of other metals, the mechanisms of catalysis by gold could have some unique features. An increased understanding of the mechanistic relationships between homogeneous and heterogeneous catalysis by gold will be helpful in stimulating advances in both areas.

Acknowledgements

We are grateful for the provision of comments, papers, and references by Stephen Hashmi, Antonio Arcadi, Silvio Carretin and Sónia Carabineiro.

References

1. R.V. Parrish, *Gold Bull.* **31** (1998) 14.
2. C. González-Arellano, A. Corma, M. Iglesias and F. Sánchez, *Chem. Commun.* (2005) 3451.
3. A.S.K. Hashmi, *Gold Bull.* **37** (2004) 51.
4. W. Schwernberger and W. Gordon, *Chem. Zentralbl.* **106** (1935) 514.
5. E. Boring, Y.V. Geltii and C.L. Hill, *J. Am. Chem. Soc.* **123** (2001) 1625.
6. J.E. Bailie and G.J. Hutchings, *Chem. Commun.* (1999) 2151.
7. F. Shi, Y. Deng, H. Yang and T.S. Ma, *Chem. Commun.* (2001) 345.
8. A.S.K. Hashmi, *Angew. Chem. Int. Ed.* **44** (2005) 6990.
9. M.A. Cinellu, G. Minghetti, F. Cocco, S. Stoccoro, A. Zucca and M. Manassero, *Angew. Chem. Int. Ed.* **44** (2005) 6892.
10. A.W. Sromek, M. Rubina and V. Gevorgyan, *J. Am. Chem. Soc.* **127** (2005) 10500.
11. J.H. Teles, S. Brode and M. Chabanas, *Angew. Chem. Int. Ed.* **37** (1998) 1415.
12. P. Roembke, H. Schmidbaur, S. Cronje and H. Raubenheimer, *J. Molec. Catal. A: Chem.* **212** (2004) 35.

13. E. Mizushima, T. Hayashi and M. Tanaka, *Proc. Gold 2003, Vancouver, Canada,* Sept.–Oct. 2003; http://www.gold.org/discover/sci_indu/gold2003/index.html.
14. E. Mizushima, K. Sato, T. Hayashi and M. Tanaka, *Angew. Chem. Int. Ed.* **41** (2002) 4563.
15. A.S.K. Hashmi, *Gold Bull.* **36** (2003) 3; http://www.gold.org/discover/sci_indu/ GBull/index.php.
16. C.J. Jones, D. Taube, V.R. Ziatdinov, R.A. Periana, R.J. Nielsen, J. Oxgaard and W.A. Goddard, *Angew. Chem.* **116** (2004) 4726.
17. Q. Xu, Y. Imamura, M. Fujiwara and Y. Souma, *J. Org. Chem.* **62** (1997) 1594.
18. A.S.K. Hashmi, T.M. Frost and J.W. Bats, *J. Am. Chem. Soc.* **122** (2000) 11553.
19. A.S.K. Hashmi, M. Rudolph, J.P. Weyrauch, M. Wölfle, W. Frey and J.W. Bats, *Angew. Chem. Int. Ed.* **44** (2005) 2798.
20. A.S.K. Hashmi and P. Sinha, *Adv. Synth. Catal.* **346** (2004) 432.
21. Z. Shi and C. He, *J. Am. Chem. Soc.* **126** (2004) 5964.
22. A.S.K. Hashmi, J.P. Weyrauch, W. Frey and J.W. Bats, *Org. Lett.* **6** (2004) 4391.
23. C.-G. Yang and C. He, *J. Am. Chem. Soc.* **127** (2005) 6966.
24. L. Zhang and S.A. Kozmin, *J. Am. Chem. Soc.* **127** (2005) 6962.
25. Y. Fukuda, K. Utimoto and H. Nozaki, *Heterocycles* **25** (1987) 297.
26. Y. Fukuda and K. Utimoto, *Synthesis* (1991) 975.
27. A. Arcadi and G. Bianchi, *Targets in Heterocyclic Systems* **8** (2004) 82.
28. C. Wei and C.-J. Li, *J. Am. Chem. Soc.* **125** (2003) 9584.
29. A. Arcadi, G. Bianchi and F. Marinelli, *Synthesis* (2004) 610.
30. Y. Ito, M. Sawamura and T. Hayashi, *J. Am. Chem. Soc.* **108** (1986) 6405.
31. M. Sawamura and Y. Ito, in *Catalytic Asymmetric Synthesis,* I. Ojima, (ed.), First edn., VCH, Weinhem, 1993, p. 367; Second edn., Wiley-VCH, New York, 2000, p. 493.
32. M. Sawamura and Y. Ito, *Chem. Rev.* **92** (1992) 857.
33. S.D. Pastor, in *Encyclopedia of Reagents for Organic Synthesis,* L.A. Paquette, (ed.), Vol. 1, Wiley, Chichester, 1995, Vol. 1, p. 447.
34. M. Georgy, V. Boucard and J.-M. Campagne, *J. Am. Chem. Soc.* **127** (2005) 14180.
35. A.S.K. Hashmi, L. Schwartz, J.-H. Choi and T.M. Frost, *Angew. Chem. Int. Ed.* **39** (2000) 2285.
36. T. Yao, X. Zhang and R.C. Larock, *J. Am. Chem. Soc.* **126** (2004) 11164.
37. Z. Shi and C. He, *J. Am. Chem. Soc.* **126** (2004) 5964.
38. N. Asao, K. Takahashi, S. Lee, T. Kasahara and Y. Yamomoto, *J. Am. Chem. Soc.* **124** (2002) 12650.
39. J.J. Kennedy-Smith, S.T. Staben and F.D. Toste, *J. Am. Chem. Soc.* **126** (2004) 4526.
40. S.T. Staben, J.J. Kennedy-Smith and F.D. Toste, *Angew. Chem.* **116** (2004) 5464.
41. M.T. Reetz and K. Sommer, *Eur. J. Org. Chem.* **18** (2003) 3485.
42. Y. Luo and C.-J. Li, *Chem Commun.* (2004) 1930.
43. Z. Shi and C. He, *J. Org. Chem.* **69** (2004) 3669.
44. Z. Shi and C. He, *J. Am. Chem. Soc.* **126** (2004) 13596.
45. G. Dyker, E. Muth, A.S.K. Hashmi and L. Dung, *Adv. Synth. Catal.* **345** (2003) 1247.
46. X. Yao and C.-J. Li, *J. Am. Chem. Soc.* **126** (2004) 6884.
47. A.W. Parkins, in *Insights into Speciality Inorganic Chemicals,* D.T. Thompson, (ed.), RSC, Cambridge, UK, 1995.
48. C. González-Arellano, A. Corma, M. Iglesias and F. Sánchez, *Chem. Commun.* (2005) 3451.

49. K. Mallick, M.J. Witcomb, A. Dinsmore and M.S. Scurrell, *Macromol. Rapid Commun.* **26** (2005) 232.
50. S. Carretin, J. Guzman and A. Corma, *Angew. Chem. Int. Ed.* **44** (2005) 2242.
51. C. González-Arellano, A. Corma, M. Iglesias and F. Sánchez, *Chem. Commun.* (2005) 1990.

CHAPTER 13

Miscellaneous Reactions Catalysed by Gold

13.1. Introduction

In this chapter there are collected short accounts of a number of reactions that do not fit easily into the previous chapters. Section 13.2 deals with the synthesis of vinyl chloride, a reaction of interest because its discovery signalled the resurgence of interest in what gold could do as a catalyst. It is also effective in the formation of C–C bonds (Section 13.3). The reactions mentioned in the other sections serve to demonstrate vividly the versatility of gold in catalysis.

13.2. Hydrochlorination of Ethyne

The heterogeneously-catalysed reaction of ethyne with hydrogen chloride leads to the formation of chloroethene, commonly known as vinyl chloride monomer. The traditional $HgCl_2$/carbon catalysts unfortunately suffer a high rate of deactivation, so alternatives have been eagerly sought. The rate-determining step is the addition of hydrogen chloride to a complex formed by ethyne and the catalyst MCl_n:

$$(HC\equiv CH)-MCl_n + HCl \rightarrow (HC\equiv CH)-MCl_n-HCl \qquad (13.1)$$

In one of the most extensive studies of metal chloride catalysts,[1] twenty of them supported on carbon were investigated, and a correlation was proposed between their activity and the electron affinity of the metal cation divided by the metal valence. Since the correlation consisted of two straight lines, it cannot be used predictively. However, electron affinity is necessarily a one-electron process, whereas hydrochlorination is more likely to be a two-electron process, involving the 2π electrons of ethyne. Because many of the cations investigated are divalent, standard electrode potential was suggested[1] as a more suitable parameter for correlating with activity.

On this basis, Graham Hutchings predicted[2-5] that gold chloride would be the most active catalyst for this reaction, and indeed it turned out that $AuCl_3$/carbon, prepared by incipient wetness impregnation with $HAuCl_4$ solution, was three times more active than $HgCl_2$/carbon. All the other supported metal chlorides deactivated when used in a fixed-bed reactor,[6] but $AuCl_3$/carbon lost its activity much less rapidly and deactivation was minimised by the use of high loadings of gold ($\geq$1%). It was temperature-dependent, being worse at high temperatures due to coke deposition,[5] arising probably through polymerisation of ethyne and vinyl chloride. Deactivation at low temperature was shown by ^{197}Au Mössbauer spectroscopy to be due to reduction of Au^{III} to Au^0; catalysts could therefore be reactivated by off-line treatment with chlorine or hydrogen chloride.[7] However, most importantly, it was observed that when nitric oxide was co-fed the activity was significantly enhanced and restored, and if introduced from the beginning, the deactivation could be almost wholly eliminated; presumably it served to re-oxidise Au^0 to Au^{III}. There was no effect on the vinyl chloride selectivity. $HgCl_2$/carbon catalysts deactivate by loss of mercury by vaporisation; this of course cannot happen with gold. Unfortunately this promising application of a gold catalyst did not prove commercially attractive at the time of the discovery, partly because the price of gold was unusually high in the 1980s and partly because there was no requirement for increased capacity for making vinyl chloride, which is not environmentally friendly; it was therefore easier to stay with well-known existing technology, whatever its limitations, rather than to venture into the unknown. The long-term stability of the gold system also required further study.[2]

13.3. The Formation of Carbon–Carbon Bonds

The formation of new C–C bonds is an important synthetic requirement in organic chemistry. It may be accomplished either by the conjunction of two identical radicals (homocoupling) or of two dissimilar radicals (cross-coupling); the latter is the *Heck reaction* and the former the *Suzuki* (or Suzuki–Miyaura) *reaction*. The Heck reaction is indeed catalysed by metallic gold and gold-containing bimetallic particles,[8,9] and also by $AuCl_4^-$;[10] the Suzuki reaction has also been performed[11] using Au^0 and supported Au^{III}.[12] Gold particles (3 nm) made by reduction of $AuCl_4^-$ by sodium borohydride and stabilised by poly(N-vinyl-2-pyrrolidone) converted phenylboronic acid to diphenyl with 60–70% selectivity at room temperature in 24 h.[11] While *n*- and *p*-methyl groups had little effect, an *o*-substituted group effectively stopped the reaction. A more thorough investigation of this reaction on Au^{3+}/nano-CeO_2 has shown that while it did not catalyse the cross-coupling reaction of phenylboronic acid with *p*-iodo-benzophenone, it effected the homocoupling of *p*-methyl-phenylboronic acid at $333\,K$ in $15\,h$ with complete selectivity to *p*-methyl-biphenyl.[12–14] The reaction was performed with K_2CO_3, which served to neutralise the boronic acid formed and thus prevent degradation of the catalyst; it could proceed in the absence of oxygen. The reaction is represented as:

$$2PhB(OH)_2 + 2H_2O + Au^{3+} \rightarrow 2B(OH)_3 + Ph_2 + Au^+ + H_2 \qquad (13.2)$$

The formation of hydrogen was confirmed by Raman spectroscopy, but in order to be catalytic there has to be a means of re-oxidising Au^+ to Au^{3+}; this may occur as

$$Au^+ + 2Ce^{4+} \rightarrow Au^{3+} + 2Ce^{3+} \qquad (13.3)$$

the Ce^{3+} then being re-oxidised by water, with formation of hydrogen.

Potentially useful examples of gold-catalysed Heck reactions include the following. 2-Alkynyl-phenylamines reacted with α,β-enones to form C-3-alkyl-indoles in the presence of $AuCl_4^-$ at $313\,K^{10}$ (Scheme 13.1), and activated methylene groups combined with alkenes when a silver–gold catalyst was used.[8] Palladium–silver–gold particles (Pd:Ag:Au $= 1{:}1{:}1$, $4.4\,nm$) have been formed by laser irradiation of a mixture of the three separate colloids; when stabilised with *N*-methyl-2-pyrrolidone they catalysed the reaction of 3-buten-2-ol with an aromatic bromo-compound 2-bromo-6-methoxynaphthalene to form the non-steroidal anti-inflammatory drug Nabumetone (Scheme 13.2) at $413\,K$ provided sodium bicarbonate was present.[9] Nabumetone is more effective than aspirin, and is comparable to Naproxen and Indomethacin. Using a traditional Heck catalyst $(Ph_3P)_2PdCl_2$ under the same conditions gave a Nabumetone yield of only 64%, compared to >95% with the trimetallic particles.

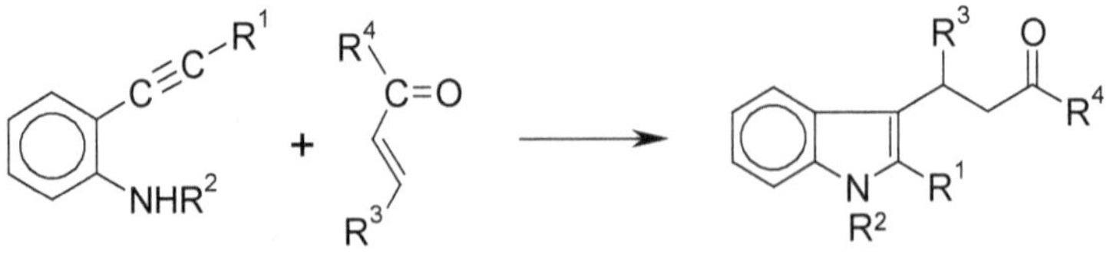

Scheme 13.1: Synthesis of C-3-alkylindoles.

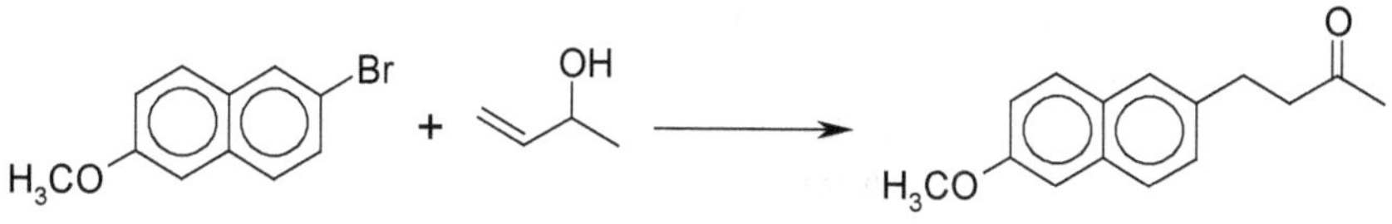

Scheme 13.2: Synthesis of nabumetone (4-(6-methoxy-2-naphthenyl)-2-butanone.

13.4. Other Reactions Catalysed by Gold

This final section gathers together a number of unrelated facets of the behaviour and potential of highly dispersed gold that are not conveniently located elsewhere in this book. They share no common features, but are briefly described here to stimulate possible future areas for development.

The amino-acid L-cysteine ($HSCH_2CHNHCOOH$) has been oligo-merised on small gold particles (3–7 nm) formed from $HAuCl_4$ by reduction with sodium borohydride.[15] L-cysteine is water soluble, and can readily bind to the gold surface through the thiolate linkage; the adsorbed layer is densely packed, and this facilitates the reaction between the amino group (–NH–) and the carboxylic acid group (–COOH) by elimination of water. This is not a polymerisation in the accepted sense of the word, but it shows that gold particles are able to adsorb and orient molecules in a way that helps them to react.

Polymer-immobilised catalysts containing gold particles smaller than 10 nm have been prepared by impregnating Merck Ion Exchanger IV with $HAuCl_4$, and have been used to synthesise ureas and carbamates by reaction of amines with carbon monoxide and oxygen;[16] so for example the oxycarbonylation of aniline proceeded at 448 K and 5 MPa pressure ($CO:O_2 = 2:1$) in 1–3 h to give yields of up to 99% of phenylmethyl-carbamate (Scheme 13.3).

In a quite different field, colloidal gold formed by reduction with cit-rate ion markedly improved the activity of silver chloride electrodes for the

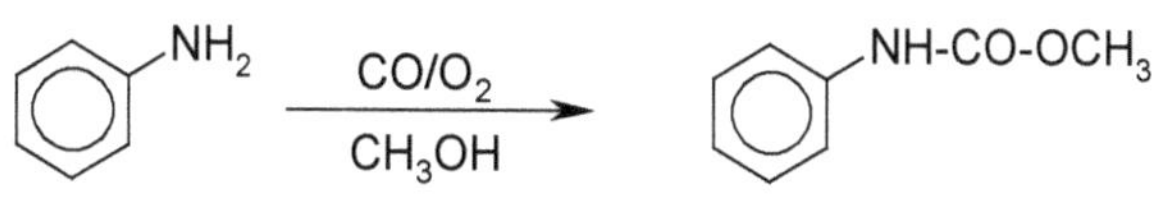

Scheme 13.3: Synthesis of phenylmethylcarbamate.

photocatalytic oxidation of water to oxygen.[17] Briefly, the light quanta convert silver chloride to silver atoms and chlorine; this reacts with water to form hypochlorous acid (HOCl) which decomposes in a reaction catalysed by Ag^+ to release oxygen. Anodic polarisation re-oxidises silver atoms, and under suitable conditions the process is catalytic and self-sustaining. Colloidal gold particles deposited on the silver chloride surface increased the photocurrent and oxygen production 3–4 fold.

We have normally declined to consider the consequences of combining gold with a catalytically active element of Group 10 where the sole function of the gold is to dilute the more active component; a great deal could be written on such systems, but, since gold has no activity of its own in these cases, it seemed inappropriate to consider them in this book. However, to indicate how the dilution effect operates, we have selected one system of great technical importance where application of the techniques of surface science have revealed exactly what happens.

We noted in Chapter 7 the requirement of the chemical industry for large quantities of pure hydrogen for processes such as ammonia synthesis and fat hardening, and that the major route followed is the *steam-reforming of alkanes*, e.g.

$$CH_4 + 2H_2O \rightarrow 4H_2 + CO_2. \tag{13.4}$$

This reaction is catalysed by supported nickel, but unfortunately parasitic reactions lead to the deposition of carbon, with consequent loss of catalytic activity. Now the dissociative chemisorption of methane into methylene groups and their subsequent polymerisation to carbon or graphite needs an ensemble of several nickel atoms, and the chance of finding such groups decreases rapidly if the surface of the nickel is coated with a small amount of an inactive metal such as gold. Some loss of initial activity may have to be tolerated, but the greater lifetime of the catalyst is a worthwhile advantage. Gold is immiscible with bulk nickel, so we are dealing with a purely surface phenomenon. The importance of nickel ensemble size was shown by studying the dissociation probability of methane on Ni(111) covered by increasing amounts of gold;[18] this fell progressively and reached zero at half monolayer coverage. DFT calculations showed that placing a gold atom next to the nickel atom on which reaction was to occur increased the activation energy by $16 \, kJ \, mol^{-1}$. The concept was tested experimentally with a $Ni/MgAl_2O_4$ catalyst containing 16% nickel, catalysing the reaction of *n*-butane with steam at 823 K; this alkane was chosen because it causes

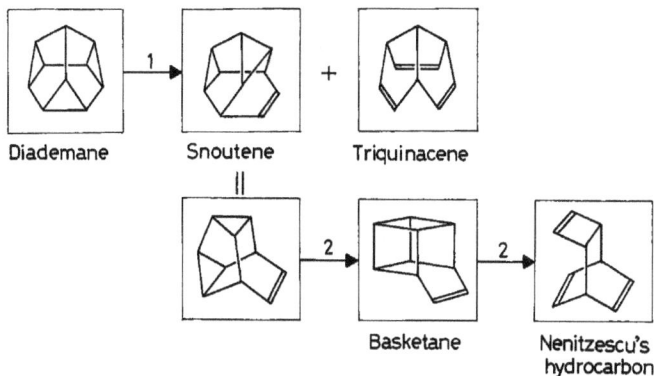

Diademane   Snoutene   Triquinacene

Basketane   Nenitzescu's hydrocarbon

Scheme 13.4: Catalysed transformations of polycyclic hydrocarbons in a gold-lined reactor; (1) 433 K; (2) 543 K.

much more rapid loss of activity than methane. While its activity fell with progressively increasing speed over 4000 min, addition of 0.3% gold stopped this entirely, and no formation of graphite could be seen. This is a powerful demonstration of the modifying influence of gold on the surface of an active catalyst, even where formation of a normal alloy cannot take place.

With another change of scene, we note the ability of gold to catalyse transformation of polycyclic hydrocarbons having fascinating structures (Scheme 13.4);[19] the reactions depicted there were catalysed either in the vapour phase using a gold-lined reactor at temperatures between 373 and 543 K and contact times of 10–20 s; they were also effected by the complex Au(DCP)$_2$Cl (DCP = dicyclopentadiene) in solution over 1 day at room temperature.

Finally, gold-plated copper turnings catalysed the fluorination of aromatic hydrocarbons at ~550 K;[20] gold has also been the means of making carbon nanotubes from ethyne (5% Au/SiO$_2$-Al$_2$O$_3$, 823 K)[21] and chloromethylsilanes.[22]

References

1. K. Shinoda, *Chem. Lett.* (1975) 219.
2. G.J. Hutchings, *Gold Bull.* **29** (1996) 123.
3. B. Nkosi, N.J. Coville and G.J. Hutchings, *Appl. Catal.* **42** (1988) 33.
4. B. Nkosi, N.J. Coville and G.J. Hutchings, *J. Chem. Soc. Chem. Commun.* (1988) 71.
5. B. Nkosi, N.J. Coville, G.J. Hutchings, N.D. Adams, J. Friedl and F. Wagner, *J. Catal.* **128** (1991) 366; B. Nkosi, N.D. Adams, N.J. Coville and G.J. Hutchings, *J. Catal.* **128** (1991) 378.

6. G.J. Hutchings, *Catal. Today* **72** (2002) 11.
7. M. Freemantle, *Chemistry in Action*, MacMillan Education, UK, 1987.
8. X. Yao and C.-J. Li, *J. Am. Chem. Soc.* **126** (2004) 2672.
9. S.-H. Tsai, Y-H. Liu, P.-L. Wu and C.-S. Yeh, *J. Mater. Chem.* **13** (2003) 978.
10. M. Alfonsi, A. Arcadi, M. Aschi, G. Bianchi and F. Marinelli, *J. Org. Chem.* **70** (2005) 2265.
11. H. Tsunoyama, H. Sakurai, N. Ichikuni, Y. Negishi and T. Tsukuda, *Langmuir* **20** (2004) 11293.
12. S. Carrettin, J. Guzman and A. Corma, *Angew. Chem. Int. Ed.* **44** (2005) 2.
13. S. Carrettin, A. Corma, M. Iglesias and F. Sánchez, *Appl. Catal. A: Gen.* **291** (2005) 247.
14. C. Gonzalez-Arellano, A. Corma, M. Igelsias and F. Sánchez, *Chem. Commun.* (2005) 190.
15. K. Naka, H. Itoh, Y. Tampo and Y. Chujo, *Langmuir* **19** (2003) 5546.
16. F. Shi and Y. Deng, *J. Catal.* **211** (2002) 548.
17. A. Currao, V.R. Reddy and G. Calzaferri, *Chem. Phys. Chem.* **5** (2004) 720.
18. F. Besenbacher, I. Chorkendorff, B.S. Clausen, B. Hamer, A.M. Molenbroek, J.K. Nørskov and I. Stensgaard, *Science* **279** (1998) 1913.
19. L.U. Meyer and A. de Meijere, *Tetrahedron Lett.* (1976) 497.
20. R.N. Haszeldine and F. Smith, *J. Chem. Soc.* (1950) 2689.
21. S.-Y. Lee, M. Yamada and M. Miyake, *Carbon* **43** (2005) 2654.
22. E.W. Krahe and E.G. Rochow, *Inorg. Nucl. Chem. Lett.* **1** (1965) 117.

CHAPTER 14

Commercial Applications

14.1. Introduction

The properties of gold catalysts described in earlier chapters imply that suitable investment in tailoring these to commercial requirements will lead to many useful applications. These could include catalysts for pollution and emission control, chemical processing of bulk and speciality chemicals, clean hydrogen production for the emerging 'hydrogen economy' and fuel cells and sensors to detect poisonous or flammable gases or substances in solution.[1] Those involved in investigating the development of various applications are making progress towards viable methods of manufacturing significant quantities of catalyst[2] and devising reliable preparative methods for ensuring suitable durability of catalysts under required operating conditions.

Already, the first known practical application for gold as a catalyst component within major industrial processes is well established for the manufacture of vinyl acetate monomer (VAM) (Section 8.4), and a pilot plant has been built for the production of methyl glycolate (Section 8.3.5). Investigation of the use of gold catalysts in respirators for carbon monoxide removal[3] and other pollution control applications is well underway (Chapter 11). The use of PROX systems to purify hydrogen by selective oxidation of carbon monoxide was discussed in Chapter 7, and its practical application in supplies of hydrogen for fuel cells and chemical processing can confidently be forecast.[1]

Most of the published methods for preparing gold catalysts in small research quantities are unlikely to prove suitable for commercial applications.[1] Complete removal of precious metal from the liquid phase is desirable when using solution methods: deposition–precipitation (DP) techniques, whilst producing highly active catalysts, also consume large quantities of water and the cost of treatment of wastewater is an expensive additional process. Other preparation methods such as appropriate modifications of impregnation via incipient wetness techniques are more likely to be suitable for commercial production if they lead to reproducible, stable and active catalysts and reduce the resulting quantities of wastewater. However, financially viable methods for catalyst manufacture are now available,[2] using both solution and metal vapour deposition approaches, and 20 kg quantities of supported gold catalysts are now available. Availability of a greater variety of soluble gold complexes would facilitate more rapid advances in the development of homogeneously catalysed gold processes.

One of the potential advantages that the use of gold catalysts offers compared with other precious metal catalysts is lower cost and greater price stability, gold being substantially cheaper (on a weight for weight basis) and considerably more plentiful than platinum. Whilst the gold mining companies are undoubtedly keen to see substantial new gold demand generated through new commercial uses of gold as a catalyst, it is not envisaged that this additional demand will result in strains on gold supply and hence upward pressure on the gold price. This is an important consideration for industrialists who wish to see stable costs in their manufacturing processes. The mild conditions under which gold catalysts are active make them attractive for sensor applications. Based on the current research efforts being devoted to gold catalysis, there is cause for optimism that many new practical applications for gold-based catalysts could emerge over the next decade.

The chemistry of the various reactions catalysed by gold was described in the earlier chapters. Here we will focus on the characteristics and requirements for commercial applications.

14.2. Pollution and Emission Control Technologies

14.2.1. Air cleaning

Gold catalysts are highly active for the oxidation of many components in ambient air at low temperatures, particularly carbon monoxide and nitrogen-containing malodorous compounds such as trimethylamine. This ability offers scope for applications related to air quality improvement and control of smells, be they in buildings, transport or other related applications such as gas masks. This a very important issue and the scope here for commercial applications is very large and significant patents have been published.[4,5] It is particularly advantageous that gold's catalytic activity is often promoted by moisture. The major remaining technical hurdle to overcome before widespread application of these technologies is the prevention of deactivation of gold catalysts caused by the accumulation of contaminants on catalyst surfaces or sintering of the active components. Nonetheless, prototype products that use gold catalysts for low temperature air quality control are now appearing in the public domain. It is believed that the market for this type of product will grow rapidly in the coming years.

Air-cleaning devices are needed for respiratory protection (gas masks) and for removing carbon monoxide and trace amounts of VOCs and ozone from ambient air indoor office space (due to smoking, etc.) and in submarines or space crafts on long missions.[6–20] Gold catalysts are promising materials to use in devices for removing carbon monoxide, and their use in respirators by its conversion to carbon dioxide is being explored.[3,21] Union Chemical Laboratories in Taiwan have developed masks for fire fighters which last for up to 100 h and which operate at room temperature: the gold particle size is 2 nm on oxide supports. Studies on carbon monoxide oxidation over gold catalysts (supported on Fe_2O_3 and TiO_2) in real air confirm that they are useful for the removal of carbon monoxide from both low (10–100 ppm) and high concentrations (10 000 ppm).[22,23]

Au/Fe_2O_3 is also a catalyst for ozone decomposition and simultaneous elimination of ozone and carbon monoxide at any ratio in the presence of oxygen at ambient temperature.[24] This catalyst is suitable for use in severe conditions such as relatively high ozone concentration, and large space velocity. It also shows a high room temperature activity and good resistance to moisture.

Incinerator exhaust gases can contain a huge variety of pollutants, such as dioxins, VOCs, hydrocarbons, nitrogen oxides, carbon oxides and amine derivatives. Due to the variety, it would be unreasonable to expect a single-component catalyst to achieve high catalytic performance for all of the exhaust gas pollutants. However, it has been shown that if several supported single noble metal catalysts are encouraged to work in synergy, the overall catalytic activity can be greatly improved.[25] Consequently, a ternary component noble metal catalyst (consisting of gold Au/Fe_2O_3–Pt/SnO_2–Ir/La_2O_3) achieved good results for purifying typical exhaust gases emanating from incinerators even at 423 K.[25,26]

Odour-producing compounds, such as trimethylamine, can be oxidatively decomposed over $Au/NiFe_2O_4$ at temperatures below 373 K[6,9] (see Chapter 11). Au/Fe_2O_3, supported on a zeolite wash-coated honeycomb, has been used commercially as a deodorizer in Japanese toilets since 1992.[6,26]

14.2.2. Autocatalysts

Air pollution generated from mobile sources is the greatest challenge in
the pollution control field. In the last 60 years the world vehicle fleet has
increased from about 40 million vehicles to over 700 million, and this figure
is projected to increase to 920 million by the year 2010.[27] The three major
pollutants emitted by internal combustion engines are carbon monoxide,
non-methane hydrocarbons and nitrogen oxides[28] (see Chapter 11). Environmental legislation governing the emission of these three types of gas
is becoming increasingly stringent. Catalysts that are capable of removing these pollutants simultaneously are referred to as three-way catalysts
(TWCs) and the design of these catalyst systems is continually evolving to
meet lower emission requirements. Typically, these catalysts must operate
in the presence of 10% water and 10–60 ppm sulfur dioxide at temperatures ranging from 623 to 1273 K and space velocities ranging from 10 000
to over $100\,000\,h^{-1}$ for the duration of 100 000 miles of operation.[28,29] All
commercial TWCs in use at present are based on platinum, palladium and
rhodium on a support comprised of zirconia-stabilized ceria, zirconia and
α-alumina. Additives include barium oxide and zinc oxide. These PGM-
based catalysts perform the task of emission control very well and many
aspects of this technology are very well established, but there are areas,
such as low light off and catalysts for diesel exhaust, where gold could play
a role in the future (see Chapter 11).

Catalyst efficiency is usually evaluated under simulated driving conditions using the standardized federal test procedure (FTP). A key problem
identified in the FTP is the liberation of unburned non-methane hydrocarbons during the cold start mode of the test when the catalyst monolith
is at ambient temperature. As a consequence, the catalyst does not reach
the hydrocarbon light-off temperature of about 573 K until approximately
2 min after the start of the test. During this delay up to 50% of the total
unburned hydrocarbons are emitted. Additionally, when the engine operates under prolonged idling conditions, the temperature at the inlet to the
converter is typically 553 K which is an accepted monolith temperature
for a catalyst converter mounted approximately 80 cm from the exhaust
manifold of a spark ignition engine of 1.8 l displacement.[28,29] This temperature is again below the light-off for hydrocarbons.

A gold catalyst with low temperature activity towards carbon monoxide and hydrocarbon oxidation could be suitable to combat cold start-up
emission problems and removal of nitrogen oxides from lean-burn gasoline

and diesel engines.[30,31] The justification for developing gold-based technologies is both their promising technical performance and the relatively stable price and greater availability of gold compared with the platinum group metals (Section 14.6); there are, however some technical difficulties to be overcome before gold catalysts can be successfully applied in the automotive sphere: these include attaining higher durability and poison resistance qualities than have been achieved to date (see below).

A gold-based material has been formulated for use as a three-way catalyst in gasoline and diesel applications.[28] This catalyst, developed at Anglo American Research Laboratories in South Africa, consisted of 1% Au supported on zirconia-stabilized-CeO_2, ZrO_2 and TiO_2, and contained 1% CoO_x, 0.1% Rh, 2% ZnO, and 2% BaO as promoters. The catalytically active gold–cobalt oxide clusters were 40–140 nm in size. This catalyst was tested under conditions that simulated the exhaust gases of gasoline and diesel automobiles and survived 773 K for 157 h, with some deactivation (see Section 11.2.7).

A significant hurdle for the gold-based TWC is the high operating temperature requirements imposed by gasoline engines. Typically, a catalyst must be able to withstand a temperature of 1373 K for at least 12 h. The gold-based TWC cannot survive under such conditions and it is accepted that gold will not be able to match the high temperature performance of the PGM-based TWCs. However, a relatively simple system in which PGM- and gold-based catalysts operate in parallel or sequentially can be envisaged, where the gold catalyst is in use at low temperatures but is by-passed in favour of the PGM catalyst at higher temperatures, or only sees the gold catalysts when at relatively low temperatures in a second exhaust box. In this way, maximum conversion activity can be maintained both at low temperatures using the gold catalyst and at high temperatures using the PGM catalyst.[28]

The formation of ionic gold trapped in an oxide lattice is thought to be responsible for the stability of some Toyota catalysts: there was no reduction in T_{50}% conversion for propene after treatment at 1073 K for 5 h (Table 14.1). A standard Au/Al_2O_3 catalyst under the same conditions suffered significant degradation.[32]

Under diesel conditions, carbon monoxide and hydrocarbon oxidation is favoured. Under the highly oxidising conditions encountered in the diesel gas stream, reduction of nitric oxide is not expected. A nitric oxide conversion window is observed at temperatures between 493 and 623 with a T_{50} value of 523 K. However, large NO absorption bands are observed at temperatures above and below the conversion window.[28]

Table 14.1: Catalyst for purifying an exhaust gas (Toyota).[32]

Catalyst composition		Temp. at 50% C_3H_6 conversion (K)[a]	
Chemical formula	Au content (wt.%)	Initial	After durability test[b]
$Au_2Sr_5O_8$	0.4	618	619
$La_2Au_{0.5}Li_{0.5}O_4$	0.2	617	618
Au/Al_2O_3	2	651	706

[a]Evaluation conditions: CO, 1000 ppm; C_3H_6, 670 ppmC; NO, 250 ppm; O_2, 7.3%; H_2, 5%; balance N_2 at 150 000 h^{-1}.
[b]Durability test conditions: CO, 1000 ppm; C_3H_6, 670 ppmC; NO, 500 ppm; O_2, 6.5%; CO_2, 10%; H_2O 10%; balance N_2 at gas temperature of 1073 K for 5 h.

Increased durability is also claimed for gold catalysts prepared by direct anionic exchange on an alumina support.[33] The durability of the catalyst was strongly improved by the complete removal of chloride using an ammonia washing procedure (*but this brings with it risks of explosions if fulminating gold is formed from reaction between soluble gold and ammonia, see Section 4.1.3*). This catalyst, tested in various reactions of saturated and unsaturated hydrocarbons from C_1 to C_3 and the oxidation of carbon monoxide, revealed a good activity, which is in an appropriate range of temperature for treatment of automotive exhaust, and longer durability tests may demonstrate further promise. Impregnation has been used for making Au/Al_2O_3, with washing with ammonia to remove chloride (*but see Section 4.1.3 for dangers involved in using this procedure*):[34] the reactivity for carbon monoxide oxidation at room temperature was comparable with catalysts prepared by DP. This 1% Au/Al_2O_3 contained 2 nm particles and was stable to hydrothermal sintering in 10 mol% steam at 873 K for 100 h. This could have important implications for their future use in autocatalyst and other pollution control applications.

14.2.3. Catalytic wet air oxidation (CWAO)

The efficient and environmentally acceptable processing of wastewater is of important industrial and environmental concern. One technique of growing interest is the wet oxidation process (Chapter 11), where the oxidation of organic compounds in an aqueous solution or in suspension by means of oxygen or air takes place at elevated temperatures (453–588 K) and pressures (2–15 MPa).[35] The organic material present is first converted into simpler organic compounds, which are then further oxidized to carbon dioxide and water. Catalysts provide the possibility of using milder conditions and most 'catalytic wet air oxidation' (CWAO) processes studied to date have been based on platinum and palladium catalysts deposited on titania or titania–zirconia. Gold catalysts could prove to be advantageous and the recent work at the Institut de Recherche sur la Catalyse, Lyon,[36] where preliminary results on the CWAO of succinic acid as a representative organic compound using Au/TiO_2 at 463 K and 50 bar air pressure, show encouraging performance for gold catalysts (Section 11.5).

Trichloroethene (TCE) is one of the most common organic pollutants found in groundwater, deriving from its use as a solvent to degrease metals and electronic parts in the automotive, metals and electronic industries, but it is a harmful environmental pollutant. Pd/Al_2O_3 catalysts have been used to dehydrochlorinate TCE, but recent work has shown that palladium supported on Au/Al_2O_3 is much more active than palladium, Pd/Al_2O_3, or palladium black, the rates being 943, 62, 12.2 and $0.421 \cdot g_{Pd}^{-1} \min^{-1}$, respectively.[37] The gold nanoparticles partially covered with Pd gave the highest activities.

14.2.4. Mercury oxidation in coal-fired power stations

Control of mercury, which has been linked to Alzheimer's disease and autism, is the subject of planned legislation by the US Environmental Protection Agency (EPA). The EPA will impose limits on mercury emissions from coal-fired boilers in the utilities industry. Current mercury control techniques used in the industry include the use of flue-gas desulfurization (FGD) units and, as a result of mercury measurements around these units, it is known that oxidized and not elemental mercury is removed by the FGDs. Thus, one method to increase mercury removal by FGD units is to introduce a catalyst to enhance the oxidation of mercury. Mercury measurement[38] led to the discovery that a gold-coated sand sample in a simulated flue-gas environment absorbed elemental mercury until an equilibrium was established and desorption of oxidized mercury began. Individual components of the simulated flue-gas were evaluated for their effect on the oxidation of mercury. Of the components present, nitrogen dioxide and hydrogen chloride

were primarily responsible for the mercury oxidation over gold; thus, it is
not yet clear whether gold is acting through a truly catalytic mechanism in
this instance, but it was the most active of the catalyst materials evaluated.

14.3. Chemical Processing

14.3.1. Vinyl acetate synthesis

A gold–palladium catalyst which includes potassium acetate is very well
established for the production of vinyl acetate monomer (VAM) from
ethene, acetic acid and oxygen in selectivities as high as 96% (see Section 8.4). VAM is an important intermediate used in the production of
polyvinyl acetate, polyvinyl butyral and a variety of other polymers, and the
gold-catalysed process followed many years of industrially focused research
and patent activity in a number of large industrial companies:[39–43]

$$\text{ethene} + \text{HO} \overset{}{\underset{}{\diagdown}} \overset{O_2}{\longrightarrow} \text{vinyl acetate} + H_2O \qquad (14.1)$$

acetic acid   vinyl acetate

The role of gold has now emerged as having greater significance than was
realised at the outset of these operations. Most of the commercial processes
are fixed-bed, but at the end of 2001, BP commissioned the brand new
plant in Hull, UK. This is the world's first fluidised-bed process for VAM,
while 80% of today's VAM plants worldwide are more than 20 years old and
use a fixed-bed process.[44]

BP Chemicals have developed this cost-saving route that allowed process
simplification and intensification, requiring only a single reactor compared
with the two reactors usually needed in the fixed-bed process. Hull in UK
was an optimum location for the commercial scale plant. Acetic acid was
already produced there and the ethene supply from Teesside was approved
for construction along with an air separation unit to produce oxygen. In a
fixed-bed reactor, the catalyst which promotes the reaction was in the form
of spheres which are packed into tubes. The reaction gases pass through the
tubes and around the catalyst particles in the spaces between the spheres,
without moving them. In a fluidised-bed reactor, however, the catalyst is
in the form of a fine powder, and as the gases flow upwards through the
reactor they blow the fine catalyst around. This gives much better mixing
and contact between the gases and the catalyst, improving heat transfer and

allowing the catalyst to be removed and replenished without having to shut down the reactor. Also fluidised-bed reactors are cheaper and easier to build (the decision to go to a fluidised bed process saved 30% in capital costs).

Moving from a fixed to a fluidised-bed operation also required a new catalyst, and the one selected was a supported gold–palladium system in the form of very fine spheres, prepared in collaboration with Johnson Matthey. Hence, gold-based catalysts are being used for this new fluidised-bed process, and are well established in fixed-bed processes for the large-scale manufacture of VAM.

14.3.2. Vinyl chloride

The manufacture of polyvinyl chloride is still very important commercially[8,45–47] and the synthesis of the monomer is therefore an important step in this synthesis (see Section 13.2):

$$HC\equiv CH + HCl \rightarrow H_2C=CH—Cl \rightarrow PVC. \quad (14.2)$$

The first significant practical demonstration of the commercial relevance of catalysis by gold was by Graham Hutchings, then working in South Africa: gold catalysts supported on activated carbon were found to be about three times more active than commercial mercuric chloride catalysts for vinyl chloride production and to deactivate much less rapidly than other supported metal catalysts. Deactivation can be minimized if high loadings of gold are used. Also, gold catalysts could be reactivated by treatment off-line with hydrogen chloride or chlorine, and by co-feeding nitric oxide with the reactants from the start of the reaction, deactivation could be virtually eliminated.[48] Gold is thus the catalyst of choice for this reaction (see Section 13.2).[6–8,45,49,50] One tonne of gold would be needed for a typical plant envisaged at that time and this was thought to be too expensive in the 1980s when the price of gold was around three times its present value: at today's gold price, the use of gold in this process would surely be viable, especially when the recovery and recycle of the gold is taken into account. This process may yet have the potential to be applied in developing countries if market demand for PVC is sustainable.

14.3.3. Production of Nylon precursors

An important recent development is the demonstration[51,52] that gold catalysts can be used in a solventless liquid-phase system to oxidize cyclohexane to cyclohexanol and cyclohexanone using oxygen. Almost all the cyclohexane produced (4.4 million tonnes per annum, and expected to grow at ca. 3%) is converted to cyclohexanol and cyclohexanone, the intermediates in the production of caprolactam and adipic acid, used in the manufacture of Nylon-6 and Nylon-66 polymers, respectively. The present commercial process for cyclohexane oxidation is carried out at around $423\,\mathrm{K}$ and $1–2\,\mathrm{MPa}$ over a catalyst such as cobalt naphthenate with ca 4% conversion and 70–85% selectivity to cyclohexanol and cyclohexanone:

$$(14.3)$$

The large demand for these products and the high energy demands for the present process could provide an opening for a more effective catalyst. These papers[51,52] are interesting both for their comparatively high conversion rates of ca. 15% and high selectivities to cyclohexanol and cyclohexanone with TONs of up to $3000\,\mathrm{h^{-1}}$, and their use of a zeolite catalyst. In addition, the reaction occurs under environmentally benign conditions involving oxygen as the oxidant in a solvent-free system. The catalyst also seems durable, at least within the limits tried so far. The catalysts used were ca. 1% Au on ZSM-5. A Solutia Inc patent describing similar technology has also been published.[53] Further investigations of this reaction[54] using organically modified mesoporous silicas as supports have given higher conversions (32%) and modified selectivities (see Section 8.2.3).

14.3.4. Methyl glycolate

Nippon Shokubai in Japan has announced the development of a gold catalyst that enables the methyl ester of glycolic acid to be made directly from ethylene glycol:[55,56]

$$(14.4)$$

A 50 tonne per annum pilot plant was commissioned and there are plans to build a larger scale plant in about three years' time. The proprietary catalyst contains highly dispersed gold supported on a metal oxide such as

Au/TiO$_2$–SiO$_2$, and is used with oxygen and under the conditions indicated in the reaction Scheme 14.4. Methyl glycolate can be used as a solvent for semiconductor manufacturing processes, as a building block for cosmetics and as a cleaner for boilers and metals. The Nippon Shokubai announcement also indicates that the catalyst technology will be used for other syntheses involving one-step esterification of carboxylic acids and lactones. One of their patents claims its use for the synthesis of methyl methacrylate.[57]

14.3.5. Selective oxidation of sugars

Gold and Au/C catalysts can be used to oxidize D-glucose to D-gluconic acid (Section 8.3.2),[58-61] and Au/C catalyst is a valid alternative to most of the investigated multimetallic catalysts based on palladium and/or platinum. Moreover, gold has the unique property of operating without the external control of pH, thus ensuring total conversion at all pH values, and total selectivity to gluconic acid. It has recently been demonstrated[62] that Au/TiO$_2$ can also be an efficient (>99% selectivity) and durable (17 runs with no loss of activity) catalyst for this conversion. Gluconic acid is an important food and beverage additive, and is also used as a cleansing agent, and made on the 60 000 tonnes per annum scale, so there may be further opportunities for gold in the food industry:

$$\text{Glucose} \xrightarrow{\text{oxidation}} \text{Gluconic acid} \tag{14.5}$$

Oxidation of lactose and maltose with Au/TiO$_2$ catalysts has been reported to give close to 100% selectivity to lactobionic acid and maltobionic acid, respectively[63] which have potential uses in the pharmaceutical and detergent industries, as well as in food. Studies of the catalytic conversion of glucose by hydrogenation and oxidation to produce sorbitol and gluconic acid respectively have also been reported.[64] Sorbitol is also manufactured on a 60 000 tonnes per annum scale.

14.3.6. Propene oxide

Current commercial production of methyloxirane (propene oxide), used extensively in the production of polyurethanes, is usually based on a chlorohydrin process. However, the direct gas-phase synthesis of methyloxirane from propene using molecular oxygen in the presence of hydrogen, offers the opportunity to eliminate chlorine from the production process, as well as reduce water consumption and salt by-products:

$$\text{propene} + O_2 + H_2 \longrightarrow \text{methyloxirane} + H_2O \tag{14.6}$$

Early results gave selectivities above 99% at low conversions using a 1% Au/TiO$_2$ catalyst system at 323 K, when both oxygen and hydrogen are present in the feed gas (H$_2$:O$_2$: propene : Ar = 10: 10: 10: 70 vol.%)[9] but more recent results give improved conversions whilst maintaining acceptably high selectivity,[65] but improved durability is now required. Patents for direct production of methyloxirane using gold catalysts have been filed by a number of major companies including Bayer, Dow and Nippon Shokubai,[66-68] indicating significant industrial interest in this application, and pilot plants are understood to be operating within the industry. Bayer researchers have claimed an 8% yield of methyl oxirane, with 95% selectivity. Dow report 92 mol% selectivity at 0.36 mol% propene conversion and a production rate of 8.3 g methyloxirane per kg$_{cat}$ h^{-1} at 433 K using a feed stream of 20% propene, 10% hydrogen, 10% oxygen and the remainder helium over 2 g 0.5% Au on a titania/silica support which had been calcined at 823 K; the flow rate was 160 cm^3 min^{-1} and alkene GHSV 480 h^{-1} at atmospheric pressure. Methyloxirane yields of 9% have now been obtained when using a silylation treatment and alkaline earth metal salts as promoters. Table 14.2 summarizes some of these results.[69,70] It was concluded that gold particle–support interaction is required together with careful selection of the titania–silica support and control of the gold-particle size. The use of

Table 14.2: Selectivities and yields (%) for propene epoxidation using gold on titanosilicate catalysts.[70]

Au (%)	C$_3$H$_6$ conv. (%)	H$_2$ conv. (%)	S (%)	Yield
0.42	3.4	35	85.4	4.0
0.6	4.5	32.1	79.6	5.6
0.007	1.0	3.5	100	1.0

TiO$_2$–SiO$_2$ support, space velocity 4000 h^{-1} cm^3 g$_{cat}^{-1}$.
Ar/C$_3$H$_6$/H$_2$/O$_2$ = 70/10/10/10, temperature 423 K, Ti/Si = 2/100: prepared by sol-gel method; catalysts calcined at 573 K.

pH 7.0 in the DP method was also recommended, together with calcination at 573 K.

14.3.7. Hydrogen peroxide

The market for hydrogen peroxide is very large (ca. 1.9×10^6 tonnes per annum) and is rising by ca. 10% per annum, due in part to it being viewed as an environmentally friendly alternative to chlorine. There is therefore a big incentive to enable hydrogen peroxide to be synthesized where it is to be used, and thus avoid the heavy transport costs for this hazardous material. Also it is currently only economic to produce it on a large scale, using the sequential hydrogenation and oxidation of alkyl anthraquinone, whereas it is often required on a much smaller scale.[71] As discussed in Section 8.5, theoretical calculations[72] and experimental results have both shown that formation of H_2O_2 from hydrogen and oxygen is favoured over gold surfaces. Au/Al_2O_3 catalysts are particularly effective for this reaction and $Au–Pd/Al_2O_3$ have now been shown to provide a significant improvement over the palladium catalysts used in industry to date.[73]

14.3.8. Hydrotreating distillates

There is continuing environmental pressure on the refining industry to decrease progressively the levels of sulfur and aromatics in gasoline and diesel distillate fuels. The present commercial dual-stage system which uses a nickel–molybdenum or cobalt–molybdenum catalyst, followed by a platinum catalyst, may not exhibit sufficient activity to achieve the final levels of sulfur and aromatics saturation required. Recent results have shown, for the first time, the possibility of using gold as a component of a hydrodesulfurization and aromatic dehydrogenation catalyst. $Au–Pd/SiO_2$ catalysts are surprisingly more active (by a factor >6) in the hydrodesulfurization of dibenzothiophene than pure palladium catalysts.[74] Such enhanced activity was explained in terms of the well-known affinity of gold for sulfur, which activates the breakage of the C–S bond without forming stable inactive sulfides.

14.3.9. Selective hydrogenation

Selective hydrogenation to remove dienes and alkynes from alkene streams is needed to prevent poisoning of the polymerization catalysts. Supported gold catalysts offer interesting potential since they can selectively hydrogenate dienes in the presence of monoenes,[75,76] and catalyse the conversion of alkynes to alkenes.[77,78] A recent paper[79] shows that hydrogen is dissociatively adsorbed on the gold particles in Au/Al_2O_3. Gold could also be associated with other hydrogenation metals, such as platinum or palladium to modify the chemisorption, activity and/or selectivity.

Some time ago the use of Au/SiO_2 or Au/Al_2O_3 for the hydrogenation of canola oil was reported.[80] It was shown that the complete reduction of linolenic acid could be achieved at a lower *trans*-isomer content in the products than that obtained using the American Oil Chemists standard nickel catalysts. Nickel catalysts have of course been used for over a century for the hardening of natural oils. Could it be that gold catalysts, using the much more advanced methods of preparation available today, will have a future role in this application? If so, this would have the appeal that any gold residues in the products would be completely harmless since gold is environmentally benign.

14.4. Fuel Cells and the Hydrogen Economy

Fuel cells are energy generators which provide a combination of high efficiency and with low pollutant emissions, based on the direct electrochemical oxidation of hydrogen or hydrocarbon fuels to provide an electric current, and at the same time forming water[30,31,45,81-87] which can be a useful by-product in space travel and submarines. Globally, fuel cells have been developed for a wide range of applications including stationary generators, cars and buses, industrial vehicles and small portable power supplies for mobile telephones and laptop computers, each of which represents a substantial market. After decades of development and large-scale trials, fuel cell generators are now entering commercial service to provide electricity and heat for individual homes, or district schemes. In addition, virtually all the world's major vehicle manufacturers are demonstrating electrically propelled passenger vehicles or municipal buses.[86]

The application of gold as an electrocatalytic component within the fuel cell itself has to date been limited primarily to the historical use of a gold–platinum electrocatalyst for oxygen reduction in the Space Shuttle/Orbiter alkaline fuel cells (AFC)[88] and the recent use of gold for borohydride oxidation in the direct borohydride alkaline fuel cell (DBAFC).[89,90] Electrocatalysts with lower cost, improved carbon monoxide tolerance and higher performance are needed for the membrane/electrode assemblies of other low temperature fuel cells (polymer electrolyte membrane, PEM and direct methanol, DM). With gold presently approximately half the cost of platinum on a weight-for-weight basis, research programmes are evaluating gold as a potential electrocatalyst component, particularly as part of a bimetallic system with platinum group metals. Results on gold and gold–platinum particles as potential fuel cell electrocatalysts, focussing for example on refining the synthesis, assembling and thermal treatment of shell-capped gold and gold–platinum particles in the 2–5 nm size range, and comparing the electrocatalytic oxygen reduction reaction (ORR) and methanol oxidation reaction (MOR) activities of the gold and gold–platinum nanoparticle catalysts with commercially available Pt/C and Pt-Ru/C catalysts. The gold–platinum catalysts with Au:Pt > 7:3 and 10–25% metal loading exhibited at least comparable, and in some cases much higher, catalytic activities than platinum (ORR) and platinum–ruthenium catalysts (MOR) in alkaline electrolytes.[91–93]

The micro fuel cell developed at QinetiQ, Europe's largest science and technology organisation, is one of the first uses of a carbon-supported gold electrocatalyst for fuel cell applications.[94] Over the last few years QinetiQ has been developing direct liquid fuel cells based on alkaline electrolytes. As well as having improved electrocatalysis and lower methanol permeation rates than methanol fuel cells, the alkaline environment stabilises several alternative fuels which decompose in neutral or acidic media. In particular, fuel cells based on sodium borohydride (DBAFC) have been shown to have good energy storage and activity. QinetiQ have now developed a prototype based on this fuel and are developing their system as a potential replacement for rechargeable batteries for mobile and portable electronics. It is possible that the DBAFC in this type of application would be rechargeable simply by replenishing the discharged fuel solution.

The QinetiQ micro fuel cell (Figure 14.1) consists of a cylindrical unit cell in which the interior of the cell is utilised as a fuel reservoir. Fuel is fed to the anode catalyst via holes in the inner cylinder which also acts as the anode current collector. Gold is used as the anode catalyst in QinetiQ's DBAFC due to complications in the anode reaction, since borohydride fuel decomposes to produce hydrogen in the presence of platinum or ruthenium. Gold is beneficial because it is relatively unreactive to this decomposition reaction while being suitably active to the desired borohydride oxidation reaction. The QinetiQ fuel cell uses a 40% Au/C (Vulcan XC72) catalyst manufactured by E-Tek Inc, USA.

Figure 14.1: QinetiQ's tubular direct borohydride alkaline fuel cell (DBAFC), reproduced courtesy of QinetiQ Ltd.

Recent developments in supported gold catalysts make them suitable for a number of potential enabling applications required for the application of fuel cell technology.[6,86,95-97] These include generation of hydrogen by reforming hydrocarbons and use in the water-gas shift (WGS) (Chapter 10). Another use for gold catalysts is the purification of the resulting gas mixture by selective oxidation of carbon monoxide (PROX) (Chapter 7).

There could be opportunities to exploit the properties of gold and gold–platinum group metals catalysts, particularly for low temperature proton exchange membrane (PEM) fuel cells, which are one of the most widely developed forms of generator and also the most susceptible to poisoning by impurities in the hydrogen.[86] Platinum group metals are currently used for both anode and cathode catalysts in most commercial PEM fuel cells, but the tendency for poisoning of platinum by carbon monoxide contaminant in the hydrogen remains a significant technical issue.[30,31] Work is now emerging[91,98,99] on investigation of the potential benefits of alternative gold–platinum group metal bifunctional catalysts. This is based on the premise that the alloying of gold with platinum in the nanosize range could generate unique bifunctional catalytic properties with platinum acting as the catalyst for the main fuel cell reaction and gold assisting via the removal of carbon monoxide under the low operating temperatures of the fuel cell.

This approach would also potentially have two other advantages over current technology. First, if platinum loadings could be reduced, it could produce a welcome reduction in the capital cost of fuel cell catalysts. In recent years, prices for platinum have significantly exceeded those for gold (see Section 14.6).[85] Cost remains a significant barrier to more widespread commercialisation of fuel cell technology and a reduction in catalyst cost is identified as a key objective throughout the industry. The second advantage from incorporation of gold into fuel cells would be the useful enhancement in electrical conductivity that could be derived, due to the lower electrical resistivity of gold compared with platinum.[30,31,85] There are patents on the use of gold–platinum particles as fuel cell catalysts.[100,101]

Electricity has been produced by catalytic oxidation of carbon monoxide using gold catalysts at room temperature, using gold nanotubes in polycarbonate membranes.[102,103] These gold nanotubes exhibit catalytic activity for carbon monoxide oxidation at room temperature, and this activity is enhanced by the presence of liquid water, promoted by increasing the pH of the solution, and increased using hydrogen peroxide as the oxidizing agent. The rate can also be increased by depositing KOH within these nanotubes. These rates are comparable with those found in heterogeneous catalysis studies with gold nanoparticles on oxide supports, which suggests that the high activity of the oxide-supported catalysts may be related to the promotional effect of hydroxyl groups. The observed rates are faster than for conventional processes operating at $500 \, K$ or higher for the conversion of carbon monoxide with water to produce hydrogen and carbon dioxide through the WGS. The elimination of the need for WGS means that there is no need to transport and vaporise liquid water in the production of energy for portable applications. The process can use carbon monoxide-containing gas streams from the catalytic reforming of hydrocarbons to produce an aqueous solution of reduced polyoxymetallate compounds that can be used to generate power. The reduced polyoxymetallate can be reoxidized in fuel cells that contain simple carbon anodes.

14.5. Sensors

The need for air-quality monitoring demands development of sensors which are selective for detection of individual pollutant gases. Gas sensors based on gold have been developed for detecting a number of gases, including carbon monoxide and nitrogen oxides. The use of gold is also particularly promising for monitoring components of body liquids but these are based on colour change and are not included here.[104]

Sensors for *carbon monoxide* detection are well established, using Au/α–Fe_2O_3[105–107] with gold particle sizes between 3.2 and 8.8 nm. Particles smaller than 5 nm have been used on the oxides of zinc, copper, nickel and cobalt.[108-111] Composite films are used for optical carbon monoxide sensors, as well as an Au/La_2O_3–SnO_2 ceramic.[112,113] The target of the last example was to develop a steadily operating and widely available carbon monoxide gas sensor, since conventional sensors had a poor selectivity to ethanol vapour, which co-exists very often in kitchens, causing false alarms; so ethanol absorbents, such as activated carbon were also incorporated. Results showed that the sensitivity to carbon monoxide was more than ten times higher than to hydrogen, methane, iso-butane and ethene.

There are a number of gold sensors for NO_x detection, e.g. porous silicon has been activated for this purpose by sputtering gold onto the surface.[114] It was found that this catalyst is suitable for sensing nitrogen oxides with negligible influence by interfering gases such as carbon monoxide, methane or methanol, but humidity appreciably affected the response.

Tungsten trioxide thin films activated by gold layers have also been used for NO_x detection.[115] This material possesses excellent sensitivity towards nitric oxide and nitrogen dioxide. An automobile exhaust gas NO_x sensor that uses a gold–platinum alloy electrode has also been reported.[116]

Sensors for nitrogen dioxide detection use a Au/PVC composite prepared by dispersion of a fine gold powder in a highly plasticised PVC matrix containing a hydrophobic electrolyte (Au/PVC electrode)[117] which could be used continuously for monitoring the gaseous environment when the nitrogen dioxide content varied only slowly.

Gold nanoparticles can be used in conjunction with coenzymes for the continuous monitoring of *glucose*.[118] Results of *in vitro* experiments show that this glucose sensor has a short response time, high sensitivity and good linearity. This is a demonstration that aqueous colloidal gold particles can enhance the activity of aqueous enzymes.

Amongst other developments, a gold-oxide composite has been developed as a sensor for *hydrocarbon* monitoring in automobile exhaust gases,[119,120] and Au–In_2O_3 ceramics have been used for detection of *ammonia* and other reducing gases.[121]

14.6. Some Economic Considerations

Whilst the cost of a material used as a catalyst is a consideration for some industrial end-users, the most important motivating factor for new catalyst development from a chemical company's point of view is often selectivity. Catalyst cost is not as significant an issue as the operating efficiency of a multi-million dollar plant. In essence, the intrinsic value of precious metal catalysts is less important than their cost effectiveness. This reflects their often increased activity and durability in many reactions compared with alternative base metal catalysts, and of course, precious metals are economically recycled.

However, catalyst cost and availability is an issue in some significant applications such as fuel cells. It is relevant therefore to compare the prices of the precious metals. Since gold is mined in far greater quantities than platinum or palladium, its price has historically been more stable (Figure 14.2) than these metals and industrialists prefer stable prices.

In addition, there is also an improved recognition that gold is not expensive compared with platinum. Due to the considerably greater availability of gold, any significant new demand for gold as a catalyst is unlikely to impact on gold price to any significant extent, unlike the PGMs where demand can outstrip supply with a consequent impact on price, as seen in recent years for palladium. This is an important economic factor in the choice of technology, particularly in emerging industries like the fuel cell industry.

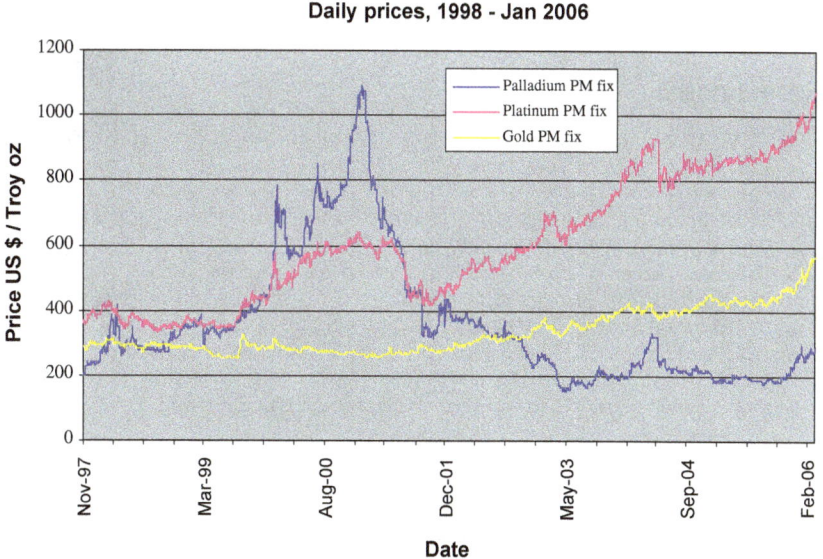

Figure 14.2: Precious metals prices 1998–2006.[1]

14.7. Future Prospects

We have tried to highlight some of the major commercial opportunities to apply heterogeneous gold catalysis in commercial applications and indicate areas that merit particular attention. However, it is not and cannot be an exhaustive list of possible application areas, since new reactions that are catalysed by gold continue to be discovered. Overall, it is believed that there is cause for great optimism that many new applications for gold catalysis could emerge over the next decade. To turn this expectation into reality requires researchers to be proactive in exploiting promising research results in both the gas and liquid phases and effectively communicating these to industry. In addition, potential end-user industries and catalyst manufacturers need to carefully and fully consider the potential business opportunities that gold catalysts undoubtedly offer. The series of recent international gold catalysis conferences are proving to be a key forum for encouraging exploitation in this regard[122–125] enabling an exchange of ideas and opportunities between industry and academia. All involved in this exciting new field need to carefully consider more ways to increase the durability of catalysts under representative operating conditions and to tailor the emerging commercially viable manufacturing methods of catalyst preparation to well-defined requirements.

References

1. C.W. Corti, R.J. Holliday and D.T. Thompson, *Appl. Catal. A: Gen.* **291** (2005) 253.
2. G.C. Bond and D.T. Thompson, *Appl. Catal. A: Gen.* **302** (2006) 1.
3. M. Cohen, *Mining Weekly* **11**(7) (2005) 5 (February 25–March 3, 2005).
4. Japanese Patent, 4281846, AIST, 1992.
5. Japanese Patent, 5115748, Matsushita Electrical Industries, 1993.
6. M. Haruta, *Chem. Record* **3** (2003) 75.
7. M. Haruta, *CATTECH* **6** (2002) 102.
8. G.C. Bond and D.T. Thompson, *Cat. Rev.-Sci. Eng.* **41** (1999) 319.
9. M. Haruta, *Catal. Today* **36** (1997) 153.
10. D.A. Bulushev, L. Kiwi-Minsker, I. Yuranov, E.I. Suvorova, P.A. Buffat and A. Renken, *J. Catal.* **210** (2002) 149.
11. H. Liu, A.I. Kozlov, A.P. Kozlova, T. Shido, K. Asakura and Y. Iwasawa, *J. Catal.* **185** (1999) 252.
12. M. Haruta and M. Daté, *Appl. Catal. A: Gen.* **222** (2001) 427.
13. G.C. Bond and D.T. Thompson, *Gold Bull.* **33** (2000) 41; G.J. Hutchings, M.R.H. Siddiqui, A. Burrows, C.J. Kiely and R. Whyman, *J. Chem. Soc. Faraday Trans.* **93** (1997) 187.
14. M.B. Cortie and E. van der Lingen, *Mater. Forum* **26** (2002) 1.

15. G.U. Kulkarni, C.P. Vinod and C.N.R. Rao, in *Surface Chemistry and Catalysis*, A. Carley, P.R. Davies, G.J. Hutchings and M.S. Spencer, (eds.), Kluwer/Plenum, New York, 2002.

16. G.B. Hoflund, S.D. Gardner, D.R. Schryer, B.T. Upchurch and E.J. Kielin, *Appl. Catal. B: Env.* **6** (1995) 117.

17. W.E. Epling, G.B. Hoflund, J. Weaver, S. Tsubota and M. Haruta, *J. Phys. Chem.* **100** (1996) 9929.

18. H. Sakurai, M. Haruta and S. Tsubota, *Proc. GOLD 2003*, Vancouver, Canada, Sept.–Oct. 2003; http://www.gold.org/discover/sci_indu/gold2003/index.html.

19. G.Y. Wang, W.X. Zhang, H.L. Lian, D.Z. Jiang and T.H. Wu, *Appl. Catal. A: Gen.* **239** (2003) 1.

20. K. Mallick and M.S. Scurrell, *Appl. Catal. A: Gen.* **253** (2003) 527.

21. W.-H. Cheng, K.-C. Wu, M.Y. Lo and C.-H. Lee, *Catal. Today* **97** (2004) 145.

22. K.-C. Wu, Y.-L. Tung, Y.-L. Chen and Y.-W. Chen, *Appl. Catal. B: Env.* **53** (2004) 111.

23. H. Sakurai, M. Haruta and S. Tsubota, *Proc. GOLD 2003*, Vancouver, Canada, Sept.–Oct. 2003; http://www.gold.org/discover/sci_indu/gold2003/index.html.

24. Z. Hao, D. Cheng, Y. Guo and Y. Liang, *Appl. Catal. B: Env.* **33** (2001) 217.

25. M. Okumura, T. Akita, M. Haruta, X. Wang, O. Kajikawa and O. Okada, *Appl. Catal. B: Env.* **41** (2003) 43.

26. M. Haruta, *Proc. GOLD 2003*, Vancouver, Canada, Sept.–Oct. 2003; http://www.gold.org/discover/sci_indu/gold2003/index.html.

27. J. Kaspar, P. Fornasiero and N. Hickey, *Catal. Today* **77** (2003) 419.

28. J.R. Mellor, A.N. Palazov, B.S. Grigorova, J.F. Greyling, K. Reddy, M.P. Letsoalo and J.H. Marsh, *Catal. Today* **72** (2002) 145.

29. R.J. Farrauto and C.H. Bartholomew, *Fundamentals of Industrial Catalytic Processes*, Chapman & Hall, London, 1997.

30. G. Pattrick, E. van der Lingen, C.W. Corti, R.J. Holliday and D.T. Thompson, *Preprints CAPoC 6*, Brussels, October 2003, O14.

31. G. Pattrick, E. van der Lingen, C.W. Corti, R.J. Holliday and D.T. Thompson, *Topics in Catal.* **30/31** (2004) 273.

32. Toyota Patent, E.P. 1043059 A1, 2000.

33. R. Cousin, S. Ivanova, F. Ammari, C. Petit, V. Pitchon, *Proc. GOLD 2003*, Vancouver, Canada, Sept.–Oct. 2003; http://www.gold.org/discover/sci_indu/gold2003/index.html.

34. Q. Xu, K.C.C. Kharas and A. Datye, *Catal. Lett.* **85** (2003) 229.

35. F. Luck, *Catal. Today* **53** (1999) 81.

36. M. Besson, A. Kallel, P. Gallezot, R. Zanella and C. Louis, *Catal. Commun.* **4** (2003) 471.

37. M.O. Nutt, J.B. Hughes and M.S. Wong, *Envir. Sci. Technol.* **39** (2005) 1346.

38. S. Meischen, *Proc. GOLD 2003*, Vancouver, Canada, Sept.–Oct. 2003; http://www.gold.org/discover/sci_indu/gold2003/index.html.

39. W.D. Provine, P.L. Mills and J.J. Lerou, *Stud. Surf. Sci. Catal.* **101** (1996) 191; European Patent, BP 0654301, 1994.

40. M. Neurock, W.D. Provine, D.A. Dixon, G.W. Coulston, J.J. Lerou and R.A. van Santen, *Chem. Eng. Sci.* **51** (1996) 1691.

41. Sennewald, U.S. Patent 3631079 (1971).

42. W.J. Barteley, S. Jobson, G.G. Harkreader, M. Kitson and M. Lemanski, US Patent 5274181 (1993).

43. N. Macleod, J.M. Keel and R.M. Lambert, *Appl. Catal. A: Gen.* **261** (2004) 37.

44. http://www.bp.com/liveassets/bp_internet/globalbp/STAGING/global_assets/
     downloads/F/Frontiers_magazine_issue_4_Leaps_of_innovation.pdf, pp. 12–15.
45. D.T. Thompson, *Appl. Catal. A: Gen.* **243** (2003) 201.
46. D.T. Thompson, *Gold Bull.* **31** (1998) 111.
47. G.J. Hutchings, *Catal. Today* **72** (2002) 11.
48. B. Nkosi, M.D. Adams, N.J. Coville and G.J. Hutchings, *J. Catal.* **128** (1991) 378.
49. B. Nkosi, N.J. Coville, G.J. Hutchings, M.D. Adams, J. Friedl and F. Wagner,
     J. Catal. **128** (1991) 366.
50. B. Nkosi, M.D. Adams, N.J. Coville and G.J. Hutchings, *J. Catal.* **128** (1991) 378.
51. R. Zhao, D. Ji, G. Lu, G. Qian, L. Yan, X. Wang and J. Suo, *Chem. Commun.*
     (2004) 904.
52. G. Lu, R. Zhao, D. Ji, G. Qian, Y. Qi, X. Wang and J. Suo, *Catal. Lett.* **97**
     (2004) 115.
53. L.V. Pirutko, A.S. Khatitonov, M.I. Khramov and A.K. Uriate, US Patent
     2004158103 A1.
54. K. Zhu, J. Hu and R. Richards, *Catal. Lett.* **100** (2005) 195.
55. *Chem Eng (New York)* **111**(9) (September 2004) 20.
56. T. Hayashi and H. Baba, W.O. Patent 2004007422 (2004).
57. T. Hayashi and T. Inagaki, W.O. Patent 2002098558 (2002).
58. Michele Rossi, *Proc. GOLD 2003*, Vancouver, Canada; http://gold.dev.cfp.co.uk/
     discover/sci_indu/gold2003/index.html.
59. S. Biella, L. Prati and M. Rossi, *J. Catal.* **206** (2002) 242.
60. S. Biella, G.L. Castiglioni, C. Fumagalli, L. Prati and M. Rossi, *Catal. Today* **72**
     (2002) 43.
61. P. Beltrame, M. Comotti, C.D. Pina and M. Rossi, *Appl. Catal. A: Gen.* **297**
     (2006) 1.
62. A. Mirescu and U. Pruesse, *Catal. Commun.* **7** (2006) 11.
63. A. Mirescu, U. Pruesse and K.-D. Vorlop, *Proc. 13 International Congress on
     Catalysis*, Paris, July 2004; http://icc2004.catalyse.cnrs.fr/CDROM/P5-059.pdf.
64. S. Schimpf, B. Kusserow, Y. Önal and P. Claus, *Proc. 13 International Congress on
     Catalysis*, Paris, July 2004; http://icc2004.catalyse.cnrs.fr/CDROM/P5-060.pdf.
65. M. Haruta, *Gold Bull.* **37** (2004) 27.
66. WO Patent 2001158887, to Bayer AG.
67. US Patent 20040176629 to Dow.
68. US Patent 2001020105 to Nippon Shokubai.
69. M. Haruta, *Proc. 13 International Congress on Catalysis*, Paris, July 2004.
70. A.K. Sinha, S. Seelan, S. Tsubota and M. Haruta, *Topics in Catal.* **29** (2004) 95.
71. P. Landon, P.J. Collier, A.J. Papworth, C.J. Kiely and G.J. Hutchings, *Chem.
     Commun.* (2002) 2058.
72. P. Paredes Olivera, E.M. Patrito and H. Sellers, *Surf. Sci.* **313** (1994) 25.
73. J.K. Edwards, B.E. Solsona, P. Landon, A.F. Carley, A. Herzing, C.J. Kiely and
     G.J. Hutchings, *J. Catal.* **236** (2005) 69.
74. A.M. Venezia, V. La Parola, V. Nicoli and G. Daganello, *J. Catal.* **212** (2002) 56.
75. D.A. Buchanan and G. Webb, *J. Chem. Soc. Faraday Trans.* **71** (1975) 134.
76. M. Okumura, T. Akita and M. Haruta, *Catal. Today* **74** (2002) 265.
77. P.A. Sermon, G.C. Bond and P.B. Wells, *J. Chem. Soc. Faraday Trans.* **75**
     (1979) 385.
78. J. Jia, K. Haraki, J.N. Kondo and K. Tamaru, *J. Phys. Chem. B* **104** (2000) 11153.
79. E. Bus, J.T. Miller and J.A. van Bokhoven, *J. Chem Phys. B* **109** (2005) 14581.

80. L. Caceres, L.L. Diosady, W.F. Graydon and L.J. Rubin, *J. Amer. Oil Chem. Soc.* **62** (1985) 906.
81. A. Luengnaruemitchai, S. Osuwan and E. Gulari, *Int. J. Hydrogen Energy* **29** (2004) 429.
82. G. Avgouropoulos, T. Ioannides, Ch. Papadopoulou, J. Batista and S. Hocevar and H.K. Matralis, *Catal. Today* **75** (2002) 157.
83. T.V. Choudhary and D.W. Goodman, *Catal. Today* **77** (2002) 65.
84. J. Zhang, Y. Wang, B. Chen, C. Li, D. Wu and X. Wang, *Energy Conv. Management* **44** (2003) 1805.
85. D. Cameron, R. Holliday and D. Thompson, *J. Power Sources* **118** (2003) 298.
86. D.S. Cameron, *Proc. GOLD 2003*, Vancouver, Canada, Sept.–Oct. 2003; http://www.gold.org/discover/sci_indu/gold2003/index.html.
87. R.J. Farrauto, *Appl. Catal. B: Env.* **56** (2005) 1.
88. J.O'M. Bockris and A.J. Appleby, in *Assessment of Research Needs for Advanced Fuel Cells*, S. Penner, (ed.), published in *Energy* **11**(1/2) (1986) 110.
89. R.W. Reeve, I.E. Eweka and G.O. Mepsted, *Proc. Eighth Grove Fuel Cell Symposium, London*, 24–26 September, 2003, 04B.6.
90. J.B. Lakeman and K. Scott, Abstract, Conference on High Energy Density Electrochemical Power Sources, Nice, France, September 17–20, 2003.
91. C.J. Zhong and M.M. Maye, *Adv. Mater.* **13** (2001) 1507.
92. C.J. Zhong, J. Luo, M.M. Maye, L. Han and N.N. Kariuki, in *Nanotechnology in Catalysis*, B. Zhou, S. Hermans and G.A. Somorjai, (eds.), Kluwer Academic/Plenum, Chapter 11, 2003.
93. K. Matsuoka, Y. Iriyama, T. Abe and Z. Ogumi, *Proc. ECS Joint International Meeting, Tapa*, 3–8 October, 2004, Abstract 1518.
94. CatGold News, Issue No. 6, Spring 2004, and private communication.
95. D. Andreeva, *Gold Bull.* **35** (2002) 82.
96. M.M. Maye, J. Luo, L. Han, N.N. Karinki and C.-J. Zhong, *Gold Bull.* **36** (2003) 75.
97. F. Boccuzzi, A. Chiorino and M. Manzoli, *J. Power Sources* **118** (2003) 304.
98. M.M. Maye, Y. Lou and C.-J. Zhong, *Langmuir* **16** (2000) 7520.
99. J. Luo, M.M. Maye, Y. Lou, L. Han, M. Hepel and C.J. Zhong, *Catal. Today* **77** (2002) 127.
100. B.E. Hayden, C.E. Lee, C. Mormiche and D. Thompsett, W.O. Patent 021740 A1 to Johnson Matthey, 2006.
101. W.O. Patent 9424710 to Johnson Matthey (1994).
102. W.B. Kim, T. Voitl, G.J. Rodriguez-Rivera and J.A. Dumesic, *Science* **305** (2004) 1280.
103. M.A. Sanchez-Castillo, C. Couto, W.B. Kim and J.A. Dumesic, *Angew. Chem. Int. Ed.* **43** (2004) 1140.
104. C.W. Corti, R.J. Holliday and D.T. Thompson, *Gold Bull.* **35** (2002) 111.
105. N. Funazaki, A. Hemmi, S. Ito, Y. Asano, S. Yamashita, T. Kobayashi and M. Haruta, *Sens. Actuators B* **14** (1993) 536.
106. G. Neri, A. Bonavita, S. Galvagno, L. Caputi, D. Pacilè, R. Marsico and L. Papagno, *Sens. Actuators B* **80** (2001) 222.
107. G. Neri, A. Bonavita, C. Milone and S. Galvagno, *Sens. Actuators B* **93** (2003) 402.
108. F. Boccuzzi, A. Chiorino, S. Tsubota and M. Haruta, *Sens. Actuators B* **24/25** (1995) 540.
109. R.-J. Wu, C.-H. Hu, C.-T. Yeh and P.-G. Su, *Sens. Actuators B* **96** (2003) 596.
110. M. Ando, T. Kobayashi and M. Haruta, *Sens. Actuators B* **24/25** (1995) 851.
111. M. Ando, T. Kobayashi and M. Haruta, *Catal. Today* **36** (1997) 135.

112. K. Fukui and M. Nakane, *Sens. Actuators B* **24/25** (1995) 486.

113. K. Fukui and S. Nishida, *Sens. Actuators B* **45** (1997) 101.

114. C. Baratto, G. Sberveglieri, E. Comini, G. Faglia, G. Benussi, V. La Ferrara, L. Quercia, G. Di Francia, V. Guidi, D. Vincenzi, D. Boscarino and V. Rigato, *Sens. Actuators B* **68** (2000) 74.

115. M. Penza, C. Martucci and G. Cassano, *Sens. Actuators B* **50** (1998) 52.

116. D.C. Skelton, R.G. Tobin, D.K. Lambert, C.L. DiMaggio and G.B. Fisher, *Sens. Actuators B* **96** (2003) 46.

117. Z. Hoherčáková and F. Opekar, *Sens. Actuators B* **97** (2004) 379.

118. M. Pan, X. Guo, Q. Cai, G. Li and Y. Chen, *Sens. Actuators A* **108** (2003) 258.

119. J. Zosel, D. Westphal, S. Jakobs, R. Müller and U. Guth, *Solid State Ionics* **152/153** (2002) 525.

120. P. Schmidt-Zhang and U. Guth, *Sens. Actuators B* **99** (2004) 258.

121. R. Cousin, S. Ivanova, F. Ammari, C. Petit and V. Pitchon, *Proc. GOLD 2003*, Vancouver, Canada, Sept.–Oct. 2003; http://www.gold.org/discover/sci_indu/gold2003/index.html.

122. *Catalytic Gold 2001*, D.T. Thompson, (ed.), *Catal. Today* **72**(1–2) (2002).

123. D.T. Thompson, *Gold Bull.* **34** (2001) 56.

124. GOLD 2003 Keynote Papers, *Gold Bull.* **37**(1–2) (2004) http://www.goldbulletin.org.

125. GOLD 2006, http://www.gold.org.

Index

www.ingramcontent.com/pod-product-compliance
Lightning Source LLC
Chambersburg PA
CBHW060756220326
41598CB00022B/2457